Studies in Logic
Logic and Argumentation
Volume 86

Reason to Dissent
Proceedings of the 3rd European Conference on Argumentation
Volume II

Volume 77
Argumentation and Inference. Proceedings of the 2nd European Conference on Argumentation. Volume II
Steve Oswald and Didier Maillat, eds.

Volume 68
Logic and Philosophy of Logic. Recent Trends in Latin America and Spain
Max A. Freund, Max Fernández de Castro and Marco Ruffino, eds.

Volume 79
Games Iteration Numbers. A Philosophical Introduction to Computability Theory
Luca M. Possati

Volume 80
Logics of Proofs and Justifications
Roman Kuznets and Thomas Studer

Volume 81
Factual and Plausible Reasoning
David Billington

Volume 82
Formal Logic: Classical Problems and Proofs
Luis M. Augusto

Volume 83
Reasoning: Games, Cognition, Logic
Mariusz Urbański, Tomasz Skura and Paweł Łupkowski, eds.

Volume 84
Witness Theory. Notes on λ-calculus and Logic
Adrian Rezuş

Volume 85
Reason to Dissent. Proceedings of the 3rd European Conference on Argumentation. Volume I
Catarina Dutilh Novaes, Henrike Jansen, Jan Albert van Laar and Bart Verheij, eds.

Volume 86
Reason to Dissent. Proceedings of the 3rd European Conference on Argumentation. Volume II
Catarina Dutilh Novaes, Henrike Jansen, Jan Albert van Laar and Bart Verheij, eds.

Volume 87
Reason to Dissent. Proceedings of the 3rd European Conference on Argumentation. Volume III
Catarina Dutilh Novaes, Henrike Jansen, Jan Albert van Laar and Bart Verheij, eds.

Studies in Logic Series Editor
Dov Gabbay   dov.gabbay@kcl.ac.uk

Reason to Dissent
Proceedings of the 3rd European Conference on Argumentation

Volume II

Edited by

Catarina Dutilh Novaes,

Henrike Jansen,

Jan Albert van Laar

and

Bart Verheij

© Individual author and College Publications, 2020
All rights reserved.

ISBN 978-1-84890-332-6

College Publications
Scientific Director: Dov Gabbay
Managing Director: Jane Spurr

http://www.collegepublications.co.uk

All rights reserved. No part of this publication may be reproduced, stored in a retrieval system or transmitted in any form, or by any means, electronic, mechanical, photocopying, recording or otherwise without prior permission, in writing, from the publisher.

Conference Organization

Organising committee

Catarina Dutilh Novaes (Vrije Universiteit Amsterdam)
Henrike Jansen (Leiden University)
Jan Albert Van Laar (chair) (University of Groningen)
Bart Verheij (University of Groningen)

Programme committee

Dale Hample (chair) (University of Maryland)
Henrike Jansen (secretary) (Leiden University)
Marcin Lewinski (Universidade Nova de Lisboa)
Juho Ritola (University of Turku)
Frank Zenker (Lund University)

ECA steering committee

Lilian Bermejo Luque (University of Granada)
Katarzyna Budzyńska (Warsaw University of Technology)
Henrike Jansen (Leiden University)
Marcin Koszowy (Warsaw University of Technology)
Jan Albert Van Laar (deputy chair) (University of Groningen)
Marcin Lewiński (Universidade Nova de Lisboa)
Dima Mohammed (Universidade Nova de Lisboa)
Steve Oswald (University of Fribourg)
Fabio Paglieri (chair) (ISTC-CNR)
Juho Ritola (University of Turku)
Sara Rubinelli (University of Lucerne)
Frank Zenker (University of Lund)

Scientific Panel

Mark Aakhus (Rutgers University)
Andrew Aberdein (Florida Institute of Technology)
Scott Aikin (Vanderbilt University)
Mehmet Ali Uzelgun (Universidade NOVA de Lisboa)
Ruth Amossy (Tel-Aviv University)
Corina Andone (University of Amsterdam)
Michael Baumtrog (Ryerson University)
Gregor Betz (KIT Karlsruher)
Sarah Bigi (Università Cattolica del Sacro Cuore)
Anthony Blair (University of Windsor)
Patrick Bondy (Cornell University)
Alessandra Von Burg (Wake Forest University)
John Casey (Northeastern Illinois University)
Daniel Cohen (Colby College)
Louis De Saussure (University of Neuchatel)

Marianne Doury (Paris Descartes University)
Michel Dufour (Université Sorbonne-Nouvelle)
Isabela Fairclough (University of Cantral Lancashire)
Eveline Feteris (University of Amsterdam)
Joana Garmendia (University of the Basque Country)
Bart Garssen (University of Amsterdam)
Ingeborg van der Geest (University of Amsterdam)
David Godden (Michigan State University)
G.C. Goddu (University of Richmond)
Jean Goodwin (North Carolina State University)
Floriana Grasso (University of Liverpool)
Sara Greco (Università della Svizzera Italiana)
Leo Groarke (Trent University)
Thierry Herman (Universities of Neuchatel and lausanne)
Martin Hinton (University of Lodz)
David Hitchcock (McMaster University)
Hans Hoeken (Utrecht University)
Michael Hoffmann (Georgia Institute of Technology)
Michael Hoppmann (Northeastern University)
Jos Hornikx (Radboud University)
Catherine Hundleby (University of Windsor)
Constanza Ihnen (University of Chile, Law Faculty, Institute of Argumentation)
Beth Innocenti (University of Kansas)
Jérôme Jacquin (University of Lausanne)
Henrike Jansen (Leiden University)
Amy Johnson (University of Oklahoma)
Manfred Kienpointner (University of Innsbruck)
Gabrijela Kisicek (University of Zagreb)
Susan Kline (The Ohio State University)
Bart van Klink (Vrije Universiteit Amsterdam)
Manfred Kraus (University of Tübingen)
Maarten van Leeuwen (Leiden University)
Christoph Lumer (University of Siena)
Fabrizio Macagno (Universidade NOVA de Lisboa)
Andrei Moldovan (University of Salamanca)
Paula Olmos (Universidad Autónoma de Madrid)
Rudi Palmieri (University of Liverpool)
Steven Patterson (Marygrove College)
Niki Pfeifer (Munich Center for Mathematical Philosophy)
Gilbert Plumer (Law School Admission Council) (retired)
Lotte van Poppel (University of Amsterdam)
Henry Prakken (University of Groningen)
Martin Reisigl (University of Bern)
Adam Richards (Texas Christian University)
Juho Ritola (University of Helsinki)
Andrea Rocci (Università della Svizzera Italiana)
Harvey Siegel (University of Miami)
Francisca Snoeck Henkemans (University of Amsterdam)
Manfred Stede (University of Potsdam)

Katharina Stevens (University of Lethbridge)
Christopher Tindale (University of Windsor)
Assimakis Tseronis (Örebro University)
Mariusz Urbanski (Adam Mickiewicz University)
Hans Vilhelm Hansen (University of Windsor)
Laura Vincze (University of Pisa)
Jacky Visser (University of Dundee)
Lena Wahlberg (Lund University)
Harry Weger (University of Central Florida)
David Williams (Florida Atlantic University)
Harald Wohlrapp (Universität Hamburg)
Carly Woods (University of Maryland)
Yun Xie (Sun Yat-sen University)
Minghui Xiong (Sun Yat-sen University)
Igor Z. Zagar (Educational Research Institute)
Marta Zampa (Zurich University of Applied Sciences)
David Zarefsky (Northwestern University)
Gábor Zemplén (Eötvös Loránd University)

Preface

After two successful editions held in Lisbon in 2015 and Fribourg in 2017, ECA was hosted in 2019 by the Faculty of Philosophy of the University of Groningen, on 24-27 June 2019. These three volumes contain the Proceedings of this third edition of the conference series, whose special theme was Reason to Dissent.

The European Conference on Argumentation (ECA) is a pan-European biennial initiative aiming to consolidate and advance various strands of research on argumentation and reasoning by gathering scholars from a range of disciplines. While based in Europe, ECA involves and encourages participation by argumentation scholars from all over the world; it welcomes submissions linked to argumentation studies in general, in addition to those tackling the conference theme. The 2019 Groningen edition focused on dissent. The goal was to inquire into the virtues and vices of dissent, criticism, disagreement, objections, and controversy in light of legitimizing policy decisions, justifying beliefs, proving theorems, defending standpoints, or strengthening informed consent. It is well known that dissent may hinder the cooperation and reciprocity required for reason-based deliberation and decision-making. But then again, dissent also produces the kind of scrutiny and criticism required for reliable and robust outcomes. How much dissent does an argumentative practice require? What kinds of dissent should we promote, or discourage? How to deal with dissent virtuously? How to exploit dissent in artificial arguers? How has dissent been conceptualized in the history of rhetoric, dialectic and logic? The papers in these three volumes discuss these and other questions pertaining to argumentation and dissent (among other themes).

ECA 2019 had 224 participants and 188 paper presentations, a clear indication that ECA continues to fulfill its role as a key platform of scholarly exchange in the field. These three volumes reflect the current state of the art in argumentation scholarship in general.

The proceedings contain papers that were accepted based on abstract submissions; each submission was thoroughly evaluated by three reviewers of our scientific board—for a full list of ECA committees, see www.ecargument.org. Volume I gathers 25 long papers and associated commentaries, together with 9 papers presented in the thematic panels that were held during ECA2019. Volumes II and III gather 69 regular papers that were presented during the conference.

Many people have contributed to the success of ECA 2019, and for the completion of the Proceedings. First of all, we must thank all members of our Scientific Panel and of our Programme Committee, thanks to whom we were able to select papers of the highest quality. In Groningen, thanks to those who provided organizational support, in particular the team of student assistants (especially Johan Rodenburg) who ensured that the conference was a pleasant experience to all participants. Our heartfelt thanks go to Jelmer van der Linde and Annet Onnes, who accomplished the gigantic task of putting all the papers together into these three volumes, and assisted us throughout in the process of producing the Proceedings. Thanks also to the European Research Council for generously supporting the production of the Proceedings by means of grant ERC-17-CoG 771074 for the project 'The Social Epistemology of Argumentation' (PI C. Dutilh Novaes).

The next edition of ECA will take place in Rome in 2021, and we look forward to seeing the ECA community gathering again for another successful event.

Catarina Dutilh Novaes, Henrike Jansen, Jan Albert van Laar, Bart Verheij

Keynote Speakers

Critical thinking as discourse

Deanna Kuhn – Columbia University

Less than it is an individual ability or skill, critical thinking is a dialogic practice people engage in and commit to, initially interactively and then in interiorized form with the other only implicit. An argument depends for its meaning on how others respond (Gergen, 2015). In advancing arguments, well-practiced thinkers anticipate their defeasibility as a consequence of others' objections, in addition envisioning their own potential rebuttals. Whether in external or interiorized form, the dialogic process creates something new, while itself undergoing development.

This perspective may be useful in sharpening definition of the construct of critical thinking and in so doing help to bring together the largely separate strands of work examining it as a theoretical construct, a measurable skill, and an educational objective. Implications for education follow. How might critical thinking as a shared practice be engaged in within educational settings in ways that will best support its development? One step is to privilege frequent practice of direct peer-to-peer discourse. A second is to take advantage of the leveraging power of dialog as a bridge to individual argument – one affording students' argumentative writing a well-envisioned audience and purpose. Illustrations of this bridging power are presented. Finally, implications for assessment of critical thinking are noted and a case made for the value of people's committing to a high standard of critical thinking as a shared and interactive practice.

Revisiting Apologie de la polémique: about some "felicity conditions" allowing for coexistence in dissent

Ruth Amossy – Tel-Aviv University

In my book entitled Apologie de la polémique (2014), I claimed that polemical discourse fulfils various social functions, among which "coexistence in dissensus" seems the most important. It means not only that disagreement is the basis of life in society, and the principle on which argumentation as a common, rational search for the reasonable, is built. It also signifies that agreement cannot always be reached in democratic societies recognizing the importance of diversity and difference, so that disagreement has to be managed through verbal confrontations, namely, agonistic discussions and polemical exchanges. It thus appears that the latter, though generally blamed for its radicalization and polarization, plays an important role in the public sphere. Among others, public polemics helps opposite parties to voice conflicting opinions and fight for antagonistic solutions without recurring to arms. To use Chantal Mouffe's words, it transforms "enemies" to be destroyed into "adversaries" who have a right to speak. Beside other social functions discussed in the book, polemics authorizes what the French call a "vivre-ensemble" – the possibility for people who do not share the same opinions, if not the same premises, to share the same national space and live together without outbursts of violence.

However, the emphasis on dissent and its polemical management is not without raising multiple questions concerning the conditions of possibility and the limits of the so-called coexistence in dissent. Obviously, the use of polemical discourse is not enough to prevent citizens from physically fighting each other and even, sometimes, to dispel the specter of civil war. Outbursts of violence against refugees regularly occur in Germany where the polemical discussion is vivid. In France, the polemical exchanges on Emmanuel Macrons' reforms and the authorized street demonstrations did not prevent urban violence. Even if polemical campaign discourse is tolerated, it did not prevent armed confrontations in certain African countries such as Ivory Coast. What, then, are the "felicity conditions" needed in order for public polemics to secure a peaceful "living together" in the framework of persistent and sometimes deep disagreements that can hardly be avoided in the democratic space? My contention is that to answer this question, it is necessary to explore polemical confrontations in their institutional framework, and to examine the functioning of polemical discourse in relation to the political, forensic and cultural factors that determine (at least partly) its degree of success. After synthetizing the finding of my first research into dissent and its polemical management, I will try – on the basis of a few case studies – to gather some of the "felicity conditions" necessary to make coexistence in dissent possible.

Dissent needed: argumentation for AI and law applications

Katie Atkinson – University of Liverpool

As technological advances in artificial intelligence are being turned into deployed products, societal questions are being raised about the need for AI tools to be able to explain their decisions to humans. This need becomes even more pressing when AI technologies are applied in domains where critical decisions are made that can result in a significant effect upon individuals or groups in society. One such domain is law, where there is a thriving market developing in support tools for assisting with a variety of legal tasks carried out within law firms and the wider legal sector. Law is a domain rich in argumentation and support tools that are used to aid legal decision making should similarly be able to explain why a particular outcome of a decision has been reached, and not an alternative outcome. Dissent needs to be captured and revealed within AI reasoners to ensure that the decision space is explored from different perspectives, if AI tools are to be deployed effectively to assist with legal reasoning tasks. In this talk I will discuss a body of work on computational models of argument for legal reasoning and show how dissent features within this work to promote scrutability of AI decision making.

Contents

Conference Organization v

Preface ix

Keynote Speakers xi

Regular papers   1

What Optimistic Responses to Deep Disagreement get Right (and
   Wrong) 3
   Scott F. Aikin

Is There a Crisis of Dissent and Disagreement in Argument Theory? . 15
   Mark Alfino

Epistemic Tolerance 27
   Dominik Balg

Structuring Controversy: The Dialectic of Disagreement 41
   Sharon Bailin, Mark Battersby

The Role of Reasoning and Arguing in Youth Democratic Participation
   in Canada 55
   Michael D. Baumtrog

Pictures and Reasoning: Visual Arguments and Objections 65
   Angelina Bobrova

Issues Leading to Argumentative Discussions During Family Mealtime
   Conversations 79
   Antonio Bova

Value-based Argumentation and the Transition to Low Carbon
   Economy in Turkey and Portugal: Values, Uncertainty and Actions   93
   Huthaifah Busuulwa

Dissent: Considering Culture and Personality 107
   Linda Carozza

Heroic Argumentation: On Heroes, Heroism, and Glory in Arguments 117
John Casey, Daniel H. Cohen

Relationships between Narrative and Argumentation. In Defence of a Functional Account 129
Guillermo Sierra Catalán

Interactive Discourse Features Supporting Aversive and Existential Acknowledgment In Legislative and Ceremonial Government Apologies 143
Martha S. Cheng

Devil's Advocates are the Angels of Argumentation 161
Katharina Stevens, Daniel H. Cohen

The Semantic Mechanisms Underlying Disagreement. An Argumentative Semantics Approach 175
Ana-Maria Cozma

Strategic Maneuvering in Implicit an Pseudo-explicit Advertising Discussion 191
Hédi Virág Csordás

Reasons for Rational Disagreement from Dialectics The Van Inwagen Cases 205
Istvan Danka

Arguing Brexit on Twitter: A Corpus Linguistic Study 217
Natalie Dykes, Philipp Heinrich, Stefan Evert

Protagoras' Principles, Disagreement and the Possibility of Error . . 231
Michel Dufour

Changing Minds through Argumentation: Black Pete as a Case Study . 243
Catarina Dutilh Novaes, Emily Sullivan, Thirza Lagewaard, Mark Alfano

The Pragma-dialectical View of Comparison Argumentation 261
Frans H. van Eemeren, Bart Garssen

An Indexical Characterization of Disagreement Based on Possible Worlds Semantics 275
Léa Farine

The Visual Rhetoric of Iconic Photographs as Topoi in Editorial Cartoons: an Argumentative Analysis 287
Eveline Feteris

Conceptual Analysis of an Argumentation: using Argumentation Schemes and the Toulmin Model 303
Marie Garnier, Patrick Saint-Dizier

Representing Second-order Arguments with Adpositional
   Argumentation (AdArg) 315
   Federico Gobbo, Jean H.M. Wagemans

Argumentation, Dissent, and Luck 327
   Job de Grefte

Argument Scheme Theory 341
   Hans V. Hansen

Resolution of Deep Disagreement: Not Simply Consensus 357
   Leah Henderson

Appeals to Popularity: Roles and Functions of 'Everyone knows X' .. 369
   Thierry Herman

Towards a Theory of Informal Argument Semantics 379
   Martin Hinton

Stereotyping 393
   David Hitchcock

Changing Norms of Argumentation 407
   Beth Innocenti

"If You Are A Scientist You Cannot Stop Such A Thing": Scientific Assent
   and Dissent in the Manhattan Project 419
   David Erland Isaksen

Formal Specifications for Dialogue Games in Multi-party Healthcare
   Coaching 433
   Mathilde Janier, Alison Pease, Mark Snaith, Dominic de Franco

How Courts Should Respond to the Stories Defendants Tell: A Bayesian
   Account of a Dutch Ruling 453
   Hylke Jellema

Regular papers

What Optimistic Responses to Deep Disagreement get Right (and Wrong)

SCOTT F. AIKIN
Vanderbilt University
scott.f.aikin@vanderbilt.edu

In this paper, I argue for three theses. First, that the problem of Deep Disagreement is usefully understood as an instance of the skeptical Problem of the Criterion. Second, there are structural similarities between proposed optimistic answers to deep disagreement and the problem of the criterion. Third, in light of these similarities, there are both good and bad consequences for proposed solutions to the problem of deep disagreement.

KEYWORDS: deep disagreement, scepticism, problem of the criterion

1. INTRODUCTION

The fact of disagreement is inescapable in social life. Who would deny it (and thereby disagree)? Some disagreements are tractable and even resolvable, but others are long-standing, wide-ranging, and seem irresolvable. The challenge of these disagreements of significant depth is that of identifying what rational procedure generally, and what tool of argument specifically, can break the logjam. Despite the fact that the special technical notion of *deep disagreements* has been around only since Robert Fogelin's 1985 essay, the form of the problem has been at work in the longstanding skeptical Problem of the Criterion since its statement in the late ancient period in Sextus Empiricus's *Outlines of Pyrrhonism*. I think it is instructive to view the problem of deep disagreement as an instance of the problem of the criterion, and so I will lay out the structural similarities between the two problems and turn then to show why this analogy is useful. In particular, the isomorphism between the two problems pays off when evaluating proposed solutions to the problems.

    The problem of the criterion admits of two broad classes of solution, aptly identified by Roderick Chisholm as *Particularism* and *Methodism*. Both forms of solution have their own appeal, but both suffer from the same structural epistemic problem: they beg the question, and so do not fully answer the challenge of the problem of the criterion. The same, as I see it, goes for optimistic replies to the problem of deep disagreement. (For the distinction between optimistic and pessimistic solutions, see Godden and Brenner 2010, and Aikin 2019). The optimistic solutions I will survey are Peter Davson-Galle's "primitive epistemic assent" (1992: 150) and Vesel Memedi's "third party" mediators (2007: 5). In both cases, there are appealing features to the optimistic replies, but here are significant drawbacks to them as replies, too. Particularly, that they, like their cousins purportedly solving the problem of the criterion, beg the question, given the description of what deep disagreements are.

2. THE PROBLEM OF THE CRITERION

The problem of the criterion is an ancient skeptical trope that can be posed with the following challenge: how can we sort true from false propositions, without presuming we have already sorted them appropriately? Here's how the challenge works, as the presumption is revealed with the following circle of reasons, or more evocatively termed by the Pyrrhonists, *the wheel* (*diallelus*):

> Q1: How do I know if these propositions (or presentations) are true (or false)?
> A1: Because they are assigned a value by a reliable or good criterion.
> Q2: How do I identify a good or reliable criterion?
> A2: A good or reliable criterion correctly sorts true from false propositions.

The problem is that in order to give A2, we must have answered Q1, which begins the series of questions and answers all over again. Hence, we see *the wheel* of critical reflection that constitutes the problem of the criterion. As Sextus Empiricus states the matter:

> [I]n order to decide the dispute which has arisen about the criterion, we must possess an accepted criterion by which we shall be able to judge the dispute; and in order to possess an accepted criterion, the dispute about the criterion must first be decided (PH II.20)

Roderick Chisholm, in his 1973 Aquinas Lecture, *The Problem of the Criterion*, comments on the history and structure of the problem:

> And now, you see, we are on the wheel. First we want to find out which are the good beliefs and which are the bad ones. To find this out, we have to have some way – some method – of deciding which ones are the good ones and which are the bad ones. But there are good and bad methods …. And so we now have a new problem: How are we to decide which are the good methods and which are the bad ones? (1974: 10).

The basic structure of the problem is that one must procedurally have two things before each other – one must possess the truths to identify criteria and criteria to possess truths. And without one or the other, one has neither. (See Amico 1993: 93, Cling 2014: 165, and Aikin forthcoming on the structure of the problem of the criterion being one of a conflict between what must be *epistemically prior*.)

Solutions to the problem of the criterion generally come in two forms – either one starts with a set of truths and generates the criteria for truth from what these truths share in common, or one begins with criteria and generates truths from their application. That is, given the two critical questions and two answers (Q1 and A1, or Q2 and A2), one could start with either complex. Chisholm identifies the set of questions as two pairs of questions:

> (A) "*What* do we know? What is the *extent* of our knowledge?"
> (B) "How are we to decide *whether* we know? What are the *criteria* for knowledge? (1973: 12)

The solutions, or better, *approaches*, to the Problem of the Criterion are to answer one of the question-complexes first, and then turn to answering the other in light of how one had answered the first. Chisholm dubs the two strategies 'Methodists,' who begin with questions of criteria, and 'Particularists,' who being with the question of instances.

> I suggest, for the moment, we use the expressions "Methodists" and "Particularists." By "Methodists," I mean… those who think they have an answer to B, and who then, in terms of it, work out their answer to A. And by "Particularists" I mean those who have it the other way around. (1973: 15)

Methodism comes in many forms in epistemology. Empiricism, for example, is a form of Methodism – experience provides us with truths, it is our criterion. Rationalism, too, is a form of Methodism, as the dictates of reason serve as the condition for accepting a content as true. Particularism, beginning with a set of truisms, is strongly associated with Common Sense traditions. One has a raft of commitments that have primarily the common thread that, to begin, is the simple fact that they are dictates of common sense – they are the kinds of things that strike us as silly or even absurd to doubt. One then designs the rules for reasoning and our criteria for knowledge around these obvious cases.

The trouble, of course, is that these strategies break the rule set forth with the initial statement of the challenge of identifying a criterion and its associated truths – we must do so without presuming that we already have an answer. We must do so without begging the question. If truths and criteria for truths are mutually epistemically prior to each other, then beginning with either will be epistemically out of order. Chisholm, in stating the options as he has (and even in stating his case for Particularism) concedes that:

> What few philosophers have had is the courage to recognize is this: *We can deal with the problem only by begging the question.* It seems to me that, if we do recognize this fact, as we should, then it is unseemly for us to pretend that it isn't so (1973: 37, emphasis added)

The lesson, as I take it, is that solutions to the problem of the criterion are less means of showing that the problem is not a problem but more ways of managing the problem. In the case of Methodism and Particularism, the answers to the challenge beg the question, and so are not solutions, given the structure of the problem, but they are ways forward for other programs of cognitive management, such as making one's beliefs more systematic and consistent. But, again, because of the problem of begging the question, they do not *solve* the problem of the criterion.

3. DEEP DISAGREEMENTS

A useful strategy for articulating what deep disagreements are is to start with a contrast. *Normal* disagreements are those wherein the two disagreeing parties nevertheless agree on some background matters – in particular, they agree on some broad set of propositions bearing on the item of disagreement and they share a number of salient epistemic resources. So, though two people may disagree, following Fogelin's famous examples, about the best path for their errands, if they agree about the geography and traffic patterns of the city, their disagreement is normal and so tractable. And if two people disagree about who was the batting champion in the baseball league for a particular year, if they agree that looking up the answer in the league's book of records will settle the matter, then they have a normal and tractable disagreement. In short, normal disagreements are those wherein the disagreeing parties share enough in common that an argument can, in principle, resolve the issue. For sure, it's still possible for a party to remain a holdout against the prevailing reasons, but this is a different kind of problem from deep disagreements.

When disagreements are deep, the parties do not share the relevant background commitments that allow simple reasoning to resolution. Nor do they share deference to the same epistemic resources. In cases of deep disagreement, the parties share no overlapping commitments or resources. Fogelin holds that, consequently, argument is not possible in these instances. He reasons that "the possibility of a genuine argumentative exchange depends … on the fact that we together share many things" (1985: 6). In turn, since those things are not shared in deep disagreements, the disagreements "cannot be resolved through the use of argument, for they undercut the conditions essential to arguing" (1985: 8).

The analogy between the problem of the criterion and that of deep disagreement should be clear. The background procedural rule for deep disagreement is what I have elsewhere termed the requirement of *dialecticality*, namely that a premise or resource must be acceptable to one's interlocutor for it to play a legitimately resolving role in an argument (Aikin 2018: 175 and forthcoming). It is analogous to the rule of support for epistemically prior commitments for the problem of the criterion – if one's justifying reasons depend on an unjustified commitment or step, then everything downstream from that unjustified component is undercut.

Given the structural similarity between the problem of the criterion and deep disagreements, the classes of proposed solutions to deep disagreements will have similar problems of question-begging that those to the problem of the criterion had. Methodist solutions to the problem of the criterion start with sources of or procedures for producing commitments. But, as we'd seen with how deep disagreements are described, there are no shared salient cognitive resources. So Methodist programs with deep disagreements cannot, given the structure of the problem, be solutions. And the same goes for Particularist solutions – finding sets of truisms from which to begin is a hopeless task for deep disagreements, since, by hypothesis, there are none in contexts of deep disagreement.

4. DAVSON-GALLE AND PRIMITIVE EPISTEMIC ASSENT

Peter Davson-Galle's 1992 essay, "Arguing, Arguments, and Deep Disagreements" has two theses. The first is that Fogelin's irresolvability thesis is defensible against Andrew Lugg's 1986 criticism that a practical perspective on argument resolves deep disagreements. Lugg's argument was that, in essence, the overlap of commitments required for resolving disagreements is itself a *product of* argument, instead of a *precondition for* it. The details of Davson-Galle's defense need not detain us here, since the target for evaluation is his second thesis, that "rational persuasion might be more powerful in other ways that Fogelin allows" (1992: 1954). So, though Fogelin's argumentative pessimism is defensible against one critical challenge from Lugg, it is open to a different form of optimistic challenge.

Davson-Galle pursues two lines of argument to substantiate the second thesis. The first is a negative line – that one can "rationally persuade someone of the error of his/her ways by tracing distasteful consequences …. One is, so to speak, pointing out an inconsistency in the set of propositions s/he subscribes to" (1992: 154). This negative line establishes that *something's gone wrong,* but not what the fix is.

The second line of argument Davson-Galle pursues is to note that there are instances of "primitive epistemic assent" around which arguments occur and from which they may begin. And further, they may be events that bring arguments to abrupt ends. Davson-Galle explains what these acts entail with a disagreement between a hypothetical *Jack* and *Jill*:

> What can Jack to? He might be able to create conditions for an act of primitive epistemic assent somehow; that is, create circumstances in which a proposition is warranted to/by Jill, but not in virtue of any argument or the provision of reasons (1992: 150)

The thought is that there are *other ways* we change our minds about things other than being on the receiving end of arguments from others. We may reflect, or feel, or have experiences. Any of these non-argumentative cognitive moments can produce a change in view and thereby a resolution. Davson-Galle provides an explanation for how this happens:

> It might be construed as akin to conversion … or it might be construed as akin to persuading someone to accept it's raining by opening a blind so that he can see that it is so. Either way … what one is doing is putting the other cognitive agent into a situation where a primitive epistemic assent occurs: assent is given without reasons for the assent (1992: 150)

The key, Davson-Galle holds, is that Fogelin's model for deep disagreements depends too much on *antecedent acceptances* for rational resolutions – one can use primitive epistemic resources to resolve disputes that need no background acceptances. (Along these lines, Wang

2018 argues that non-argumentative strategies are the only way forward at this stage.)

It is clear that Davson-Galle's solution to the problem of deep disagreement is a form of Methodism – that two methods may be used for resolution – finding internal contradictions and inducing primitive epistemic assents. Davson-Galle concedes that the former has significant limitations in that it, again, does not propose a truth on the other side of the inconsistency, and it also depends on one's interlocutor sharing some "canons of logic" (1992: 154). The latter, however, Davson-Galle seems to hold has limits only as far as what can be primitively epistemically presented.

First, it is worth noting that what Davson-Galle's proposal has done is significantly shrink the domain of what can count as deep disagreements. If, by hypothesis, the disputants do not share either salient commitments or cognitive resources in deep disagreements, then finding instances of resolutions consequent of discovered internal contradictions or primitive epistemic assents is also finding instances that are not deep disagreements. That does not mean that they are not resolutions, but it does mean that they are not resolutions to deep disagreements.

The crucial insight is that if the disagreements are *deep* in the way that Fogelin's program has stipulated, then it is not just a non-overlap of commitments but epistemic resources, too, for resolution. So internal *reductio* depends on shared resources for deriving the contradiction, and so also does primitive epistemic assent. Given the demands of deep disagreements, these resolutions, if successful in these fashions, demonstrate the disagreement is not deep. And if the disagreement is truly deep, these strategies will, given the dialiecticality requirement, beg the question.

5. MEMEDI AND THIRD-PARTY MEDIATION

Vesel Memedi argues in "Resolving Deep Disagrement" that at least some deep disagreemetns "can be rationally resolved by introducing the concept of 'third party' to those particular discourses" (2007: 1). Memedi's primary example is that of the conflict between Macedonian governmental forces and Albanian armed groups living in Macedonia in 2001. The narratives of the conflict's history and what the respective sides are trying to accomplish varied greatly between the two parties. The Madedonians accused the Albanians of trying to create, as Memedi reports, "a 'Greater Albanian' state," but the Albanians reported their motives only as to attain "greater rights for Albanians in Madedonia" (2007: 6). When the debate proceeded along these lines, Memedi holds, there was an incommensurability of reasons, and so an argumentative impasse.

The crucial element to Memdi's case study was a third party to the critical discussion. Memedi observes that the audience the two primary disagreeing sides appealed to was one outside the debate:

> [T]here is a presence of another audience that I think plays a crucial role in reconstructing better the discussion between Madedonian and Albanian language media. This role is played by 'the international community' (2007: 7)

The upshot is that the way to break the logjam of the deep disagreement, Memedi reasons, is to find a *judge* incorporated into the discourse. Memedi's two criteria for these third party judges are that (1) they must be "capable of being influenced," and (2) they must "have the capability to act as 'mediators of change,'" and in particular, this agency means that they are "more powerful" than the sides being mediated (2007: 9).

One question, of course, is how the fact that the mediator is more powerful than the singular parties is relevant to the *rational* resolution to the debate. That one party can convince a stronger party to take their side does not count as any more rational a resolution than that he party has stronger allies or that one party is itself simply stronger than the opposition. I presume that the issue of identifying stronger parties as mediators is an acknowledgement of the realism of disagreements – that rational resolution is by itself not sufficient for lasting resolutions, but some plausible threat of sanctions is required too. Regardless, the important element to Memdi's third party mediator is implicit, at least in the case study of the Albanian-Macedonian conflict, that both parties to the dispute argumentatively appeal to the same third party. That is, despite the apparent depth of the disagreement between the two parties, they agree on a mediating source of resolution.

It is not difficult to see Memedi's program as a form of deep disagreement Methodism – as Davson-Galle's program had been earlier – one that proposes a procedure for producing resolving reasons for the dispute. The problem, as seen before, is that if the sides *do agree* on the mediating party, then they share a cognitive resource, and so do not have a deep disagreement, properly described. If they are deeply disagreeing, then they will not have a mutually agreed upon mediator.

Seeing deep disagreement as an instance of the problem of the criterion is useful in this regard, since the mediator strategy is one of the methods the ancients had considered when addressing the problem of the criterion. Sextus Empiricus went out of his way to argue that it begs the question given the structure of the problem. "[S]ince there exists great difference among men, let the dogmatists first agree together that this is the particular man to whom we must attend, and then and only then, let them bid us also to yield him our assent. ... For if they declare that we must believe the sage, we shall ask them, 'what sage?' .... [T]hey will be unable to return us a unanimous answer" (PH II.38). And so, insofar as there is first-order deep disagreement about the facts of the case, there will be second-order disagreement about who is the right resource for accurate judgment. Again there may be agreement about who is the *strongest* and who can most effectively enforce a decision, but this is not the challenge of the criterion or of deep disagreement.

Memedi qualifies his solution to the problem of deep disagreement by noting that the third-party mediation strategy is highly contingent, so his conclusion is "modest" (2007: 10). His qualification is that his solution is indexed only to cases where there is a deep disagreement and the two parties address to a moveable and stronger third party. Only under those conditions is there hope for this kind of rational resolution to deep disagreements. The results of the case study are not universal for deep disagreement, but are restricted to these conditions. "There is no intention on my part to generalize these two criteria to other types of discourse," Memdi clarifies (2007: 9).

6. THE GOOD AND BAD NEWS

Let me start with the good news. What optimistic programs with deep disagreement, those on analogy with the problem of the criterion with formally Methodist approaches in particular, get right is that there are many ways that what look initially like deep disagreements are, in fact, not *absolutely deep* disagreements. To appreciate this point, a distinction is in order. *Depth* is a gradable concept, so disagreements may be of various depths, some more deep than others (see Duran 2016, Wang 2018, and Aikin 2019). Disagreements are deeper in terms of how many dialectical steps must be traversed to address the issue, and those that are absolutely deep have no upper limit on the steps necessary. So with some disagreements, only one argument is necessary, and with others, many back-and-forths are required. The former are not deep at all, and the latter are of degreed depth. But, again, given Fogelin's description of deep disagreement, there are others without any number of steps to get to resolution – any given argument will occasion more challenges. And so, with these, we are thrown upon the skeptic's wheel.

What deep disagreement optimism gets right, then, is that there are instances wherein we have disagreements of depth, but we may yet find new, unanticipated, epistemic and argumentative resources that contribute to the resolution of the dispute. In Davson-Galle's case,

11

'primitive epistemic assent' is a shared cognitive resource that can break the logjam of conflicting reasons. But, if we can do this and appeal to these sources of belief, it shows that though the disagreement may have depth, it is not absolutely deep. The same goes for Memedi's program of third-party mediation. The disagreement on the first level may be deep in that when the parties address each other, they cannot provide any reasons that meet the requirements of dialecticality. But if they happen to share regard for a third party to judge the dispute, they in turn have a shared cognitive resource. And so, though the dispute has a degree of depth, it is not absolutely deep.

Consequently, the good news is that optimistic programs with deep disagreement give us hope and methods for proceeding in the face of what often seems to be deep disagreement. The hope is that not all disagreements that are deep are absolutely deep, and the methods provided are those of revealing ways forward in cases wherein we think we face a deep disagreement.

The bad news is that these are not solutions to the problem of deep disagreement, so described. If the disagreements are deep, or as I've termed them for clarity's sake with the gradable concept of depth, *absolutely deep*, then these solutions will share the same problem that all Methodist solutions to the problem of the criterion have had – they will beg the question, and so are not really *solutions*.

What deep disagreement optimism of this particular Methodist form amounts to, then, is a program of showing that the domain of deep disagreement recedes when we apply our critical skills to the disagreements before us. Argument's reach is broader than the pessimistic view that many who read Fogelin take – disagreements may be of depth, but there are many ways that we may nevertheless get to the bottom of things in them. What deep disagreement optimisms, at least of the form evaluated here, get wrong, however, is that these are not solutions to the problem of deep disagreement. This is because conditions for optimism identified in them are conditions that simply don't obtain in instances of deep disagreement, properly described. The problem, as I take it, has a particular dialogical structure, and given that structure, the approaches proposed cannot be rationally satisfactory solutions. If the conditions do obtain, then the optimistic views are about disagreements that are not absolutely deep. Thereby, they are not solutions to the problem of deep disagreement, but more incremental restrictions of the domain for the problem.

In a way, this conclusion should be a happy one for both deep disagreement optimists and for deep disagreement pessimists. The pessimists are right that the problem, properly described, is rationally insoluble in ways analogous to the longstanding skeptical challenge of the problem of the criterion. But the result is also happy for the optimists, because we see the extension of the domain for deep disagreements shrink. What loomed as a worrisome global problem for argumentation is by piecemeal theoretical work, being progressively pushed back. And this occasions a question: *are there any really absolutely deep disagreements?* And if they are really only theoretical possibilities, not regular and inescapable social realities, does it matter?

REFERENCES

Aikin, S. (forthcoming). Deep Disagreement and the Problem of the Criterion. *Topoi.*
Aikin, S. (2019). Deep Disagreement, the Dark Enlightenment, and the Rhetoric of the Red Pill. *The Journal of Applied Philosophy,* 36(3), 420-435.
Aikin, S. (2018). Dialecticality and Deep Disagreement. *Symposion,* 5(2), 173-179.
Amico, R. (1993). *The Problem of the Criterion.* Lanham: Rowman and Littlefield.
Chisholm, R. (1973). *The Problem of the Criterion.* Milwaukee: Marquette University Press.
Cling, A. (2014.) The Epistemic Regress Problem, The Problem of the Criterion, and the Value of Reasons. *Metaphilosophy,* 45(2), 161-171.
Davson-Galle, P. (1992). Arguing, Arguments, and Deep Disagreements. *Informal Logic,* 14(2&3), 147-156.
Duran, C. (2016). Levels of Depth in Deep Disagreement. OSSA Conference Archive, OSSA 11. 109. https://scholar.uwindsor.ca/ossaarchive/OSSA11/papersandcommentaries/109/
Fogelin, R. (1985). The Logic of Deep Disagreements. *Informal Logic,* 7(1), 3-11.
Lugg, A. (1986). Deep Disagreement and Informal Logic: No Cause for Alarm. *Informal Logic,* 8(1), 47-51.
Memedi, V. (2007). Resolving Deep Disagreement. OSSA Conference Archive, OSSA 7. 108. https://scholar.uwindsor.ca/ossaarchive/OSSA7/papersandcommentaries/108/
Sextus Empiricus (1990). *Outlines of Pyrrhonism.* Trans. R.G. Bury. Amherst;: Prometheus Books. (Referenced as PH Book#, Page#).
Wang, C. (2018). Beyond Argument: A Hegelian Approach to Deep Disagreement. *Symposion,* 5(2), 181-195.
Zarefsky, D. (2010). The Appeal for Transcendence: A Possible Response to Cases of Deep Disagreement. In F. van Eemeren, B. Garssen, D. Godden, and G. Mitchell (eds.) *ISSA 2010 Proceedings.* http://rozenbergquarterly.com/issa-proceedings-2010-the-appeal-for-transcendence-a-possible-response-to-cases-of-deep-disagreement/

Is there a crisis of dissent and disagreement in argument theory?

MARK ALFINO
Department of Philosophy, Gonzaga University, USA
alfino@gonzaga.edu

Abstract: Is there a crisis of dissent and disagreement in argument theory? This paper explores this question by considering recent evidence from moral and political psychology, including theories such as social intuitionism, which suggest that our epistemic judgements in moral and political matters are persistently influenced by relatively fixed emotional responses and by stable aspects of personality. This suggests that some cases of political dissent appear to be argumentative when in fact they may be some form of *ad hominem* or objection to one's opponent's identity. This poses a challenge to our ordinary intuition (supported in social epistemology) that disagreements are in principle resolvable, and it suggests the need for new strategies for interpreting political disagreement and engaging in dissent.

KEY WORDS: argument theory, disagreement, social epistemology

1. INTRODUCTION

There is nothing new in claiming that there are limits to rational discourse, that rhetoric and persuasion, much less propaganda, can short-circuit reason, that reason is a mask for power, or that all viewpoints are relative to presuppositions. These challenges to reason span the history of Western philosophy from Plato's Gorgias to Wittgenstein's On Certainty and postmodernism. Without dismissing them, I would like to focus on a much more contemporary body of evidence and reflection on rational disagreement which, I believe, poses new and more fundamental challenges to our usual ways of thinking about and theorizing about argumentation, especially in the areas of moral and political discourse. Recent evidence from neuropsychology, moral psychology, and political science undermines key assumptions in argument theory and social epistemology.

It might be helpful to put the argument itself forward succinctly so we can see what might be at stake if this thesis is true. Three of the key Enlightenment assumptions about reason that contemporary research undermines are the idea that reason and our epistemic faculties are universal, that they can transcend personal or subjective identity, and that they are transparent to us. The transparency of reason can mean a variety of things, such as that we have accurate introspective access to our own mental states and processes, that we know when we are engaging in argument, and that we are not fundamentally deceived about the relationship of reasoning to doxastic commitment and behaviour.

But what if these assumptions are not true? What if we have enduring orientations to basic issues affecting moral and political life, orientations that are not just biases, but part of our identity, part of the way we process our social and moral experience? What if introspection does not clearly reveal the role of these orientations and the extent to which they can change? If so, then there may be a point, especially in strong and polemical dissent and political argumentation, in which we think we are having an argument with someone, but we are really making an unreasonable argument *ad hominem*. This would be an instance of what I call "argument illusion" and its prevalence and the difficulty of determining when it takes place together pose a serious challenge, if not crisis, for traditional epistemically oriented theories of argument, especially those which do not distinguish moral and political epistemology from other contexts.

Perhaps "crisis" talk is even more appropriate when we think about the effect of argument illusion on other Enlightenment ideas such as freedom of speech. On classical Millian grounds, for example, we ought to hold very positive attitudes toward polemical political speech, as well as political dissent. For Mill, optimism about extended disagreement and dissent is connected to a faith that free and open discourse will in fact promote the "livelier impression of truth," due to its collision with error. The modern commitment to freedom of speech and intellectual freedom is partly based on the assumption that ours and our opponent's views are ultimately comparable and not products of diverse moral matrices or orientations to basic social dilemmas. If contemporary evidence undermines this assumption, then it should also lead us to wonder about our faith that moral and political disagreement is always in principle resolvable, especially through free and open discussion. From here you should be able to see the edge of the cliff.

2. THE 'SPACE OF REASON' AND 'ARGUMENT ILLUSION'

In order to distinguish contemporary challenges to the power of reason from the more traditional ones, I will need to focus our attention on our tacit view of reason in argumentation, informed by the Enlightenment, and to review recent empirical evidence which, I think, poses challenges to this view of reason and to contemporary epistemology and argument theory.

Let's use the phrase "space of reason" to refer to views of reason that hold that rational discourse provides a general set of tools which can, in principle, transcend personal bias and prejudice, as well as the contingencies (personal, biographical, and sociological) that produce irrational commitments to beliefs. It does this in part by asserting a strong separation between reason and emotion. The "space of reason" is also a pragmatic orientation that speakers in a culture of free speech typically take toward each other. Speakers create the "space of reason" pragmatically, by signalling a willingness to hear and be open to influence by others' views, by mutual commitment to principles of logic and rules for the evaluation of evidence and inference, and for conducting argumentation. Typical among these rules is the prohibition against illicit appeals and arguments, such as varieties of ad hominem argumentation.

There are many places in Enlightenment thought to anchor this concept historically, but Mill's view of our encounter with truth and falsity in *On Liberty* makes a useful *locus classicus* for the concept because Mill, like many Enlightenment thinkers, saw the potential for reason, operating in a cultural ethos of liberty of thought and discussion, to critique unjust conditions such as the subjugation of women, and to promote other progressive ideals. Quoting another author in *On Liberty*, Mill rails against the "deep slumber of decided opinion" (49). More than others, he had a strong intuition that there could be a virtuous "social epistemology" to intellectual freedom. Mill's view also foregrounds the practical social and political stakes for our discussion. If we cannot enter a "space of reason" in discourse, then at least some of the justifications for protecting liberty of thought and discussion and for building a political theory of democracy based on the "open society" are called into question. I will return to this practical problem at the end of the paper, as we do have resources for addressing it. However, the focus here is on the implications of emerging views of reason for argument theory.

In contemporary social epistemology one finds a more precise version of this concept of a "space of reason" in the idea that there is, in principle, "no rational disagreement," at least among epistemic peers. Social epistemology emerges in the latter half of the 20th century in part as a response to debunking theories of reason from Kuhn, Foucault, Derrida, Lyotard, Latour and others who posed various challenges to the possibility that discourse can ultimately be oriented toward truth. Social epistemology has the virtue of moving beyond the traditional model of an isolated knower to consider how testimony and social practices can be rationally assessed and, more importantly, how doxastic attitudes can be revised and updated, perhaps in Bayesian fashion, to reduce or eliminate rational disagreement. While the literature of social epistemology also includes deep critiques of the claim that there is "no rational disagreement," it is safe to say that the dominant voices in the field have developed a more sophisticated model than classical epistemology of the possibilities for assessing the epistemic quality of social and group beliefs, as well as social institutions. Thus, social epistemology gives us a well-elaborated view of the space of reason and strong support for the idea that, when discourse among epistemic peers is properly constrained, epistemic relativism can be eliminated. As Feldman put it in 2007 in his "uniqueness thesis:" "This is the idea that a body of evidence justifies at most one proposition out of a competing set of propositions..." (148). If the uniqueness thesis is true, then it follows that epistemic peers facing disagreement (one believing p and the other believing ~p) must be open to revising their commitments to p or ~p. In principle, there is no scenario in which belief revision stops without agreement. The contingent circumstances of actual social disagreement do not count against this claim. They only signify that the process of inquiry has not reached its inevitable conclusion.

There is a counter-literature within social epistemology which claims that there can indeed be reasonable disagreement. Fogelin and others have followed this line in *Informal Logic*, drawing on Wittgenstein's view in On Certainty that raising doubt about some propositions depends upon some propositions not in doubt (Fogelin, 2005). Michael Hoffman, also writing in *Informal Logic*, makes a case for belief relativism based on the cognitive situation and belief system of the interlocutor (Hoffman 2005). In the literature of social epistemology in general, "nonconformists" argue that there are at least some circumstances in which it is rational to continue to believe p despite the fact that one's epistemic peer believes ~p. An excellent treatment along these lines comes from Christian Kock, "Norms of Legitimate Dissensus," who comes closest to my approach by focusing on the unique epistemic challenges of moral and political discourse (2007).

Social epistemology provides a very clear elaboration of the concept of a "space of reason" in which dissent and disagreement is always provisional, always a contingent fact about the short term rather than the long run. But belief in a "space of reason" can also be justified by our practical experience. We are well advised to be open to belief revision whether or not Feldman's uniqueness thesis is true. When you are in the presence of epistemic peers, much less epistemic superiors, and you disagree, you should reconsider your doxastic attitudes toward propositions you claim to be true. Epistemic virtues and vices can be explicated, in part, by how we engage or resist belief revision in the face of conflicting testimony and argumentation. Contemporary scientific accounts of reason do not undermine this pragmatic understanding, but do suggest that we are often deceived about the role of reasoning in guiding belief commitment and action. They also call into question strong claims in social epistemology, such as the belief that there is no rational disagreement.

Contemporary evidence about reason, dissent, and disagreement should also lead us to doubt other aspects of our experience of argumentation, such as the idea that we always know when we are engaged in argumentation. Nothing might seem more transparent to us than this. Just as I typically know when I am "looking" at something, I know when I am engaged in argumentation. But if we can have perceptual illusions, perhaps there is also "argument illusion." Simple cases of argument illusion might be discovered retrospectively, as when we thought we were experiencing argumentation and later found good evidence that that was not the case. We might overhear an argument in the next room only to look in and see that actors were rehearsing for a play. Just as you may learn that the person who swore at you really had Tourette's syndrome, you may learn something about your interlocutor that leads you to conclude that the "argument" you just thought you had with this person might not deserve the name. Maybe your interlocutor is a certain kind of conspiracy theorist, with a warp in their space of reason. Or maybe you missed the fact that he is suffering from a mental illness that distorts his use of reason. Maybe the illness is subclinical. Maybe the exchange was mediated by keyboards and your interlocutor was a computer program. Finally, consider deliberately deceptive interlocutors, who might just be playing with you. You might think you were having a robust argumentative exchange in these cases only to conclude, on reflection, that it was not really an argument.

Argument illusion occurs when we mistakenly believe that the normal conditions for argumentation are present. It is easy to notice in cases such as these, where there is a significant failure of ideal conditions, especially of sincerity or capacity, but what if these are the easy cases? What if we have enduring differences in how we view the challenges of social life, differences that are consequential for an argumentative exchange, but not really open to revision? Then, at some point in an argumentative discussion, you may have the illusion that you are advocating good reasons for belief change when in fact you are just expressing a real difference in orientation and approach, a difference that is much more like a personality trait than a commitment resulting from reasoning.

3. SOME RESEARCH

New evidence on the nature of reasoning comes to us over the last forty years from many disciplines, especially neuro-psychology, psychology, moral psychology, biology, and behavioural economics. These and other fields are contributing to a broad collection of relatively stable findings, but also competing theories that take a naturalistic approach to understanding reason. We find, for example, considerable evidence from cognitive psychology reminding us of Hume's model of reason and consciousness in which reason is, if not a slave of the passions, deeply entwined with emotional processing. Motivated reasoning is well-understood now and suggests that we are not always aware of the drivers of our argumentative behaviours. We are often acting and arguing to reduce cognitive dissonance. Damasio's famous study of impaired reasoning in patients with emotional dysfunction due to lesions suggests that, at least in cases of pragmatic reasoning about values and life planning, the space of reason metaphor is mistaken (Damasio, 2005). Effective reasoning about pragmatic matters is not corrupted by emotion, but supported by emotional inference processes. Emotional processing and emotional cognition are increasingly seen on a continuum of inferential processes that include self-conscious and reflective reasoning.

Evolutionary and naturalistic accounts of reason have built upon and produced results compatible with Mercier and Sperber's famous essay advocating an "argumentative theory of reason," which suggests that reason is originally a persuasive faculty which facilitates the exchange of reasons rather than a tool of inquiry (Mercier and Sperber, 2011). Our ability to use reason for inquiry might then be seen as a special case of reasoning. Reasoning about social and pragmatic issues may sometimes be improved by the kind of reasoning which abstracts from emotion, but it does not follow that social and pragmatic deliberation can or should isolate itself from engaging social intuitions.

Indeed, social intuitionists, such as Jonathan Haidt, suggest that, at least in matters related to personal and political morality, self-conscious reasoning is often the "tail that wags the dog" after many relatively automatic processes provide us inferences about the matter at hand and establish credibility and affiliation among social agents (Haidt, 2011). While Haidt holds to a different distinction between reason and intuition than the argumentative theory, both theories suggest reason evolved more to be our "inner lawyer" than our personal private detective. For both theories, reason plays a much more *ex post facto* role in our moral and social life than we typically believe.

The other side of Haidt's theoretical project, which he developed in collaboration with many other researchers, is "moral foundations theory" (MFT). MFT is based in empirical surveys using questionnaires that purport to elicit our baseline approaches to five or six fundamental problems of social life for partially social creatures such as we are. These include at least Care, Fairness, Loyalty, Authority, and Sanctity, but maybe Liberty as well. They map onto both original and current problems of social life such as care of children and the vulnerable, commitments to coalitions, acceptance of hierarchy, etc. For example, an original trigger for the "care" foundation would be neglect of a child, whereas a current trigger, for some, might include animal suffering or care of the environment. The important result of this research for our purposes is that the questionnaires do not contain reference to political views or matters yet they appear to accurately predict political orientation and strength of orientation. In combination with a wide range of other evidence in social psychology and behavioural economics, this empirical research suggests that there is a distinctive pattern to the way we are "triggered" by situations involving fundamental social dilemmas. We are not determined by this pattern, but we are somewhat blind to it in social interactions.

Research on cognitive bias and implicit associations adds a second wave of evidence to challenge the Enlightenment model of a pure "space of reason." Contemporary neuropsychological models of bias suggest that mental life is continually engaged in expressing and overriding automatic inferences that may represent biases. With computer-based research these latencies and responses are measurable. The Implicit Association Test purports to measure these response differences for a wide range of biases such as ethnicity, gender, age, and weight. Research projects such as those surrounding the Implicit Association Test have their critics. We may not have a validated theory about how biases get established or how they relate to biologically instantiated traits. The IAT, however, is just one research program among many which document our "cognitive opacity". Our access to mental processes through introspection is much more limited than we believe, and we often engage in confabulation in accounting for our mental life (Nisbett and Wilson, 1977; Wilson, 2002). As Mercier and Sperber argue in their recent book, we are often deceived about whether the reasons we provide to explain our actions actually guided those actions or were fabricated after the fact (Mercier and Sperber, 2018). If we cannot reliably know about or control the pre-conscious and emotional processes which affect our reasoning, then it is hard to see how the space of reason can transcend these influences.

A third wave of evidence comes from recent work in political science, especially from Hibbing et. al, *Predisposed: Liberals, Conservatives and the Biology of Political Difference*. Like Jonathan Haidt's research connecting political orientation to relatively stable moral foundations such as care, fairness, loyalty, etc., Hibbing and others have shown persuasively that people have relatively stable predispositions that predict their responses to "bedrock social dilemmas" about how "society works best" (44). These include intuitions about the need for shared social values, treatment of outsiders, authority, leadership, etc. Like Haidt, Hibbing et. al. have been able to predict political orientation from questionnaires involving non-political questions. Their account builds on existing literature about authoritarian personalities, but develops a more sophisticated account of the relationship between the relatively invariable problems of human social life and the variability of both our nature and the actual political positions and opinions we hold over time. When this research is seen in relationship to a larger body of research on the biological aspects of personality, the picture that emerges is clear: "Predispositions...can be thought of as biologically and psychologically instantiated defaults that, absent new information or conscious overriding, govern responses to given stimuli" (24).

4. INTERPRETATION

This research takes us beyond simple reductive or deterministic models of behaviour, but also limits our ability to assess the epistemic predispositions that we do have. Our personalities have biologically instantiated traits, but experience and our interpretations of our experience also shape the character and expression of our social and political orientations. We can become aware of some of our predispositions, but they are also part of the way we see and process our experience. There is no question of purging ourselves from the cognitive-emotive intuitions that help us think about moral and political matters.

We can certainly still learn to see the world the way people different from us do, and sometimes this leads us to change our beliefs or meta-cognitive practices, but there is no basis for a cognitive evaluation of the predispositions that shape our moral matrices or our orientations to "bedrock social dilemmas," except perhaps at the extremes. An extreme liberal concern for harm or an extreme conservative trigger for threat detection can be evaluated as dysfunctional, just as an obsessive-compulsive disorder might be, but, to play on Feldman's uniqueness thesis, there is no unique doxastic attitude that follows from this evaluation. There is no epistemically privileged or justified set of predispositions for approaching bedrock social dilemmas. The predispositions that inform our persistent orientations toward moral and political life are a "population phenomenon." Like height and eye colour, they vary predictably in any population of humans. The idea that our judgements of moral and political matters are pervasively influenced by relatively fixed commitments and biologically instantiated traits undermines the uniqueness thesis.

Two implications and a paradox follow from this evidence: First, there will always be some beliefs about moral and political matters that cannot be evaluated epistemically since they will be the result of an indeterminate mix of predisposition and cognition. Pragmatically, it means that "argument illusion" is a persistent feature of moral and political argumentation. There is, then, a basis for rational disagreement, but not because the positions at odds are both known to be rational, but because neither can be known to be rational. Second, we can never be sure that we are not committing the fallacy of *ad hominem*. At some point in a political discussion you may indeed find yourself arguing that your interlocutor should not see or process their experience in the way that they do, but your warrant for this claim will be, in some cases, simply that you see the world the way that you do. In other words, you will be implying that your opponent should not be the person that they are.

The space of reason is less transparent and less criterion-based than it formerly appeared. We mistake our biases for reasoned positions, we reason *ad hoc* and often to reduce dissonance, and we are largely unaware of many of the extrinsic and irrelevant things that influence our judgements, like hunger and odours. Even though we know that our views are shaped by a consistent and relatively fixed set of orientations, we experience them as the result of careful reflection and weighing of reasons and evidence. This gives rise to a paradox in our experience of moral and political argumentation. On the one hand, we experience our moral and political commitments as epistemic products, the result of truth seeking behaviour and practice. When we advocate our views argumentatively, we intend others to take our reasoning and evidence as a basis for belief change. On the other hand, the picture of reasoning emerging from contemporary research suggests that every population of humans has a distribution of diverse but largely overlapping perspectives for understanding and making inferences about moral and social life. Personal experience supports the view from research that we largely encounter people with consistent moral and political outlooks. Paradoxically, we experience political discourse and dissent as though this were not true, as though they might change to our orientation, even though we have good evidence (both third person and personal) that dissent is often the product of persistent orientations and automatic inferences retrospectively rationalized.

In spite of this, we are not completely without criteria for assessing the reasonableness of someone's moral matrix. In some cases, maybe with highly partisan brains, you might be able to show, like a good cognitive behavioural therapist, that someone's belief about a political issue is at odds with other beliefs the person holds or is demonstrably dysfunctional. So, for example, a highly partisan conservative may have an extreme trigger for threat detection which leads him to favour an approach to immigration or health care that entails grave harm to innocent people. Assuming he also holds some form of a non-harm principle, there may be a process of rational persuasion that leads to a moderation of views. Likewise, highly partisan liberals are often dumbfounded when pushed on how to control immigration. Their harm avoidance intuitions often colour their assessment of threats from immigration. Sometimes noticing that we are dumbfounded produces belief change. So political discourse might still play a robust role in helping us think about how to connect our moral matrices to specific and changing social and political policies. But the result of this process is often a refinement of this connection, rather than a change of orientation. As Hibbing, et al. point out, when conservatives in the US stopped being isolationists after Pearl Harbor, they did not stop being conservatives.

The evidence emerging from sciences that have been studying reason and reasoning does not support the claim that all argumentation on moral and political matters involves argument illusion, but rather a weaker sceptical conclusion: We cannot know with certainty when we are arguing against an individual's default modes of processing bedrock social dilemmas, but we can be relatively certain that this will occur in the ordinary course of moral and political discussion. Likewise, we can never be sure that our arguments are legitimate proposals for belief change versus illegitimate demands that our interlocutor become a different sort of person.

I think it follows directly that this poses a crisis for traditional theories of dissent and disagreement about political and moral issues. Such theories suppose that the space of reason is equally transparent and open to argumentative and epistemic processes no matter what the topic is. It may be true that a researchers' personality is irrelevant to the assessment of the effectiveness of a drug in a clinical trial, yet it remains relevant in discussing whether to socialize health care or close the borders.

The hard problem here is that there is no obvious way to account for these biases of orientation epistemically, at least not once we have excluded the extremes. You might try to treat the problem analogously to problems of measurement and estimation. Maybe we can just throw out the extremely partisan views as outliers and then "average" the rest. That probably does work in some cases, but if there really is an existential threat to a socio-political group from an external foe, such averaging might be catastrophic. Further, some political and moral decisions simply do not offer compromise solutions. You either make buildings accessible to the disabled or you do not. You either get the lead out of gas or you do not. It's not clear, for example, how to accommodate social conservative intuitions about abortion or gay marriage and still do justice to liberal (and libertarian conservative) claims about the harms of constraining liberty of action.

Even if we cannot always treat the crisis as a problem of measurement or estimation, there may be other solutions. Knowing about the biological basis of our socio-political orientation may not change the way we look at issues, but it may change the way we look at each other, especially during polemical and heated discussions. We may need to cultivate new sensitivities and discussion virtues to account for the possibility that the *ad hominem* is endemic to moral and political discourse. At the same time, we may need to qualify the otherwise good pragmatic advice from social epistemologists that we ought to always to update our doxastic attitudes in light of contrary testimony from epistemic peers. In some cases, we need to conclude that an apparently argumentative discussion is really an illusion and that the paradox of moral experience is leading us to mistake something about our identity for an epistemic appeal. Like Mill, the social epistemologist might encourage us to keep arguing, but a new kind of argument theory might also tell us when to stop arguing at signs of argument illusion. We might then approach some impasses in social and political life more like negotiations than epistemic inquiries, more like an accommodation than a truth-seeking process.

5. CONCLUSION

In conclusion, new models of reason and reasoning challenge some of the assumptions behind Enlightenment theories of reason, theories which also underlie beliefs about the importance of free speech and democracy as institutions supporting a social epistemology oriented toward truth. Recent challenges to freedom of speech on US campuses tellingly rely on claims about threats to identity and emotional safety which, even if exaggerated or used strategically, may not be easily dismissed in light of this evidence. Likewise, new forms of manipulation of thought through propaganda and social media may derive their effectiveness from facts about our persistent biases and vulnerabilities. Mill and others thought that the right response to dissent and discord in the public square was to rededicate ourselves to liberty of thought and discussion. The idea that the solution to distorted and manipulative speech is more speech is harder to maintain in light of these challenges to Enlightenment optimism about the space of reason.

REFERENCES

Damasio, Antonio. (2005). Descartes' error. New York: Penguin.
Feldman, R. 2007. Reasonable religious disagreements. In L. Antony (Ed.), *Philosophers without gods*, (194-214). Oxford: Oxford UP.
Feldman, R. and T. Warfield, Eds. (2010). Disagreement. Oxford: Ox-ford UP.
Fogelin, R. J. (2005). The logic of deep disagreements. *Informal Logic* 25(1), 3-11.
Haidt, J. (2001). The emotional dog and its rational tail: A social intui-tionist approach to moral judgement. *Psychological Review*, 108(4), 814-834.
Haidt, J., Ed. (2012). *The righteous mind: Why good people are divided by politics and religion.* New York, Pantheon.

Hibbing, J. R., et al. (2013). *Predisposed: Liberals, conservatives, and the biology of political differences*. Routledge.
Hoffman, M. (2005). Limits of truth: Exploring epistemological ap-proaches to argumentation. *Informal logic,* 25(3), 245-260.
Kock, C. (2007). Norms of legitimate dissensus. *Informal Logic,* 27(2), 179-196.
Mercier, H. and Dan Sperber. (2011). Why do humans reason? Argu-ments for an argumentative theory. *Brain and behavioral sciences,* 32, 57-111.
———. (2019). *The enigma of reason*. Harvard UP.
Mill, J. S. (1991). *On liberty and other essays*. Oxford UP.
Nisbett, R. E., & Wilson, T. D. (1977). Telling more than we can know: Verbal reports on mental processes. *Psychological review*, 84(3), 231-259.
Smith, K. B., et al. (2011). Linking genetics and political attitudes: Re-conceptualizing political ideology. *Political psychology*, 32(3), 369-397.
Wilson, T. D. (2002). Strangers to ourselves. Cambridge, Massachusetts, Belknap/Harvard UP.

Epistemic Tolerance

DOMINIK BALG
Department of Philosophy, University of Cologne, Germany
dominik.balg@posteo.de

> When it comes to political, religious or ethical issues, many people consider a tolerant "live and let live"-attitude to be the best reaction to disagreement. However, the current debate about the epistemic significance of disagreement within social epistemology gave rise to certain worries about the epistemic rationality of tolerance. Setting aside those already extensively discussed worries, I would like to focus on the instrumental rationality of a tolerant attitude with respect to our epistemic goals.
>
> KEYWORDS: disagreement, fallibility, humility, open-mindedness, permissivism, rationality, relativism, tolerance, toleration

1. INTRODUCTION

In this paper, I will discuss a certain 'live and let live'-attitude towards recognized disagreement that might be called *epistemic tolerance*. What does it mean to react to a disagreement in an epistemically tolerant way? Tolerant people *agree to disagree* and respect each other's opinion as equally reasonable. They stick to their guns, but they don't impose their opinions on others. To put it a little more formally, a person who displays a tolerant attitude towards a recognized disagreement (i) evaluates the other person's belief as false, (ii) evaluates both her own and the other person's belief as equally reasonable, (iii) retains her own belief, and (iv) refrains from any attempt to modify the other person's belief.

Especially when it comes to political, religious, moral or scientific disputes, many people consider an epistemically tolerant attitude the best way to go. Take for example the following passage from Richard Feldman, where he recalls a situation in one of his classes:

> A few years ago I co-taught a course on 'Rationality, Relativism, and Religion' [...]. Many of the students [...] displayed a pleasantly tolerant attitude. Although [...] [they]

> disagreed with one another about many religious issues, almost all the students had a great deal of respect for the views of the others. They 'agreed to disagree' and concluded that 'reasonable people can disagree' about the issues under discussion. (Feldman 2007, p. 194)

Later on, Feldman describes his student's attitude in a little more detail:

> Thinking someone else has a false belief is consistent with having any of a number of other favorable attitudes toward that person and that belief. You can think that the person is reasonable, even if mistaken. And this seems to be what my students thought: while they had their own beliefs, the others had reasonable beliefs as well. I think that the attitude that my students displayed is widespread. It is not unusual for a public discussion of a controversial issue to end with the parties to the dispute agreeing that this is a topic about which reasonable people can disagree. (Feldman 2007, p. 200)

The attitude that Feldman ascribes to his students seems to be exactly the attitude that is picked out by my initial characterization of epistemic tolerance. Furthermore, I agree with Feldman that this attitude is widespread. One reason for the prima facie attractiveness of epistemic tolerance is that it seems to avoid both skepticism and dogmatism. It allows us to stick to our guns, while also leaving room for respecting conflicting opinions as equally valuable. Given its remarkable popularity, a critical assessment of a tolerant reaction to recognized disagreement is directly relevant to our epistemic practice.

In what follows, I will argue that tolerance cannot be an epistemically adequate reaction towards a recognized disagreement. In section 2, I will argue that although there are some legitimate worries about the epistemic rationality of a tolerant attitude, there will be many situations where reacting tolerantly towards a recognized disagreement is epistemically rational. In section 3, I will argue that there is nevertheless a fundamental argument to be made against a tolerant attitude. More specifically, I will argue that in order to come to a complete epistemic assessment of a tolerant reaction towards disagreement, considerations concerning its instrumental rationality with respect to our epistemic goals need to be taken into account. Once the dimension of instrumental rationality enters the picture, it becomes clear why epistemic tolerance cannot be an epistemically adequate reaction towards disagreement.

2. THE EPISTEMIC RATIONALITY OF TOLERANCE

In this section, I will argue that at least from the perspective of epistemic rationality, a tolerant reaction towards disagreement is not necessarily problematic. This claim has bite, for although many people find an epistemically tolerant attitude initially attractive, the recent debate within social epistemology about the rational reaction to recognized disagreement has raised some fundamental worries about the epistemic rationality of a tolerant stance. For example, one worry is that evaluating one's own belief and another person's conflicting belief as equally reasonable is only rational against the background of relativistic theories of epistemic justification or permissivistic accounts of evidential support relations (Feldman 2007). Another worry is that retaining one's own belief in the face of disagreement is epistemically irrational. On the one hand, many authors have argued that sticking to one's guns is epistemically irrational in disagreement situations where one respects the other person as one's epistemic peer (see Christensen 2009 for an overview) or as an epistemic authority (Constantin and Grundmann 2018; Zagzebski 2012), but also in situations where one is unsure about the other person's epistemic status (Hallsson and Kappel 2018; King 2012; McGrath 2009) and even in some situations where one considers oneself as epistemically superior (Priest 2016). On the other hand, an epistemically tolerant attitude might be irrational because of an *intrinsic rational tension between its components*. More specifically, the worry is that it is epistemically irrational to retain one's own belief while at the same time evaluating another person's conflicting belief as equally reasonable (Feldman 2007).

While surely important and interesting, I don't think that these worries suffice to establish a fundamental argument against the epistemic adequacy of a tolerant attitude. In fact, it is pretty easy to come up with cases where evaluating both one's own belief and another person's conflicting belief as equally reasonable while retaining one's own belief is perfectly rational. Consider for example the following case: Lea and Nick are two detectives investigating a murder. They have been working together for many years and respect each other as reliable and competent colleagues. After carefully evaluating the evidence, they both suspend judgement on who the killer is. Then they receive a phone call from the victim's butler, who claims that the gardener did it. Having no special reason to distrust the butler, Lea forms the belief that the gardener did it. Nick, however, knows that the butler is a notorious liar and remains agnostic. At the same time, he knows that Lea isn't aware of the butler's tendency to lie and thus considers her belief as reasonable. Because he respects Lea's belief as reasonable, he refrains from modifying it and decides not to tell her about the butler's unreliability.

In this case, it should be clear that Nick is perfectly rational in retaining his own belief in face of the disagreement with Lea. Nevertheless, it also seems perfectly rational for Nick to evaluate Lea's belief as reasonable. Moreover, there is some substantial sense in which Nick's and Lea's beliefs are epistemically on a par: For example, they are both formed on the basis of a thorough and careful evaluation of the available evidence, and both are caused by reliable belief-forming processes. Given that, it even seems to be rational for Nick to evaluate Lea's and his own belief in some substantive respects as *equally* reasonable. If this is correct, then at least from the perspective of epistemic rationality, there will be no fundamental argument to be made against a tolerant attitude towards disagreement. Retaining one's own belief while respecting another person's conflicting belief as equally reasonable is not necessarily irrational.

However, this line of thought might lead to the following worry: Even if we accept that Nick's reaction is epistemically rational, his positive epistemic evaluation of Lea's belief seems to be too weak to plausibly constitute the proper basis for a genuinely tolerant attitude. Although the idea that a tolerant attitude is essentially based upon an ambivalent normative evaluation is widely accepted, it should also be clear that not *every* ambivalent normative evaluation rationalizes a tolerant attitude (Forst 2013, 2017; King 1998). And while Nick's reaction is based on *some* ambivalent epistemic evaluation, he is obviously only rational in retaining his original belief due to a *significant evidential asymmetry* between him and Lea. So the worry is that as long as my conception of epistemic tolerance allows for cases in which the positive epistemic evaluation of a conflicting belief is harmless enough to not render retaining one's own belief epistemically irrational, it will be too weak to only capture cases where it is natural to speak of tolerance, and the cases that philosophers are primarily interested in when they talk about tolerance as a specific intellectual attitude towards persistent disagreements over moral, political or religious questions.

While it is true that my conception allows for cases with some epistemic asymmetries, I don't see why this should be a problematic feature. In fact, it seems independently plausible that a tolerant attitude is compatible with significant asymmetries - several authors within practical philosophy have explicitly argued that at least for some instances of tolerance, such an asymmetry is even constitutive.[1] And the

---

[1] For example, in his essay *Answering the Question: What Is Enlightenment?*, Immanuel Kant already called tolerance a "presumptuous title" (Kant 1991, p. 58). Taking up Kant's criticism, Rainer Forst has developed a much discussed "permission conception" of tolerance, according to which toleration consists in a unilateral relation between an authority and an inferior party (Forst 2013, 2017). Forst is convinced that the permission conception of toleration is not

same seems to be true for the theoretical domain. Consider for example demands for tolerance with respect to scientific communities. Many philosophers of science believe that scientific communities would benefit from a tolerant behavior of their individual members. However, the idea is not that scientists should be tolerant towards conflicting theories that are equally well supported by the available evidence. The idea is rather that scientists should be tolerant towards theories that are *only weakly supported* by the available evidence – because they could easily turn out to be better than expected (Chang 2012, Šešelja et al. 2015).

Given that, the sole fact that there is a significant epistemic asymmetry in cases like the one of Nick and Lea doesn't necessarily speak against them being genuine instances of a tolerant attitude. Nevertheless, it is right that those are not the cases that philosophers are primarily interested in when they think of tolerance as an intellectual attitude towards conflicting opinions. I think that at this point, it will be helpful to distinguish between *appropriate* and *inappropriate* instances of tolerance. While Nick's attitude towards Lea's belief arguably is a genuine instance of tolerance, it is clearly inadequate. At the same time, it is also clear that Nick's attitude is not epistemically irrational. So why is it so problematic? How can we distinguish appropriate from inappropriate instances of epistemic tolerance? And are there even any appropriate instances of epistemic tolerance? To answer these questions, a critical discussion of epistemic tolerance has to go beyond the assessment of its epistemic rationality.

3. THE INSTRUMENTAL RATIONALITY OF TOLERANCE

Given the initial characterization of epistemic tolerance, it shouldn't come as a surprise that evaluating the epistemic rationality of a tolerant attitude won't suffice to come to a complete assessment of its adequacy. Reacting tolerantly towards a recognized disagreement doesn't just mean to stick to one's guns – it also means to refrain from any attempt to modify the other person's belief on the basis of a specific epistemic evaluation of that belief. A tolerant reaction doesn't just consist in a specific doxastic response to recognized disagreement, but also in a certain *behavior towards conflicting beliefs of others*. Given that, a critical discussion of an epistemically tolerant attitude needs to take into

---

just based on a rich philosophical tradition, but still informs our understanding of the term to a considerable extent. I think that this diagnosis is correct. For example, only against the background of a conception of toleration that allows for significant asymmetries, it gets clear why many sexual, religious or political minorities take offense at being tolerated.

account considerations concerning its instrumental rationality with respect to epistemic goals.²

The idea behind demands for epistemic tolerance seems to be that from an epistemic point of view there are circumstances under which it is advisable to refrain from modifying other people's beliefs, although we consider them to be false. More specifically, the idea is that we shouldn't try to modify conflicting beliefs of others if those beliefs are reasonable. This suggests that in cases where the conflicting beliefs of others are unreasonable, we should try to modify them. Tolerating a conflicting belief doesn't just mean to refrain from modifying that belief *and* to evaluate it as reasonable – it means to refrain from modifying that belief *because* it is reasonable. This specific relation between tolerating something and interfering with it is no peculiarity of an epistemically tolerant attitude, but a structural feature of toleration in general. To see this, it is helpful to consider a case of practical tolerance. Suppose Sam is a sexist journalist who deeply hates all women. Determined to convince as many people as possible of the legitimacy of male supremacy, he frequently publishes articles in which he argues for his androcentric world view. However, none of his readers find his arguments convincing. Quite the contrary, his offensive articles lead many people to reflect on their sexist prejudices and to actively engage in feminist advocacy.

In this situation, it might be plausible to argue that tolerating Sam's behavior could be morally adequate.³ However, it should be clear that tolerating Sam's behavior would only be a morally adequate reaction, if in general it was adequate to interfere with sexist behavior. Tolerating Sam's behavior doesn't just mean to refrain from interfering with it and to appreciate its desirable consequences – it means to refrain from interfering with it because it has desirable consequences. So one core idea behind a tolerant attitude is that there are specific positive evaluations of other persons' objectionable beliefs or actions that make it rational to refrain from interfering with them (Forst 2013,

---

² The idea that considerations concerning instrumental rationality with respect to epistemic goals need to be taken into account in order to come to a complete assessment of our intellectual conduct is not a new one. For example, Thomas Kelly has argued that theoretical rationality is a 'hybrid' virtue that involves sensitivity to both epistemic and instrumental reasons (Kelly 2003). To use Kelly's terminology, we can say that a tolerant attitude towards disagreement will only be theoretically rational if it is both epistemically rational and instrumentally rational with respect to our epistemic goals.

³ Whether tolerance really would be an appropriate reaction in this case is, of course, a difficult ethical question. All I am assuming here is that it is at least not absurd or obviously misguided to consider tolerating Sam's behavior as potentially appropriate.

ch. 1). And this idea seems to imply that in general we should try to interfere with other persons objectionable beliefs and actions.

Accordingly, to establish the instrumental rationality of epistemic tolerance, two claims need to be defended: The first claim is that in general, it is instrumentally rational with respect to our epistemic goals to modify conflicting beliefs of others. The second claim is that it is irrational with respect to our epistemic goals to modify conflicting beliefs of others that are reasonable.[4] In this section, I will argue that while the first claim is plausible to at least some degree, the second claim is clearly wrong.

Let's begin with the first claim. Why should it in general be instrumentally rational with respect to our epistemic goals to modify conflicting beliefs of others? At first glance, there seems to be a straightforward explanation. If minimizing falsehood in a large body of beliefs is a core feature of our epistemic goals, it will be instrumentally rational with respect to these goals to modify any belief we consider to be false – regardless of whether or not it is part of our own belief system or not.

However, this line of thought will be highly controversial and is likely to be rejected by most epistemologists. The way that it is usually interpreted, the goal of getting at the truth and avoiding falsehood is an individualistic goal - instead of interfering with other people's beliefs, we should try to maximize truth and minimize error within our own belief system. Maybe a proponent of epistemic tolerance could try to shift the burden of proof here and argue that although most epistemologists actually accept that there is an epistemically important difference between our own false beliefs and false beliefs of others, it is not clear at all why there should be such a difference. And as long as there are no convincing arguments for a fundamental distinction

---

[4] At this point, a little more needs to be said about exactly what it means to modify 'conflicting beliefs of others'. One natural worry is that as long as trying to modify other people's beliefs only means to engage in argumentative exchange, it is hard to see how this can be a problematic activity at all. However, what I have in mind when I talk about the modification of other people's beliefs is rather an intentional attempt to get other people to believe certain things. Engaging in argumentative exchange and presenting reasons for one's own view is certainly one possible means to make other people believe something, but of course there will also be other ways of doing this. Furthermore, many people frequently engage in argumentative exchange without thereby intending to convince others of their own position – for example, to simply gather new evidence or to come to a better understanding of conflicting standpoints.
Given this, it is also clear that a tolerant attitude isn't necessarily incompatible with engaging in argumentative exchange – at least as long as one doesn't do so with the intention to modify the other person's belief.

between the epistemic significance of falsity within our own belief system and falsity in other people's belief systems, so the idea, a social interpretation is the most natural way to understand our epistemic goals. However, the plausibility of such a move is at least questionable. It seems that as long as a core part of the argument for the instrumental rationality of epistemic tolerance rests on such a controversial assumption as a social interpretation of epistemic goals, the prospects of success look pretty poor.

Luckily, there is an alternative route the proponent of a tolerant stance could choose. Even if our epistemic goals are purely egocentric, to the effect that every person should only aim at the truth and avoid falsehood with respect to her own belief system, there is still some room to argue that in general we have good reasons to modify conflicting beliefs of others. To see why, one just needs to consider the social aspects of our epistemic reality. In forming beliefs about the world, every person heavily relies on the beliefs of others. Given our fundamental epistemic dependency on other people, it is clear why we should care about our epistemic environment. By improving the quality of our epistemic environment, we can effectively increase the chance of forming more true than false beliefs in the future (Werning 2009). And modifying false beliefs of others, so the idea, is one obvious way to improve the epistemic quality of our environment - since the beliefs of other people are part of countless reasoning and communication processes that are impossible to track, even beliefs we know to be false can have epistemically infectious effects that can easily be harmful to our own belief system.

Of course, there are many situations in which I know that another person's false belief won't have any negative impact on my own belief system. But given the high degree of communicative interaction in our globalized world, one could argue, those will be cases where I have *positive reasons* to believe that the other person's belief is epistemically harmless. In the absence of such reasons, it seems that the prima facie rational thing to do is still to try to modify the other person's conflicting belief. A similar point can be made with respect to cases where we don't have the time or the cognitive resources to modify conflicting beliefs of others. Also in those cases, one could argue, the falsity of the other person's belief still constitutes a prima facie reason for modifying it that is only outweighed by specific pragmatic considerations. To defend the instrumental rationality of a tolerant attitude, one doesn't have to argue that we should try to modify *all* conflicting beliefs that are not reasonable – all that needs to be established is that we have prima facie reasons to modify conflicting beliefs of others. And as we have seen, even against the background of an individualistic interpretation of our epistemic goals there are still some interesting arguments for the proponent of a tolerant attitude to support this claim. Although there is a lot more to say about those arguments, and although the above considerations do in no way suffice to establish this claim, I will accept it in the following for the sake of argument.

Suppose that it is in general instrumentally rational with respect to our epistemic goals to modify the conflicting beliefs of others - why should it be irrational in cases where the other person's belief is reasonable? The underlying idea behind this second claim seems to be that there is an intimate connection between tolerance and fallibility. This idea has some philosophical tradition. In 'On Liberty', John Stuart Mill already argued that one main reason for tolerating conflicting beliefs is that – given our fundamental fallibility - they could always turn out to be true (Mill 2001, p. 19).[5] So the thought behind demands for epistemic tolerance is that although it is generally rational to modify false beliefs of others, we should at least refrain from modifying those conflicting beliefs of others we consider to be reasonable - modifying reasonable beliefs of others just because they conflict with our own fallible opinion would be a form of intellectual hubris.

How convincing is this line of thought? On the one hand, it seems plausible that - given our fundamental fallibility - we should be very careful in our attempts to modify conflicting beliefs of others. Just because we believe that someone has a false belief, that doesn't mean we are in fact right. Especially under hostile epistemic conditions, intellectual virtues like humility and open-mindedness are indispensable for an epistemically responsible way of dealing with conflicting opinions of others. So instead of prematurely dismissing differing beliefs, we should try to take them seriously and make up our mind as impartially as possible.

On the other hand, this does in no way mean that we should react tolerantly towards disagreement. A tolerant person refrains from modifying those beliefs of others she considers to be reasonable. But as the considerations from section 2 suggest, a tolerant attitude is only epistemically rational in cases with significant evidential asymmetries.

---

[5] It is important to note that Mill presupposes a slightly different conception of toleration when talking about tolerance towards other person's conflicting beliefs. More specifically, Mill doesn't think that we should only tolerate conflicting beliefs if they are reasonable. According to Mill, we should tolerate *all* conflicting beliefs, because all conflicting beliefs could easily turn out to be true. Furthermore, Mill thinks that we should even tolerate conflicting beliefs we know to be false, since sincerely engaging with conflicting opinions is necessary to fully understand and appreciate one's own insights. Nevertheless, the argument from fallibility plays an important role in Mill's theory of toleration and can directly be applied to the conception of epistemic tolerance that is relevant here.

And in those cases, it seems very unlikely that the tolerated beliefs could surprisingly turn out to be true, even if they are reasonable. Take the case of Nick and Lea: Given that Lea's belief is formed on the basis of unreliable testimony, the fact that it is reasonable doesn't make it any more likely to be true. This also explains why it wouldn't be appropriate for Nick to tolerate Lea's belief – given that it is most likely to be false, he should simply inform Lea of the butler's tendency to lie.

However, a proponent of epistemic tolerance could just agree and still argue that there are some specific circumstances under which distinguishing between reasonable and unreasonable beliefs is epistemically significant. In fact, she could rightly point out that she never claimed that we should refrain from modifying *all* conflicting beliefs that are reasonable. It seems to be obvious that sometimes reasonable beliefs of others give us reasons to change our own beliefs, and that sometimes we should try to change the beliefs of others even if they are reasonable. Given that, plausible demands of epistemic tolerance are always restricted to specific circumstances or domains. The idea is not that we should always tolerate conflicting beliefs that are reasonable, but rather that there are specific circumstances under which we should tolerate conflicting beliefs that are reasonable. But what are those circumstances supposed to be?

One idea is that there may be circumstances under which it is appropriate to tolerate reasonable beliefs because, although there seems to be a significant evidential asymmetry in face of which the conflicting belief is sufficiently likely to be false, there is also a certain chance that this asymmetry turns out to be illusory. For example, consider the following case: Tim and Mary are two philosophers who respect each other as thoughtful and open-minded colleagues. While Tim believes that antinatalism is true, Mary believes that antinatalism is false. Having discussed the issue for a while, both have presented various arguments and considerations for their respective views. Mary, however, sticks to her guns. For her, antinatalism is just obviously false. At the same time, she is well aware that there is no way of sharing this sense of obviousness with Tim. Nevertheless, she decides to refrain from trying to modify Tim's belief, because she knows that he might have his own sense of obviousness.

This case, the idea goes, is a good example for an appropriate instance of epistemic tolerance. First of all, Mary seems to be justified in retaining her own belief because she has private evidence that clearly speaks for its truth. At the same time, there is also a sense in which Tim's belief is reasonable. He has carefully engaged with all the arguments Mary put forward to support her view, but after thorough investigation, he still didn't find them very convincing. Finally, Mary's decision to refrain from further attempts to modify Tim's belief also seems to be justified. Given that Tim's belief could be partly based on private evidence as well, it might turn out to be better supported than expected.

If this line of thought was convincing, a tolerant reaction towards another person's conflicting belief would be epistemically adequate given that (i) there are good reasons to believe that there is a significant evidential asymmetry in face of which retaining one's own belief is epistemically rational (ii) there are also good reasons to believe that the alleged evidential asymmetry might break down. It is important to note that the possibility of such a constellation doesn't necessarily presuppose the existence of private evidence. Even if there isn't anything like private evidence, there will be cases that satisfy the two conditions specified above.[6] For example, we can imagine a situation that resembles the one of Nick and Lea. Suppose that Nick justifiedly believes that Lea's belief is based on misleading evidence, but that he is also aware that his belief that the butler is a liar could easily turn out to be false – maybe because it is only based on comparatively weak evidence. In this case, a tolerant attitude towards Lea's belief might also be appropriate. Nevertheless, it should be clear that the above conditions are still extremely specific. Even if it was epistemically adequate to display a tolerant attitude under those conditions, general demands for epistemic tolerance in entire domains like philosophy or politics would probably be inappropriate.

However, it seems that even under those very specific circumstances the distinction between reasonable and unreasonable beliefs that is characteristic of a tolerant attitude is still arbitrary. To see this, just consider the following sequel to Mary's case. Suppose that, after talking to Tim, Mary meets Gary. Being the member of a radical religious cult, Gary has been brainwashed into believing that antinatalism is true. In this case, it is not clear at all why Mary should try to modify Gary's belief, but not Tim's. From Mary's perspective, Gary's belief could just as easily turn out to be probably true as Tim's. But in contrast to Tim's belief, Gary's belief isn't reasonable at all. So, even if it is instrumentally rational to refrain from modifying Tim's belief because it could easily turn out to be probably true, this has nothing to do with the fact that it is reasonable. Given that, the initial worry remains. It seems that the cases where a tolerant attitude towards disagreement is not epistemically irrational are exactly those cases in which the reasonableness of the other person's conflicting belief is not a good

---

[6] Nevertheless, it shouldn't come as a surprise that demands for epistemic tolerance seem to be especially common in exactly those domains in which philosophers have usually argued for the possibility of private evidence, like for example ethics or religion (see e.g. Feldman 2007, Rosen 2001, van Inwagen 1996).

basis for deciding whether to modify it or not. Or to put it differently: A tolerant reaction is only epistemically rational in cases where it is instrumentally irrational.

4. CONCLUSION

If my arguments are convincing, a tolerant attitude towards recognized disagreement is necessarily epistemically inadequate. Given the remarkable popularity of epistemic tolerance, this result is highly relevant to our intellectual practice. However, to fully understand the epistemic problem with a tolerant attitude, it is not enough to discuss its epistemic rationality - as I have argued in section 2, there are cases where reacting tolerantly towards recognized disagreement is epistemically rational. In section 3, I have suggested that to come to a complete epistemic evaluation of epistemic tolerance, considerations concerning instrumental rationality with respect to epistemic goals need to be taken into account. Once the dimension of instrumental rationality is included, it gets clear why a tolerant attitude is so problematic. The problem with a tolerant stance towards disagreement is not that it is epistemically irrational, but rather that in those situations where it is epistemically rational, it doesn't provide a good basis to decide which conflicting beliefs of others we should try to modify. So to establish a convincing argument for the epistemic inadequacy of a tolerant attitude, considerations concerning its instrumental rationality with respect to epistemic goals have to be taken into account.

ACKNOWLEDGEMENTS: I would like to thank Jan Constantin, Thomas Grundmann and Steffen Koch for helpful comments on earlier versions of this paper. I would also like to thank the participants of the ECA Groningen 2019 for a vivid and fruitful discussion.

REFERENCES

Chang, H. (2012). *Is Water H2O? Evidence, Realism and Pluralism*. Dordrecht: Springer.
Christensen, D. (2009). Disagreement as Evidence: The Epistemology of Controversy. *Philosophy Compass*, 4(5), 756–767.
Constantin, J., & Grundmann, T. (2018). Epistemic authority: Preemption through source sensitive defeat. *Synthese*, 82, 1–22.
Feldman, R. (2007). Reasonable Religious Disagreements. In L. M. Antony (ed.), *Philosophers Without Gods. Meditations on Atheism and the Secular Life* (pp. 194-214). Oxford: Oxford University Press.

Forst, R. (2013). *Toleration in conflict. Past and present.* Cambridge: Cambridge University Press.
Forst, R. (2017). Toleration. In E. N. Zalta (ed.), *The Stanford Encyclopedia of Philosophy* (Fall 2017 Edition).
Hallsson, B. G., & Kappel, K. (2018). Disagreement and the division of epistemic labor. *Synthese*, 29, 97–122.
Kant, I. (1991). An Answer to the Question: "What is Enlightenment?". In H. Reiss (ed.), *Kant: Political Writings* (pp. 54-60). Cambridge: Cambridge University Press.
Kelly, T. (2003). Epistemic rationality as instrumental rationality: a critique. *Philosophy and Phenomenological Research*, 66(3), 612-640.
King, N. (2012). Disagreement. What's the Problem? or A Good Peer is Hard to Find. *Philosophy and Phenomenological Research*, 85(2), 249–272.
King, P. T. (1998). *Toleration.* London: Frank Cass.
McGrath, S. (2009). Moral disagreement and moral expertise. In R. Shafer-Landau (ed.), *Oxford Studies in Metaethics* (pp. 87-108). Oxford: Oxford University Press.
Mill, J. S. (2001). *On Liberty.* Kitchener: Batoche Books.
Priest, M. (2016). Inferior Disagreement. *Acta Analytica*, 31(3), 263–283.
Rosen, G. (2001). Nominalism, Naturalism, Philosophical Relativism. *Noûs*, 35(s15), 69–91.
Šešelja, D.; Straßer, C.; Wieland, J. W. (2015). Withstanding Tensions. Scientific Disagreement and Epistemic Tolerance. In E. Ippoliti (ed.), *Heuristic Reasoning* (pp. 113-146). Cham: Springer.
van Inwagen, P. (1996). It Is Wrong, Everywhere, Always, for Anyone, to Believe Anything upon Insufficient Evidence. In J. Jordan (ed.), *Faith, freedom, and rationality. Philosophy of religion today* (pp. 137-154). Lanham: Rowman & Littlefield.
Werning, M. (2009). The Evolutionary and Social Preference for Knowledge. How to solve Meno's Problem within Reliabilism. *Grazer Philosophische Studien*, 79(1), 137–156.
Zagzebski, L. T. (2012). *Epistemic authority. A theory of trust, authority, and autonomy in belief.* Oxford: Oxford University Press.

Structuring controversy: The dialectic of disagreement

SHARON BAILIN
Faculty of Education, Simon Fraser University, Vancouver Canada
bailin@sfu.ca

MARK BATTERSBY
Department of Philosophy, Capilano University, Vancouver Canada
mbattersby@criticalinquirygroup.com

> There is considerable evidence that the consideration of alternative views and opposing arguments is crucial for coming to reasoned judgments. Yet disagreement and controversy may result in animosity, adversariality, and polarization. This paper addresses the issue of how to incorporate disagreement into critical thinking instruction in a way that results in productive interaction and robust outcomes.
>
> KEYWORDS: adversariality, alternative views, controversy, dialectical inquiry, polarization, reasoned judgment

1. INTRODUCTION

There is considerable evidence that the consideration of alternative views and opposing arguments is crucial for coming to reasoned judgments. Yet disagreement and controversy may result in animosity, adversariality, and polarization. This paper addresses the issue of how to incorporate disagreement into critical thinking instruction in a way that results in productive interaction and robust outcomes.

2. EPISTEMIC PROBLEMS OF ONE-SIDEDNESS

A significant obstacle to arriving at reasoned judgments is posed by the failure to seriously consider views and arguments which conflict with the position which one holds. One major cause is confirmation bias, the common tendency to primarily seek evidence in support of one's existing views. This problem is exacerbated by the one-sideness of the claims and arguments to which people are often exposed. The effects of search engines whose personalization algorithms direct people to views which are similar to their own (Pariser, 2011), the tendency of social media users to follow those who hold similar views (Halberstam & Knight, 2014), and the increasing geographic homogeneity of political beliefs (Aisch, 2018) all works against an exposure to alternative views and opposing arguments.

Such limited exposure to alternative views and opposing arguments has epistemic consequences. A number of authors have shown how significant errors of reasoning can be attributed to a lack of understanding of other positions and the failure to pursue alternative lines of reasoning (e.g., Finocchiaro's historical study of scientific reasoning (1994); Perkins's experimental investigations (1989; Perkins et al., 1983).

There is considerable current research in cognitive psychology which supports these conclusions. This research has demonstrated the ubiquity of myside bias, involving a failure to consider alternatives and to fairly and adequately evaluate arguments with which one disagrees (Perkins, 1989; Perkins & Tishman, 2001). It appears that people are generally much better at evaluating the arguments of others than they are at evaluating their own reasoning. They tend, for example, to have a limited ability to generate counter-arguments and counter-examples to views they hold (Mercier, 2016; Mercier & Sperber, 2017).

Such a consideration of alternatives is crucial for coming to a reasoned judgment because fully evaluating a view is a comparative enterprise requiring the weighing of evidence and arguments for and against the various alternative views (Kuhn, 1991; Bailin & Battersby, 2016, 2018b). Evaluation, as Kuhn argues, is meaningful only in a framework of comparison (Kuhn, 1991, pp. 266-267). In this context, the generation of counter-examples and counter-arguments plays a crucial role in the evaluating of one's own views in comparison with alternative views (Kuhn, 1991). As Kuhn puts it: "Paradoxically, to know that a theory is correct entails the ability to envision and address claims that it may not be" (p. 171).

3. EPISTEMIC BENEFITS OF CONTROVERSY

There are strategies that have shown some success in countering myside bias involving explicitly encouraging individuals to take others' perspectives (Galinsky & Ku, 2004) and to consider alternatives and counterarguments (Anderson, 1982; Hirt & Markman, 1995; Lord, Lepper, & Preston, 1984).

Another approach which can be effective is the actual exposure to conflicting views that can be facilitated in the context of deliberation within a group. There is a considerable body of research indicating that group deliberation can be superior to individual reasoning in many contexts, including political and economic forecasting (Mellers et al.,

2014), jury deliberations (Ellsworth, 1989), political deliberations (Fishkin, 2009; Mercier & Landmore, 2012) and scientific investigations (Dunbar, 1997).

In all these cases, the primary factor contributing to the effectiveness of group reasoning appears to be the confrontation of conflicting views. The existence of disagreement can counteract confirmation bias (Druckman, 2004; Schulz-Hardt et al., 2000) and can help people to see both sides of an issue (Kuhn & Crowell, 2011), acknowledge counter-arguments (Mercier & Sperber, 2017, p. 298; Kuhn, Shaw, & Felton, 1997), and make better judgments and decisions (Mercier & Sperber, 2017, p. 298).

4. RISKS OF CONTROVERSY

4.1 Problems of adversariality

Although engaging in controversy has important epistemic benefits, it also presents risks. One of these is adversariality. In one sense, controversy is, by its nature, adversarial in that it involves a confrontation of opposing views (Govier, 1999). And we are in agreement with Govier that the existence of controversy, in this sense, is a healthy thing. Getting the strongest arguments on various sides of an issue on the table for consideration is crucial for the comparative evaluation of arguments about controversial issues.

Problems arise when the confrontation between views is seen as a confrontation between arguers. According to Cohen, the dominant model of argumentation (DAM) frames argumentation as essentially an adversarial enterprise in which arguers are opponents or enemies in a battle to win (Cohen, 2015). The interlocutors are seen in roles of opponent and proponent with the goal of prevailing in the argument. Aikin (in a 2011 paper) supports this oppositional framing, maintaining that we argue with others because we believe that our views are correct and theirs are not and that those who disagree with our views are wrong and need correction (Aikin, 2011). Similarly, Govier (1999) states:

> Insofar as we are engaged in a controversy, we will be arguing with others who disagree with us and are, in that sense at least, our opponents or antagonists (p. 247).

Such an oppositional framing can be problematic in terms of the modes of discourse it encourages. Numerous theorists have criticized the dominance of battle metaphors in argumentation and the type of aggressive discourse which it can engender (Lakoff & Johnson, 1980; Moulton, 1989; Ayim, 1991; Cohen, 1995; Govier, 1999; Rooney, 2010).

These modes of discourse can interfere with reasonable and productive interactions and with rational exchange (Hundleby, 2013, p. 240).

It has been argued, however that adversarial argumentation need not result in aggressive modes of interaction. Govier (1999) suggests that argument is not necessarily confrontational and that adversariality can be kept to a logical and polite minimum, what she calls minimal adversariality.

The proposal for minimal adversariality, although it does address the issue of aggressive language and modes of interacting, is nonetheless problematic in accepting the framing of the enterprise in terms of opponents and winning. Govier, for example, states that in argumentation, "people occupy roles which set them against each other, as adversaries or opponents" (p. 242).

This slide from "arguing *for* claims" to "arguing *against* people who disagree with those claims" is problematic (as Govier herself acknowledges places).[1] Moreover, viewing the person holding the opposing position as one's opponent introduces an unnecessary and unhelpful element of adversariality (Rooney, 2010). As Rooney states:

> [W]hy are you my "opponent" if you are providing me with further or alternative considerations in regard to X ... whether I end up agreeing with X or not-X? (p. 221)

Govier herself, in fact, recognizes the difficulty inherent in this oppositional terminology:

> Given all the positive aspects of controversy, there is an important sense in which such people are helping us by disagreeing with us. Thus we might wish to regard them as partners, not opponents (p. 254).

A related issue has to do with the effect of this contest metaphor on the goal of epistemic improvement. Some theorists (e.g., Aikin, 2011) argue that adversariality, with its accompanying desire to win, contribute to epistemic goals:

> it is in the enacting of the debates, the attempts by each side's proponents to make the best case, rebut the opponent's counter-arguments, and lay out the best criticisms of the alternatives that we gain an understanding of an issue (Aikin 2011, p. 260).

---

[1] "We can argue *for* a claim without arguing *against* a person – even in contexts where we are addressing our arguments to other persons with whom we deeply disagree" (Govier, p. 64).

The opposite result is often the case, however. The imperative to win that is inherent in adversarial argumentation may well eclipse the goal of coming to a reasoned judgment, undermining co-operation, open-mindedness, and a willingness to concede to the strongest reasons.

Such counter-productive tendencies may be tamed through adherence to appropriate dialectical norms (e.g., pragma-dialectical rules) (Aikin 2011) which ensure, among other things, that one concedes to the most defensible position. Such rules require that claims be put to the test of reason and that those which are to be accepted are those that have the strongest warrant. In other words, their justification lies in the epistemic goals of argumentation. If one's goal were simply to win an argument, then one would have no reason to concede to a more defensible position. Dialectical rules are based on an implicit recognition of the priority of the epistemic goal of reaching better justified positions over the goal of winning.

The framing of the argumentative enterprise in terms of winning and losing is, in fact, an inaccurate and misleading description, as Rooney (2010) points out. If our interlocutor offers a better argument for their position than we offer for ours, we don't lose. We actually gain. We are, epistemically speaking, better for it (pp. 121-122).

Accepting the force of this criticism of an oppositional framing, Aikin later offers a modification of Govier's minimal adversariality, proposing what he calls dialectically minimal adversariality:

> The only adversariality in this model is the matter of weighing the force of the better reasons, and so this is minimal and only dialectically adversarial. As a consequence, the force of this notion of dialectical adversariality is in the reasoned weighing of evidential considerations for and against a view (2017, p. 16).

Adversariality in this sense of the confrontation of opposing views is precisely what we are advocating. We acknowledge, with Aikin, that there are various dialectical tasks or "moves of critical probing" that must be performed and that some of them are oppositional from a dialectical perspective. In a similar vein, Stevens and Cohen (2018) argue that argumentative contexts vary and that one might choose a more adversarial role in some contexts. Roles are, however, fluid and often overlap in practice, as Cohen himself argues (Cohen, 2015). These dialectical tasks may be (and often are) performed by, shared among, and even switched between various numbers or combinations of individuals depending on the context. What matters is that the various tasks be performed. Moreover, the oppositionality entailed by these moves is ultimately "in the service of a broader *cooperative* goal of dialectical testing of reasons and acceptability" (Aikin, 2017, p. 16). The overarching goal is epistemic betterment (Stevens & Cohen, 2019). And from this perspective argumentation needs to be seen as a collaborative endeavour.

4.2 Problems of polarization

Another potential risk of controversy is polarization. Although the exposure to conflicting views can enhance the making of reasoned judgments, it does not always reap such epistemic benefits.

One obstacle is belief tenacity. Numerous studies have shown that beliefs, once formed, can survive strong counter-arguments and discrediting evidence (Jennings, Lepper, & Ross, 1981; Anderson, Lepper, & Ross, 1980, Lord, Ross, & Lepper, 1979). Moreover, the process of defending one's position against counter-arguments and counter-evidence often creates a backfire effect, with individuals becoming even more entrenched in their original positions (Sloman & Fernbach, 2017; Kahan, 2013; Bail et al., 2018).

One explanation for these tendencies is in terms of defensive biases. People tend to identify with their beliefs and so are motivated to protect their beliefs as a way of protecting their feelings of adequacy and self-worth (Cohen et al., 2007; Sherman & Cohen, 2002).

Another likely factor is cultural cognition, which involves individuals holding onto specific beliefs as a way of expressing their group identity and evaluating information in a selective pattern that reinforces their group's worldview (Kahan, 2013; Kahan, Jenkins-Smith & Braman, 2011).

The research cited earlier suggests that group deliberation is a way to overcome some of these obstacles. Yet these epistemological benefits only accrue in groups in which genuine arguments for alternative views are presented (Lunenburg, 2012; Schultz-Hardt et al., 2002; Sunstein, 2006) and in which group members feel free to express their views. Discussion groups in which critique, argumentation, and the consideration of alternatives is absent and in which members feel pressure to conform to the majority view can, in fact, have a negative impact on reasoning, amplifying errors, reinforcing existing beliefs, increasing commitment to poor decisions (Janis, 1982; Schultz-Hardt et al. 2000, 2006; Sunstein & Hastie, 2015) and sometimes even resulting in more polarized views (Sunstein 1999, p. 1).

5. DIALECTICAL INQUIRY

Our objective is to find a way to incorporate disagreement into critical thinking instruction in a way that mitigates these problems and enhances epistemic benefits. The research and arguments rehearsed above suggests that such instruction should instantiate the following features:
1. It should focus explicitly on a comparative evaluation of conflicting views.
2. It should have group deliberation as a central focus.
3. It should be framed with an inquiry orientation.

5.1 The nature of dialectical inquiry

The approach to critical thinking instruction that we propose to meet this challenge is based on dialectical inquiry (Bailin & Battersby, 2016, 2018b). In dialectical inquiry, the goal is to come to a reasoned judgment on a controversial issue and this is viewed as an essentially dialectical and collaborative process. Students work in groups to comparatively evaluate arguments on all sides of an issue rather than simply offering and defending their own arguments. Thus the exploration of conflicting views is at the centre of the inquiry process, but the process of reaching a reasoned judgment is a collaborative rather than adversarial endeavour. The focus is on the confrontation of conflicting positions without the adversariality implicit in oppositional argumentation, and the collaborative, community orientation can mitigate the type of polarization which often accompanies controversy.

5.2 Aspects of dialectical inquiry

There are particular aspects of dialectical inquiry which instantiate the desired elements.

5.2.1 Comparative evaluation of conflicting views

Aspects of the structure of the inquiry process ensure an exposure to conflicting views. These include the requirement that students research the actual arguments that have been presented on various of issues. A useful heuristic in this regard is a dialectical argument table which represents the debate on the issue, including the arguments pro and con as well as objections to the arguments and responses to the objections. Reaching a reasoned judgment takes place through a comparative evaluation of the relative strengths of the various arguments in the overall case.

5.2.2 Group deliberation

The advantages of group deliberation are facilitated in a number of ways. Students frequently engage in group interaction, discussing, questioning, challenging, and critiquing. They engage in collaborative inquiries, jointly researching, evaluating, and coming to a joint judgment. They also engage in individual inquiries in which they conduct the inquiry in stages, working in groups to get feedback and critique from peers at each stage. Strategies for further promoting the inclusion of conflicting views within the groups include creating heterogeneous groups, devil's advocacy (Schulz-Hardt, Jochims & Frey, 2002), and structured controversy (where students alternatively defend different sides of an issue and then collectively come to a reasoned judgment) (Johnson & Johnson, 1988, 2009). These strategies can help to mitigate the pitfalls of adversariality in group argumentation while ensuring that alternative views are given a full hearing.

The epistemic ends of inquiry are also fostered by the creation of a community of inquiry. This is a community which instantiates the norms of rational inquiry, promoting open-minded and fair- minded exchanges, rigorous but respectful critique, and changing one's mind when justified by the evidence and arguments. It is also a community committed to respectful treatment, meaningful participation, and productive interaction (Bailin & Battersby, 2016b). Such a community can mitigate defensive biases in that value is placed not on supporting particular views but rather on being reasonable. It can also address the challenges posed by cultural cognition by creating a community of affiliation centred on rational inquiry as an alternative to or counter-balance to one's cultural community.

5.2.3 Inquiry orientation

An important aspect of the approach involves the framing of argumentation in terms of inquiry rather than persuasion. Students need to understand that the confrontation is really between views and not between people. The epistemic goals of argumentation and the essentially collaborative nature of the enterprise are emphasized in dialectical inquiry.

6. WILL IT MAKE A DIFFERENCE?

One objection that has been raised to dialectical inquiry is that it is not really a way of dealing with controversy. To quote a reviewer, "Instead of engaging IN a controversy, students are encouraged to think ABOUT a controversy. These are different kinds of activities and it is hard to see how - at least in the wild, i.e., outside of the classroom - people could be motivated to shift from the first kind of activity to the second."

We would, however, dispute the claim that these are fundamentally different kinds of activities. Even when individuals are engaging in argumentation over a controversial issue with the goal of persuading others of the soundness of their position, they are bound by normative rules which require conceding to the most defensible position. Thus, whatever the initial intent of the argument, the arguers are essentially testing claims in the interest of arriving at the best justified position. They are, in other words, inquiring. This indicates that the dichotomized framing of the activities in terms of engaging in a controversy versus thinking about a controversy is problematic. The criticism seems to assume that engaging in a controversy involves adversarial argumentation aimed at persuasion and that the confrontation of opposing views through inquiry is not really engaging in the controversy. But dialectical inquiry is not just thinking about a controversy. It is trying to come to a reasoned judgment about the controversial issue and so is also engagement.

The dialectical inquiry approach aims at helping students learn to treat controversial issues with an inquiry orientation. They learn to reframe the process of argumentation, acquiring the habit of considering both/many sides of an issue and treating the dialogical interaction as one in which they need to be willing to be open to other views and to change their mind when warranted. We recognize that people, including our students, engage in argumentation in different contexts and for various purposes, and that sometimes the immediate purpose is to persuade. But through engaging in dialectical inquiry, they learn the need to inquire into an issue before attempting to make a case. Making a reasonable case can be seen as presenting the results of an inquiry in such a manner that the interlocutors will also come to see that the judgment is reasonable. But whatever their more proximal purposes, the aim is for students to keep in mind the overriding epistemic goals. The intention is that students learn to frame the activity through an epistemological orientation.

An important aspect of the dialectical inquiry approach is the development of a spirit of inquiry. The approach not only aims at equipping students to make reasoned judgments via critical inquiry but also puts considerable emphasis on fostering the habits of mind of the critical inquirer. This means fostering in students the disposition to believe and act on the basis of reason and the motivation to inquire in the face of disagreement.

Confronting controversy is not limited to instances of direct argumentation with others. We are frequently confronted with conflicting views and arguments and need to think them through on our own. Given that our interest is in education, our aim is not only to enhance the ability and disposition of people to make reasoned judgments when they argue with each other but also, importantly, to enhance the ability and disposition of individuals to make reasoned judgments when they think through controversial issues on their own. This includes anticipating counter-arguments, generating alternatives, and fairly evaluating all sides of an issue. These aims are addressed directly through aspects of the dialectical inquiry process (e.g., the requirement to research arguments on various sides, the pro con argument table) and also through the community of inquiry and group deliberation.

There is, in fact, considerable evidence that the epistemic benefits of group deliberation carry over to the individual context. Kuhn, for example, found that students who had engaged in argumentation with peers offered more complex arguments incorporating both sides of the issue when writing individual essays on a different topic than did students who had been reasoning on their own (Kuhn & Crowell, 2011). They also demonstrated an increased capacity to anticipate counter-arguments in contexts when an interlocutor was not present and with respect to topics beyond those discussed in the group (Mercier, 2017, p. 11). Through engaging in group argumentation, students interiorized the dynamics of argumentation and become better reasoners on their own (Mercier, 2017).

7. CONCLUSION

There is considerable evidence that the consideration of alternative views and opposing arguments is crucial for coming to reasoned judgments. Yet controversy often results in adversariality and polarization, which tend to have a negative effect on our epistemic goals. The approach to critical thinking instruction which we propose, dialectical inquiry, offers a way to incorporate the confrontation of conflicting views into critical thinking instruction in a way that minimizes adversariality and polarization. The goal is not to avoid controversy but rather to structure it to this end.

REFERENCES

Aikin, S. F. (2011). A defense of war and sports metaphors in argument. *Philosophy and Rhetoric,* 44(3), 250-272.

Aikin, S. F. (2017). Fallacy theory, the negativity problem, and minimal dialectical adversariality. *Cogency,* 9(1), 7-19.

Aisch, G. (2018). The divide between red and blue America grew even deeper in 2016. *The New York Times.* Retrieved from https://www.nytimes.com/interactive/2016/11/10/us/politics/red-blue-divide-grew-stronger-in-2016.html.

Anderson, C. A. (1982). Inoculation and counter-explanation: Debiasing techniques in the perseverance of social theories. *Social Cognition,* 1, 126–139.

Anderson, C. A., Leper, M. R., & Ross, L. (1980). Perseverance of social theories: The role of explanation in the persistence of discredited information. *Journal of Personality and Social Psychology,* 39, 1037-1049.

Ayim, M. (1991). Dominance and affiliation. *Informal Logic,* 13(2), 79-88.

Bail, C., Argyle, L., Brown, T., Bumpus, J., Chen, H., Hunzakeer, F., Lee, J., Mann, M., Merhout, F., & Volfovsky, A. (2018). Exposure to opposing views can increase political polarization: Evidence from a large-scale field experiment on social media. Retrieved from www.files.osf.io

Bailin, S., & Battersby, M. (2016). *Reason in the balance: An inquiry approach to critical thinking.* 2nd edition. Cambridge, Mass: Hackett.

Bailin, S., & Battersby, M. (2018a). DAMed if you do; DAMed if you don't: Cohen's 'Missed Opportunities.' In M. Battersby & S. Bailin, *Inquiry: A new paradigm for critical thinking*. Windsor, ON: Windsor Studies in Argumentation.

Bailin, S., & Battersby, M. (2018b). Inquiry: A dialectical approach to teaching critical thinking. In M. Battersby & S. Bailin, *Inquiry: A new paradigm for critical thinking*. Windsor, ON: Windsor Studies in Argumentation.

Cohen, D. H. (2015). Missed opportunities in argument evaluation. In B.J. Garssen, D. Godden, G. Mitchell, & A. F. Snoeck Henkemans, (Eds.), *Proceedings of ISSA 2014: Eighth Conference of the International Society for the Study of Argumentation* (pp. 257–265), Amsterdam: Sic Sat.Govier

Cohen, D.H. (1995). Argument as war … and war is hell: Philosophy, education, and metaphors for argumentation. *Informal Logic,* 17(2), 177-188.

Cohen, G. L., Bastardi, A., Sherman, D. K., McGoey, M., Hsu, L., & Ross, L. (2007). Bridging the partisan divide: Self-affirmation reduces ideological closed-mindedness and inflexibility in negotiation. *Journal of Personality and Social Psychology,* 93(3), 415-430.

Dunbar, K. (1997). How scientists think: On-line creativity and conceptual change in science. In T. B. Ward, S. M. Smith, J. Vaid, & D. Perkins (Eds.), *Creative thought: An investigation of conceptual structures and processes.* Washington, DC: American Psychological Association, 461-493.

Ellsworth, P. C. (1989). Are twelve heads better than one? *Law and Contemporary Problems,* 52(4), 205-224

Finocchiaro, M. (1994). Two empirical approaches to the study of reasoning. *Informal Logic* 16(1), 1-21.

Fishkin, J. S. (2009). *When the people speak: Deliberative democracy and public consultation.* Oxford: Oxford University Press.

Galinsky, A. D., & Ku, G. (2004). The effects of perspective-taking on prejudice: The moderating role of self-evaluation. *Personality and Social Psychology Bulletin*, 30, 594–604.

Govier, T. (1999). *The philosophy of argument.* Newport News, VA: Vale Press.

Halberstam, Y., & Knight, B. (2016). Homophily, group size, and the diffusion of political information in social networks: Evidence from Twitter. *Journal of Public Economics,* 143, 73-88.

Hirt, E., & Markman, K. (1995). Multiple explanation: A consider-an- alternative strategy for debiasing judgments. *Journal of Personality and Social Psychology, 69,* 1069–1086.

Hundleby, C. (2013). Aggression, politeness, and abstract adversaries. *Informal Logic* 33(2), 238–262.

Janis, I. L. (1982). *Groupthink* (2nd rev.). Boston, MA: Houghton Mifflin.

Jennings, D. L., Leper, M. R., & Ross, L. (1981). Persistence of impressions of personal persuasiveness: Perseverance of erroneous self-assessments outside the debriefing paradigm. *Personality and Social Psychology Bulletin, 7,* 257-263.

Johnson, D. W. & Johnson, R. T. (2009). Energizing learning: The instructional power of conflict. *Educational Researcher, 38*(1), 37–51.

Johnson, D. W. & Johnson, R. T. (1988). Critical thinking through structured controversy. *Educational Leadership, 45*(8), 58-64.

Kahan, D. M. (2013). Ideology, motivated reasoning, and cognitive reflection. *Judgment and Decision Making,* 8(4), 407-424.

Kahan, D. M., Jenkins-Smith, H., & Braman, D. (2011). Cultural cognition of scientific consensus. *Journal of Risk Research*, 14*(2),* 147-174.

Kuhn, D. (1991). *The skills of argument.* New York: Cambridge University Press.

Kuhn, D. & Crowell, A. (2011). Dialogic argumentation as a vehicle for developing young adolescents' thinking. *Psychological Science,* 22(4), 545-552

Kuhn, D., Shaw, V., & Felton, M. (1997). Effects of dyadic interaction on argumentative reasoning. *Cognition and Instruction*, 15, 287–315.

Lakoff, G., & Johnson, M. (1980). *Metaphors we live by.* Chicago: University of Chicago Press

Lord, C., Lepper, M., & Preston, E. (1984). Considering the opposite: A corrective strategy for social judgment. *Journal of Personality and Social Psychology,* 47, 1231–1243.

Lord, C. G., Ross, L., & Lepper, M.R. (1979). Biased assimilation and attitude polarization: The effects of prior theories on subsequently considered evidence. *Journal of Personality and Social Psychology*, 37(11), 2098-2109.

Lunenburg, F. (2012). Devil's advocacy and dialectical inquiry: Antidotes to groupthink. *International Journal of Scholarly Academic Intellectual Diversity*, 14(1), 1–9.

Mellers, B., Ungar, L., Baron, J., Ramos, J., Gurcay, B., Fincher, K., Scott, S. E., Moore, D., Atanasov, P., Swift, S. A., & Murray, T. (2014). Psychological strategies for winning a geopolitical forecasting tournament. *Psychological science*, 25(5), 1106-1115.

Mercier, H. (2016). The argumentative theory: Predictions and empirical evidence. *Trends in Cognitive Sciences*, 20(9), 689-700.

Mercier, H. & Landemore, H. (2012). Reasoning is for arguing: Understanding the successes and failures of deliberation. *Political Psychology*, 33(2), 243-258

Mercier, H. & Sperber, D. (2017). *The enigma of reason.* Cambridge, Mass: Harvard University Press.

Moulton, Janice. (1989). A paradigm of philosophy: The adversary method. In S. Harding & M.B. Hintikka (Eds.), *Discovering reality: Feminist perspectives on epistemology, metaphysics, methodology, and philosophy of science* (pp. 149-164), Dordrecht, Holland: D. Reidel.

Pariser, E. (2011). *The filter bubble: What the internet is hiding from you.* London: Penguin.

Perkins, D. N. (1989), Reasoning as it is and as it could be: An empirical perspective. In D. N. Topping, D. C. Crowell, & V. N. Kobayashi (Eds.), *Thinking across cultures: The third international conference on thinking.* Hillsdale, NJ: Lawrence Erlbaum Associates.

Perkins, D. N., Allen, R., & Hafner, J. (1983). Difficulties in everyday reasoning. In W. Maxwell (Ed.), *Thinking: The expanding frontier.* Philadelphia: The Franklin Institute Press.

Perkins, D. N., & Tishman, S. (2001). Dispositional aspects of intelligence. In S. Messick & J. M. Collis (Eds.), *Intelligence and personality: Bridging the gap in theory and measure- ment* (pp. 233-257). Mahwah, NJ: Lawrence erlbaum. Retrieved May 17, 2006, from http://learnweb.harvard.edu/alps/thinking/docs/Plymouth.pdf

Rooney, P. (2010). Philosophy, adversarial argumentation, and embattled reason. *Informal Logic* 30(3), 203-234.

Schulz-Hardt, S., Frey, D., Luthgens, C., & Moscovici, S. (2000). Biased information search in group decision making. *Journal of Personality and Social Psychology,* 78, 655-669.

Schulz-Hardt, S., Jochims, M., & Frey, D. (2002). Productive conflict in group decision making: Genuine and contrived dissent as strategies to counteract biased information seeking. *Organizational Behavior and Human Decision Processes,* 88, 563 – 586.

Sherman, D.K., & Cohen, G.L. (2002). Accepting threatening information: Self-affirmation and the reduction of defensive biases. *Current Directions in Psychological Science,* 11(4), 119-123.

Sloman, S., & Fernbach, P. (2017). *The knowledge illusion: Why we never think alone.* Penguin.

Stevens, K., & Cohen, D. (2019). The attraction of the ideal has no traction on the real: On adversariality and roles in argument. *Argumentation and Advocacy,* 55(1), 1-23.

Sunstein, C. R. (1999). The law of group polarization. *University of Chicago Law School, Coase-Sandor Working Paper Series in Law and Economics.*

Sunstein, C. R. (2006). *Infotopia: How many minds produce knowledge.* Oxford: Oxford University Press.

Sunstein, C.R., & Hastie, R. (2015). Garbage in, garbage out? Some micro sources of macro errors. *Journal of Institutional Economics,* 11(3), 561–583.

The Role of Reasoning and Arguing in Youth Democratic Participation in Canada

MICHAEL D. BAUMTROG
Ted Rogers School of Management, Ryerson University
<u>baumtrog@ryerson.ca</u>

In 1970 the voting age in Canada moved from 21 to 18. Since then, there have been calls to lower it further, most commonly to age 16. Against the motion, however, it has been argued that youth may lack the ability to exercise a mature and informed vote, which I take to mean a vote exercised on the basis of informed reason. This paper aims at testing the veracity of this worry.

KEYWORDS: youth, voting, democratic participation, informed reason, decision making.

1. INTRODUCTION

Youth have plenty of reasons to participate in political dissent. Like adults, cultural shifts over time may change the focus of their interests, but they never eliminate them. Today, perhaps most prominently, we see global youth action aimed at fighting the climate crisis. In the past year young advocates such as Autumn Peltier and Greta Thunberg have captured global attention for their clarity, determination, courage, and remarkable poise while advocating for climate action in the presence of some of the world's most powerful people. Beyond exceptional exemplars, however, youth are also becoming increasingly visible supporters of non-binary gender policies and protecting and promoting educational quality.

Unfortunately, however, in many countries youth are prohibited from one of the most basic forms of political recognition – voting. In Canada, my home and native land, the legal age for voting is 18. This means that 7, 176, 144 people (19.35% of the population) are restricted from voting (www.statscan.ca). Nevertheless, a country that restricts nearly 1/5 of its population from voting prides itself on providing "universal" suffrage. I argue below that although the colonial government of Canada has been expanding suffrage since its independence, at least one of the arguments against lowering the voting age, namely that youth cannot come to a mature and informed decision regarding who to vote for, is unconvincing.

The paper proceeds as follows. In the next section I will offer a very brief overview of the expansion of the right to vote in Canada, focusing on an articulation of some of the reasons youth are restricted from voting and on the claim that youth cannot come to a mature and informed decision. The third part of the paper offers a review of some of the literature focused on youth's ability to reason and argue. The fourth section demonstrates young peoples' political competency by showing how their abilities are already often employed in a number of political activities. In the conclusion, I tie together the argument against the claim that 16-year-olds cannot come to a mature and informed decision and should thereby be prohibited from voting.

2. CANADIAN SUFFRAGE – A VERY BRIEF HISTORY

On July 1, 1867, the day Canada became a country, it only included four provinces: Ontario, Quebec, New Brunswick, and Nova Scotia. At that time, "control of the federal franchise would remain a provincial matter until Parliament decided otherwise" (Elections Canada, p. 40). Though there were some individual differences, at confederation the provinces each maintained three common conditions an individual must meet to be eligible to vote - being male, having reached the age of 21, and being a British subject by birth or naturalization. By 1885 when the federal government first gained control of the franchise, additional conditions requiring the would-be voter to own property and/or meet a minimum income level were also in place. It then took until 1918 before women successfully won the right to vote and until the 1960's before Inuit and "status Indians" could effectively vote (Dabin et al. 2019).[1] In 1970, the Canada Elections Act, passed under Prime Minister Pierre Trudeau, lowed the voting age from 21 to 18 with little resistance, with calls to lower it to 16 occurring regularly ever since.

In 1989, the Royal Commission on Electoral Reform and Party Financing, better known as the "Lortie Commission",[2] was established to review the Elections Act and make recommendations. Among other items within its purview, the Lortie Commission specifically addressed the question of lowering the voting age to 16, concluding that despite evidence 16-year-olds have a significant stake in the society, can exercise a mature and informed vote, and that they generally act

---

[1] I use "effectively" because while the Inuit were legally granted the right to vote in 1950, the practice was only really made possible when ballot boxes were placed in more Inuit communities in 1962. See Leslie (2019).
[2] In honour of its chairman Pierre Lortie.

responsibly when they participate in public affairs, "[u]ltimately, any decision on the voting age involves the judgement of a society about when individuals reach maturity as citizens" and that "there remains a strong conviction that the time has not come to lower the voting age" (Lortie, 1991, pp. 48-49). This conclusion was so persuasive that it was still being appealed to in 2005 during debate of bill C261, another attempt to lower the voting age. In that same debate, arguments against the ability for 16-year-olds to exercise a mature an informed vote were also still being forwarded by the opposition. Most recently, on May 1, 2018, Green Party MP Elizabeth May introduced Bill C-401, which proposes that "Every person who is a Canadian citizen and is 16 years of age or older on polling day is qualified as an elector."

As long as calls to lower the voting age to 16 have been proposed, they have been met with vehement opposition. Wall (2014) categorises the opposition into two broad categories: a lack of capacity and the potential to cause harm. In terms of harm, allowing children the right to vote has some people worried that it may cause harm to other children, adults, and the culture more generally. In terms of capacity, Wall distinguishes arguments that children lack 1) the competency to make rational judgements, from 2) knowledge of political systems, and 3) independence from outside influence. In this paper, I am only concerned with the claim regarding competency. In political forums (for example, the debate on Bill C261) and the popular media alike (e.g. Lum, 2018, Burnett, 2017), this opposition is usually described as the inability for 16-year-olds to exercise a mature and/or informed vote. In what follows I first review some of the literature arguing for youth's competency and then look at how these abilities are put into practice in non-voting related activities every day.

3. WHAT DO THE STUDIES SAY?

It is important to first identify the target, in other words, what I mean by "competency". To do so, it may be helpful to distinguish decision-making competency from excellency. Articulating decision-making excellency would require a determination of *how well* 16-year-olds can make an informed decision. Since our concern is with competency rather than excellency, we only have the less complicated task of determining if 16-year-olds can reason *well enough* to make an informed decision, i.e., whether they meet a certain standard or threshold. It seems fitting then to make that standard the rough equivalent of the competency of the majority of the rest of the voting population.[3] In this spirit, Wall (2014,

---

[3] One could argue that we need only measure against the competency of the lowest common denominator in the voting citizenry. I specify, "majority of the

p. 110) has identified competency as, "the capacity for political reason as expressed in such abilities as public critical thinking, discourse with others, and the ability to weigh society-wide outcomes of decisions." I appreciate this characterization for its flexibility and applicability and thus take it up as the target for the remainder of the discussion.

So, what does the literature say about the ability for people 16 years of age (and under) to meet our target? In an excellent doctoral dissertation, Schär (2019) has pointed out that the literature on youth reasoning and argumentation tends to fall into two broad streams. The first stream is product focused and looks at the ability for youth to reason individually. In other words, competency is assessed by looking at the argumentative products – essays, scores on tests, etc. – that youth produce. This approach is developed most prominently by Dianna Kuhn (1991). The second stream is process focused, and more often looks at the role of reasoning and decision-making in social, or group, contexts. In other words, the argumentative behaviours of youth are monitored in social situations where people interact and argumentation emerges (or doesn't emerge) within their discussions. Both streams, however, are clear that 16-year-olds maintain at least equal competency to their older counterparts.

For example, Kuhn argues that age matters until around the age of 14, and then education takes over as the most important determinant of reasoning ability. She states,

> After ninth grade, educational level (college vs. noncollege) takes over as the factor predictive of [argumentative] performance, as found here. Young adults with at least several years of college performed significantly better than ninth-graders, while *the performance of noncollege young adults was intermediate between that of sixth- and ninth-graders.* (Emphasis added. Kuhn 1991, p. 285)

This means that if competency ought to be a measure of eligibility for voting, many noncollege adults ought to be prohibited as well, since they demonstrate performance equivalent to children aged from approximately 11 to 14 years.

The shift that Kuhn identifies as occurring around the age of 14 involves metacognitive tasks. She argues that prior to ninth grade (approximately age 14), youth can still apply theories to evidence, but are not as good, or may be unable, to conduct the metacognitive tasks of

---

rest of the voting population", however, so as not to measure the general population of 16-year-olds against the sub-set of people over the age of 18 with mental disorders and illnesses, all of whom were granted the right to vote in Canada in 1988.

"specifying forms of evidence that would show a theory to be correct or incorrect and to evaluate the bearing of forms of evidence presented by the interviewer on different causal and noncausal theories" (284).

At this very conference, Kuhn (See also Kuhn 2019, pp. 155ff.) also noted that educational intervention with students aged 11 to 13 enabled them to competently discuss and decide on questions such as "Should people be required to pay a social security tax from each paycheck that will provide money when they retire, or should people save on their own for their retirement?" and "Should a powerful nation intervene to help another nation in trouble or only focus on its own problems?" These questions are obviously political, and the transcripts Kuhn provided at clearly showed how the children met all of Wall's criteria to demonstrate political competency.

The process-oriented study of youth reasoning and argumentation is currently being developed in Switzerland by scholars such as Greco, Mehmeti, Perret-Clermont (2017), who found that students aged 8-13 discussing environmental issues were largely able to meet the demands of the ideal model of a critical discussion proposed by van Eemeren & Grootendorst (1984, 2004). As opposed to asking students to complete a test or write an essay individually, in this study children were encouraged to interact and discuss issues with each other with a teacher facilitating. The study shows that these students were able to "open new issues for a discussion; they advanced standpoints and arguments in support of their standpoints. Moreover, they were able to follow the teacher when she shifted the issue and opened new paths for their discussion" (p. 213).

Another proponent of the process approach, Hugo Mercier (2011), has argued that reasoning is innate and that it evolved to improve argumentation. As evidence for the innateness of reasoning, he points to studies demonstrating that by age 3, children have "recourse in argumentation to social rules, to the material consequences of action or the consequences for others' feelings" (182). He further highlights that children, like adults, reason and argue better when motivated to do so – e.g. when facing or anticipating disagreement. This motivation is also, like adults, closely linked to the confirmation bias:

> In another study it was found that the large majority of 9-year-olds' utterances supported their own point of view (Pontecorvo & Girardet, 1993). It is important to stress that this early emerging confirmation bias does not entail a lack of ability to attack arguments—when they are the arguments of the other party in the conflict (Howe, Rinaldi, & Jennings, 2002; Tesla & Dunn, 1992).

Like Mercier (2011), Kuhn in her more recent and more process-based work (2019, p. 154) highlights differences between individual and social argumentation. While in written essays, most evidence produced by children is in support of their own position, "an average of one third of evidence-based claims served the function of weakening the opposing position (versus under 10% in the essays of these same participants)". In her talk at this conference, Kuhn concluded that children best 1) Support own standpoints (confirmation bias), 2) find weakness in others', 3) find strength in others', and 4) see weakness in their own. She also noted that youth, like adults, reason and argue far better when they have more access to more information. Thus, all three of the studies looked at thus far recognize the ability for young children, well below the age of 16, to generate and support their standpoints and find weaknesses in others'. These results align well with the same strengths in adult reasoning performed in light of the confirmation bias (Kahneman 2011). Thus, from the two main streams of research into youth reasoning and argumentation, there does not appear to be any significant difference between how a 16-year-old and an average person of the legal voting age would form an opinion.

4. YOUTH POLITICAL ACTIVITIES

Given that the literature points to the ability for youth to reason and come to an informed decision, it is unsurprising that they also exercise these abilities in a number of forums where they demonstrate their ability to meet the aforementioned "capacity for political reason as expressed in such abilities as public critical thinking, discourse with others, and the ability to weigh society-wide outcomes of decisions."

Take, for example, the recent global marches for climate. Before, during, and after these events, youth are engaging in discourse with others, especially about the society-wide outcomes of decisions. Before marches, young people are discussing current and potential policy changes and costs. They do so at home, at school, and often on social media. The marches themselves are a manifestation of at least two conclusions drawn during these previous discussions, namely, that something must be done to solve the climate crisis and that so far there is little enough public momentum do to anything that the urgency must be expressed through public protest.

The movement has been some time in the making but has consistently been led by young people. For example, Autumn Peltier, from Wiikwemkoong First Nation on Manitoulin Island in Northern Ontario made headlines in 2016 when she confronted Prime Minster Justin Trudeau about his broken environmental promises. She has also now spoken at to the General Assembly of the United Nations in both

2018 and 2019 (Manitoulin teen, 2019). In the United States, 11-year-old Amariyanna (Mari) Copeny, gained international attention in 2016 when she received a response from American President Barack Obama regarding the water crisis in her hometown of Flint, Michigan (Wikipedia). She has been fighting for clean water since, recently running a successful GoFundMe campaign to provide water filters to the community.

More popularly, 16-year-old Greta Thunberg's School Strike for Climate, which has now evolved into a mass movement taking place around the world, has put youth political reasoning competencies on full display, with more and more youth voices hitting the airwaves on major networks as recognised knowers on expanding social and political topics. Indeed, two of the six main political parties in the 2020 Canadian election have now committed to lowering the voting age should they win in part due to this visibility.

When not striking, many secondary school students are exercising their political competency through school-based activities, such as mock voting preparation and execution (www.studentvote.ca), which are aligned with real elections. In these events, like adults, students learn about each candidate and the implications of their platforms. They compare and contrast candidate platforms, and eventually cast a mock ballot. But their participation is not limited to the municipal, provincial, or even national level. On the international scale, students also participate in political forums such as the model U.N. and NATO summits wherein they roleplay representatives from participating countries and make decisions based on complicated country profiles. It should be noted that all of the school-based activities mentioned thus far are also happening in addition to every student's participation in their mandatory civics class.

Finally, outside of school, youth often sign petitions after discussions with advocates on the street, engage in debate about political issues with their families at home, and some even join the youth wings of adult-run political parties and organizations, most of which in Canada welcome all participants 25 years of age and under, despite the legally recognized 18-year-old voting age.

5. CONCLUSION

As mentioned at the outset, I have only addressed one of several arguments forwarded for restricting 16-year-olds from voting. I believe there are other responses to the other arguments, but I leave them for another work. For now, three important lessons emerge when considering the results of the academic literature and observations regarding youth democratic participation. First, if an age had to be selected for cognitive development, 14 would make more sense as that is the age at which the last major cognitive development occurs. Second, education is much more important than age for making an informed decision. Since this is also the case for adults, we should not be surprised that it also holds for the young, and we can see it both in the academic studies as well as in the forums where youth participate in politics before being allowed to legally vote. The consequence of this result is that if one feels understandably uncomfortable with restricting the uneducated from voting, s/he ought to at least feel as uncomfortable restricting 16-year-olds. Third, like adults, youth demonstrate better reasoning and decision-making abilities when they are motivated to do so (have a stake in the game, so to speak) and have the chance to investigate and learn about a topic. When others disagree with them, or when they are operating in a political context, youth regularly demonstrate their abilities to meet Wall's conditions for political competency. This suggests that engaging, motivating, and informing young people would suffice to ensure they meet the standard of the average of their adult voting counterparts. Further, since educating the electorate is a standard goal of all liberal democracies, including the young should not require a foundational shift – it would simply require recognizing 16-year-olds as a valuable voting demographic, with their unique interests and abilities, much like all of the existing demographics already considered.

ACKNOWLEDGEMENTS: I would like to thank Michelle Baumtrog, CW, and SW for numerous conversations on the topic. I would also like to thank the employee at the Ontario Science Centre who provided me with the opportunity to vote for saving either the trees, air, water, or land at 5 years old. After a patient hour of painful deliberation, I had finally learned how difficult and important voting is.

REFERENCES

Dabin, Simon, Jean François Daoust and Martin Papillon (2019) Indigenous Peoples and Affinity Voting in Canada. *Canadian Journal of Political Science* 52(1), 39–53
van Eemeren, F. H., & R. Grootendorst. (1984). *Speech acts in argumentative discussions*. Dordrecht/Cinnaminson: Foris.
van Eemeren, F. H., & R. Grootendorst. (2004). *A systematic theory of argumentation: The pragma-dialectical approach*. Cambridge: Cambridge University Press.
Elections Canada. (2007). *A History of the Vote in Canada* (2nd Ed). *Chief Electoral Officer of Canada.* https://www.elections.ca/res/his/History-Eng_Text.pdf

Howe, N., Rinaldi, C., & Jennings, M. (2002). "No! The lambs can stay out because they got cosies": Constructive and destructive sibling conflict, pretend play, and social understanding. *Child Development*, (73), 1460–1473.
Greco, Mehmeti, Perret-Clermont. (2017). Do adult-children dialogical interactions leave space for a full development of argumentation?: A case study. *Journal of Argumentation in Context*. 6(2).193–219
Kahneman, D. (2011). *Thinking, Fast and Slow.* London: Macmillan.
Kuhn. D. (1991). *Skills of Argument*. Cambridge University Press. New York.
Kuhn, D. (2019). *Critical Thinking as Discourse.* Human Development, 62(3), 146–164. doi:10.1159/000500171
Leslie, J., Indigenous Suffrage (2019). In, *The Canadian Encyclopedia*. https://www.thecanadianencyclopedia.ca/en/article/indigenous-suffrage
Lortie, P. (Chair). (1991). *Reforming electoral democracy: Final report of the Canadian Royal Commission on Electoral Reform and Party Financing.* Ottawa: Ministry of Supply and Services Canada.
Lum, Z. (2018). Arguments Against Lowering Canada's Voting Age To 16 Are Retrieved from 1970: Expert. *Huffington Post.* https://www.huffingtonpost.ca/2018/03/19/arguments-against-lowering-canadas-voting-age-to-16-are-from-1970-expert_a_23389984/
MacDonald, Heidi. 2017. Women's Suffrage and Confederation. *Acadiensis*, 46(1), 163-176.
Macedo, A. (2011). *The Development of Children's Argument Skills*. Dissertation. University of London: Dissertation.
Manitoulin teen to address UN again on need for clean water. (2019, September 27). *Gananoque Reporter.* Retrieved from: https://www.gananoquereporter.com/
Mercier, H. (2011). Reasoning serves argumentation in children. *Cognitive Development*, 26, 177-191.
Pontecorvo, C., & Girardet, H. (1993). Arguing and reasoning in understanding historical topics. *Cognition and Instruction*, 11(3), 365–395.
Schär, R. G. (2018). *An argumentative analysis of the emergence of issues in adult-children discussions.* USI Lugano: Dissertation.
Tesla, C., & Dunn, J. (1992). Getting along or getting your own way: The development of young children's use of argument in conflicts with mother and sibling. Social Development, 1(2), 107–121.
Turcotte, M. (2015). Political participation and civic engagement of youth. Ottawa: Minister of Industry.
Wall, J. (2014). Why Children and Youth Should Have the Right to Vote: An Argument for Proxy-Claim Suffrage. *Children, Youth and Environments*, 24(1), 108-123.
Wikipedia contributors. (2019, August 22). Amariyanna Copeny. In *Wikipedia, The Free Encyclopedia*. Retrieved 23:41, September 28, 2019, from https://en.wikipedia.org/w/index.php?title=Amariyanna_Copeny&oldid=912039800

Pictures and Reasoning: visual Arguments and Objections

ANGELINA BOBROVA
Russian State University for the Humanities
angelina.bobrova@gmail.com

The paper contributes to the debates on visual argumentation. From the perspective of Peirce's Existential Graphs theory, I'll specify the logical background of visual arguments conception and demonstrate how it cooperates with arguments evaluative techniques. A particular emphasis is made on the visual ways of argument refutation. The paper ends with an example that shows the profits of visual argumentation once we need to see reasons for arguments dissents.

KEYWORDS: argument, diagrammatic reasoning, dissent, reasoning, visual argumentation, visuality.

1. INTRODUCTION

Although the debated on visual arguments are still in progress, the fact of visual argumentation is almost not disputed. Indeed, "people now live in a reality that is not merely visually permeated – it is visually mediated" (Groarke, Palczewski & Godden, 2016, p. 233). Once we admit the idea of visuality, we have to find "a legitimate place for visual elements within argumentation" (Dove, 2012, p. 223) and decide if they are a special class of arguments or not. The current paper contributes to this long-lasting discussion. It focuses on the problem of arguments refutation through visual reasoning.

The mixture of linguistic (most common) and visual levels in argumentation is not surprising. Almost all arguments are multimodal, and "some of the most used arguments are not verbal in character. They are mental or logical or cognitive operations that can be expressed verbally as well as visually. ... The verbal loses its prerogative of being the paradigm of all argumentation" (Roque, 2009, p. 9), and speaking of argumentation in terms of words and sentences (Fleming, 1996; Johnson, 2003) is not relevant. Argumentation is not only linguistics activity, and it also presumes a visual part. Meanwhile, I would only partly share

Johnson's autonomy thesis, as Godden (2013, p. 4) classifies it, due to which "visual argument is a distinct and autonomous type of argument, and is not to be treated as an extension of verbal argument" (Johnson, 2010, p. 2). Visual arguments are specific, but they are, first of all, arguments, and this fact groups them with other reasoning styles.

This paper starts by discussing the definition of visual argumentation. The third section pays attention to the concept of iconicity that specifies both why visual arguments are arguments, along with others, and why they are, nevertheless, exclusive. The fourth section briefly introduces normative frames for evaluation and refutation, while the fifth one scrutinizes an example.

2. TOWARDS VISUAL ARGUMENTS DEFINITION

Visual arguments are popular but not trivial for dealing with them. It is hard to know if a picture is an argument or not. Once it has been interpreted as an argument, it is difficult to say whether the interpretation is correct. The frames and functions of visual arguments are still vague. Alcolea-Banegas insists that they are those "arguments in which the propositions and their argumentative function are expressed visually" (Alcolea-Banegas, 2009, p. 261). Groake specifies a visual argument as "an argument conveyed by (non-verbal) visual means." He continues: "One can find visual arguments that contain no verbal elements, but most combine the visual and the verbal. In some cases, a visual argument makes the same claims both visually and verbally, reinforcing the verbal with the visual (or the visual with the verbal). More frequently, the visual and the verbal contribute different elements that combine to create an argument" (Groarke, 2003, p. I).

There are many other definitions, and all of them appraise visuality as the distinguishing feature. At the same time, they often diverge in arguments essence understanding. Some approaches speak of arguments in a static manner (argumentative schemes) while others draw attention to their dynamic core (the process of argumentation). Some solutions identify normative frames (mostly logical) whereas some others introduce the problem descriptively (rhetorical). The borders between two distinctions are somewhat vague, but they specify the core of visual arguments that puts them in a row with other styles of argumentation. Which approach is more promising in our case?

The first dichotomy appeals to static and dynamical ways of reasoning interpretation. Indeed, it can be considered as a particular structure or a procedure. However, dynamical solutions come closer to the idea of dialogs, and it is essential for argumentation. Nevertheless, despite the differences, both approaches are various sides of the same coin. Both of them work within the field of argumentation, acknowledge the role of pragmatics and stress tight connections between arguers and listeners. The most important aspect is that they both refer to argument *schemata*, i.e., premises and conclusions relation that specifies the ways of how a proponent communicates with her opponents. This distinction emphasizes the crucial part of any reasoning: it should be seen as a *sequential activity* that follows premises and conclusion frames. If we see it in a picture, the latter can be called an argument. However, the easiest way to observe schemata is provided with a normative approach that solves the normative *versus* descriptive opposition. Since "arguments are 'claim-reason complexes'" (Hitchcock, 2007) and "the methodological focus of argument identification ... must include a search for reasons" (Godden, 2013, p. 6), the work within the normative approach seems to be more productive. There are several ways in which pictorial normativity can be treated (Blair, 2015, Aberdein, 2017), but it will presume rational or logical analysis. The logic simplifies arguments evaluations as it provides precise methods for it.

The normative dynamical approach gives a fruitful perspective for visual arguments studies as visual information can be the point of logical (Hammer, 1995) or mathematical (Cellucci, 2019) interests. Some logical conceptions diagrammatize inferences, e.g., Venn or Peirce's theories. Let me cite Peirce's Existential Graphs system[1] as an example. It is less known than Venn's ideas, but this theory suits us better as it clarifies what a visual argument is. Basic units of this conception are diagrams (Figure 1) that can be roughly correlated with propositions. Peirce calls his diagrams (Figure 1) as "moving pictures of thoughts". They move[2] towards the conclusion (see modus ponens transformations as an example), and such movements introduce argumentation as a specific process of information exchange, in which premises are logically related to their conclusions.

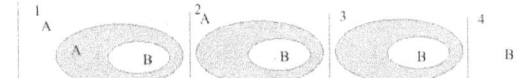

Figure 1 – Peirce's diagrams (modus ponens transformation:
"If A then B (two ovals). A. Consequently, B").

It seems that strict visual deductions discover the essence of reasoning, and this essence is worth being allied to visual arguments. When we look at a picture, these sequential relations have to be

---

[1] I will not draw correlations between Peirce's and algebraic logical theories. It is worth mentioning the papers for those who might be interested in this conception: Roberts (1973) and Zeman (1964) offer formal introductions whereas Champagne and Pietarinen (2019) provide an informal representation.
[2] The term "movement is not just about physical transportation.

observable. The conclusion has to be "made explicit, by gradually transforming the starting layout" (Champagne & Pietarinen, 2019, p. 26). Once it is done, "the visual argument risks begging the question" (Champagne & Pietarinen, 2019, p. 26).

To sum up, visual arguments are arguments if they demonstrate certain schemata. However, it does not mean that they are not specific. To see their peculiarity, we need another Peirce's idea.

3. ICONICITY

Visual and verbal arguments have similar schemata, but its transparency is different. Verbal arguments mostly demonstrate premise and conclusion procedures brighter than pictorial arguments do. Meanwhile, sometimes pictorial arguments are confident and even more persuasive than verbal. The term "persuasive" should not mislead as this paper develops a normative, which means mostly logical[3], rather than rhetorical, approach. In Peircean terms, we would say that different arguments have various levels and styles of iconicity[4]. Iconicity refers to semiotics (the study of signs and sign processes) with its most famous signs trichotomy: icons, index, and symbols. We live in a world of signs, and each sign can be scrutinized in terms of this trichotomy. Icons match signs with real conditions fixing resemblances with their objects; indexes are caused by their objects while symbols have no precise coincidence and have to be culturally learned.

It might come to mind that icons refer to visual arguments, and iconicity is a distinguishing feature of visual argumentation. However, it would be a rash decision as 'iconic' does not mean 'visual'. It is our habit to associate both words since we got used to perceiving the world visually. In the case of reasoning, iconicity is responsible for structure-preserving mappings, which means that we see an argument when we presume the premise and conclusion consequence. Logical schemata are always presented in an iconic way, but this way is not unique. There are several types of icons that are images, diagrams, or metaphors. If images presume precise resemblance, diagrams point to structural similarity whereas the resemblance in metaphors, the most distant from their object type, is getting more conditional (for details see CP 2.278-2.282). Any arguments are given iconically, but images are more iconic than logical diagrams, diagrams are more iconic than "the algebras, and the algebras are more iconic than linguistic expressions (see MS 1147, 1901)" (Pietarinen & Bellucci, 2017, p. 188).

---

[3] Logic it treated in both formal and informal ways.
[4] Stjernfelt (2011) scrutinizes this fact from the perspective of philosophy.

Iconicity let us see the common basis of all arguments. At the same time, it explains the peculiarities of pictorial argumentation. Pictures per se are icons that are modeled most closely on their objects (images). That is why they are read comfortable and fast. Visibility simplifies interpretation and objections construction as it represents the whole situation at once. At the same time, it works well only if a picture is correctly understood, *viz.* iconically designated content follows reasoning schemata that are also given in an iconic way. Otherwise, problems appear. For instance, two persons can treat the same image in opposite manners or miss the idea of argument at all. A picture works as an argument if the concept of drawing conclusions through the use of reasons is presented. There are two opportunities for how it could be done.

The first road deals with reasoning observation, *i.e.*, we see how premises are turned into the conclusion. The process can be fixed within one image (Figure 2, the 1st image[5]), or one picture can substitute the other one until the final result is reached. This procedure reminds commercials without words (Figure 2, the 2d image[6]). Besides, we can play with iconicity types. Figure 2 (the last pair) demonstrates how an image is diagrammatized (the transition from an image to a diagram). These movements turn a picture into an argument.

Figure 2 – Iconical games

The second way of schemata introduction is indirect. It relies on the fact that the majority of visual arguments are needed to be verbally specified. Verbal elements facilitate the right form of conclusion identification or, in other words, they identify the rule of how an image has to be moved and, as a consequence, read. Technically, such arguments behave like "joint arguments" (Roque, 2012, p. 283). The size of the verbal part depends on the situation. I would say that the relationship between visual and verbal components is inverse: fewer pictorial details, more words. For example, visual reasoning is unreliable for situations in

---

[5] The picture is borrowed from the Map Hause project (https://flint-culture.com/war-map-new-exhibition-book-showcase-rare-20th-century-pictorial-conflict-maps/).
[6] The pictures are borrowed from Groake (2017).

which limits are involved. If accurate information is needed, a picture will be fulfilled with a large text or even substituted with it.

To conclude, *pictures are arguments if they represent thoughts movements in the most iconic way that could match signs with real conditions by association and likeness.* This definition simplifies specifies why visual arguments should be estimated in a normative way.

4. EVALUATIVE CRITERIA OF VISUAL ARGUMENTS

Do we need new evaluative criteria for visual reasoning or one? Both alternatives are possible, but the second one looks more attractive. For instance, Gilbert proposes four argumentative modes: logical, emotional, visceral (physical) and kisceral (intuitive) (Gilbert, 1997, p. 75), when "each of the modes can define, for itself, relevance, sufficiency and acceptability" (Gilbert, 1997, p. 97). However, it is not clear how these modes are measured. At the same time, existed normative theories with its limited but well-developed and well-working methods get work done. It seems that Godden (2013, 2017) is right, and visual arguments do not require any revision. Any known approach, such as counterexample, arguments schemes and critical questions, ARS (acceptability, relevance and sufficiency) or ARG (acceptability, relevance and good grounds) and even fallacies method can be used. In fact, if visual argumentation is a part of the whole argumentation, its usual methods should be applied.

Groarke has provided general frames for the evaluation of visual arguments. He states them as follows:

> Once we have identified the structure of simple and extended visual arguments we can assess them by applying well-established theories of argument developed by logicians, rhetoricians and pragma-dialecticians. Among other things, these theories raise the questions:
> 1. whether a visual argument's premises are acceptable;
> 2. whether a visual argument's conclusion follows, deductively or inductively, from its premises;
> 3. whether a visual argument is appropriate or effective in the context of a particular audience or a particular kind of dialogue; and
> 4. whether a visual argument contains a fallacy or conforms to some standard pattern of reasoning (argument by analogy, straw man reasoning, modus ponens, and so on). (Groarke, 1996, p. 114)

These frames demand specifications, and below I will single out two solutions, namely Dove and Lake & Pickering's approaches, as examples.

Dove (2013, 2016) justifies argumentation schemes method to the needs of visuality (there is also a paper of Dove and Guarini, 2011). He argues that some subset of argumentation schemes could be of use in assessing visual arguments: "Schemes for patterns such as argument from analogy, argument from sign, argument from perception, or the various abductive schemes would also seem to be applicable with minimal modification to a variety of visual arguments" (Dove, 2016, p. 261). For example, visual analogies vividly match several situations, which is entirely predictable since analogy perfectly works with iconic representations. Dove notes that "the application of these schemes will always leave skeptics of visual argumentation with the suspicion that visual argument is simply parasitic upon verbal argumentation" (Dove, 2016, p. 261), but I think it is not the point. Reasoning structure does not belong to the level of words. It is captured iconically, and if a picture does this job, it starts representing an argument.

Lake and Pickering (1998) concentrate on procedures that presume refutations. They propose three methods, in which pictures "can refute and be refuted in a mixed-media environment." These are normative but quite informally introduced as follows:

> (1) dissection, in which an image is broken down discursively;
> (2) substitution, in which one image is replaced within a larger visual frame by a different image;
> (3) transformation, in which an image is recontextualized in a new visual frame. (Lake and Pickering, 1998, p. 79)

The scholars agree that their procedures are not the only strategies of visual refutation, but they propose an idea that can cooperate with Dove's approach.

Despite the differences, both conceptions pay attention to visuality and meet the requirements of normativity in arguments evaluation. They complement each other as they accentuate various sides of arguments analysis. If Dove's position can be identified as static (it is keen on argumentation schemes), Lake and Pickering's proposal regulates evaluative procedures dynamically (the dichotomy from the 1st section). In the next section, I will demonstrate how these approaches can be combined and effectively used as a criterion for patterns of dissents predictions.

5. EXAMPLE

The section analyzes the debates on the tallest building in Warsaw destroying. This building is the Palace of Culture and Science (the left image of Figure 3) that was built like a soviet gift and turned into political games after the collapse of the Soviet block. The question under discussion is as follows: is the building needed to be saved or destroyed? Although today the polemic is almost gone, this story provides an excellent instance of visual refutation. Besides, this example is quite famous in argumentative society as it has been analyzed at several conferences (ISSA, ECA, etc.). Today I will not scrutiny debates in general (political and ethical parts are left aside), but show how verbal argumentation can be visually declined.

Let me start with a passage from a polish newspaper. Argumentation is mostly verbal (the article is supplied with the only image of the palace), but I will use visual means for critics. Once we open a web-page, we read the following:

> A lot of people would like to completely erase socialist realist architecture from the cityscape," says Zawadzki, the architect. This includes some middle-aged adults and seniors who lived through the Soviet oppression who yearn to see the entire socialist realist style of the city replaced and Warsaw returned to a more traditionally Polish aesthetic.
> Like other socialist realist architecture spread across the former Soviet bloc, its mandate was to be "socialist in content, national in form," pairing realistic imagery with grand scale to control the public consciousness and help forge a new social order. In other words, it was designed to be an actual mechanism of tyranny.
> Some [people] have described it as ugly. But others believe it is now an integral part of the city even if it is not attractive to look at. (https://nextcity.org/features/view/the-movement-to-destroy-warsaws-tallest-building)

The passage argues against the building retention, and its arguments can be summarized as follows: (1) the building is an example of socialist realist architecture, which means (2) that it was designed to be an actual mechanism of tyranny; finally (3), it is not attractive but ugly.

To start critics, it is worth appealing to Dove's scheme from analogy and adding different similar pictures to the original photo (Figure 3). This step makes the resemblance visible.

Figure 3 – The row of similar buildings

If the pictures are similar, the initial image can be replaced by the other painting. This move reminds Lake and Pickering's procedure of transformation, in which an image is recontextualized. Let me substitute the left (original) photo with the picture from the right (Figure 3). The transformation automatically generates a counterexample that we need (Figure 4). Instead of socialist realist architecture, we have Art Deco style that ruins the first argument and questions the rest of them. The Art Deco style building can hardly be a mechanism of socialistic tyranny. It can scarcely be called 'ugly' as it is the Empire State Building in New York that was designated a National Historic Landmark in 1986 and was ranked number one on the American Institute of Architects' List of America's Favorite Architecture in 2007. Someone might say that such visuality does not provide a consistent refutation, but it, at least, undermines the confidence, which is also important.

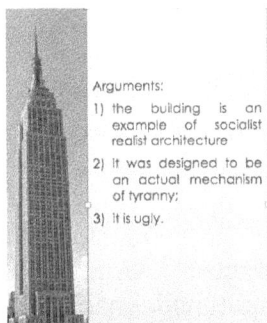

Figure 4 – Substitution and arguments

For those who are not convinced, the same resemblance can be presented dynamically (as a cartoon). We take the previous row of pictures (the first line in Figure 5), retract the color (the second line in Figure 5), and diagrammatize the images (the third line).

Figure 5 – Games with iconicity 2

We play with various types of icons and move from images toward diagrams. These movements clarify the analogy schemata as they make the resemblance quite obvious.

6. CONCLUSION

I am far from thinking that the paper makes a step towards a theory of visual argument (Birdsell & Groarke, 2007), but I guess that it clarifies three positions. First, visual arguments are arguments, but not every picture is an argument. To be recognized as such, it has to be correlated in our heads with premises and conclusions relations. It is possible if an image moves per se or is supplied with verbal elements that specify the way of transformations. Such transformations are not about the propositional nature of pictures. On the contrary, they argue for "an acknowledgment of the multimodal nature of communication in which argumentation as a social and rational activity finds its place" (Tseronis, 2013). Second, as visual arguments are arguments, they can be reconstructed, evaluated, and even abandoned if it is needed. To see that, we can use any normative conception of arguments analysis since "the cogency of the argument does not depend on what language it is stated in, just as it does not depend on whether the argument is heard, let alone understood. Rather, the cogency of an argument is a function of how well its reasons support its claim" (Godden, 2013, 2017). Third, visual argumentation is special, and the idea of signs and objects resemblance confirms it. Visual arguments can play with various icons, and this peculiarity is fruitful in cases of reasons to dissent investigations (see the example from the fifth section).

ACKNOWLEDGEMENTS: The reported study was funded by RFBR, project number 18-011-00889A.

REFERENCES

Aberdein, A. (2017). Virtuous norms for visual arguers. *Argumentation* 32(1), 1-23.
Alcolea-Banegas, J. (2009). Visual arguments in film. *Argumentation* 23(2), 259-275.
Bellucci F. & Pietarinen A-V. (2017) Two Dogmas of Diagrammatic Reasoning: a View from Existential Graphs. In K.A. Hull, R.K. Atkins (Eds.), *Peirce on Perception and Reasoning: From Icons to Logic*, (pp. 174-196). New York, NY: Routledge, 2017.
Birdsell, D. S., & Groarke L. (2007). Outlines of a theory of visual argument. *Argumentation and Advocacy* 43(3-4), 103-113.
Blair, J. A. (1996). The possibility and actuality of visual arguments. *Argumentation and Advocacy* 33(1), 23-39.
Blair, J. A. (2015). Probative norms for multimodal visual arguments. Argumentation 29(2), 217-233.
Champagne, M. & Pietarinen, A-V. (2019). Why Images Cannot be Arguments, But Moving ones Might. *Argumentation*. https://doi.org/10.1007/s10503-019-09484-0
Cellucci, C. (2019) Diagrams in Mathematics. *Foundations of Science*, 24 (3), 583-604.
Dove, I. & Guarini, M. (2011). Visual analogies and arguments. In Zenker, F. (ed.). Argumentation: Cognition and Community. *Proceedings of the 9th International Conference of the Ontario Society for the Study of Argumentation (OSSA)*, May 18-21, 2011 (pp. 1-16). Windsor, ON.
http://scholar. uwindsor.calossaarchive/OSSA9/papersandcommentaries/75
Dove, I. J. (2012). On images as evidence and arguments. In F.H. van Eemeren & B. Garssen (Eds.), Topical Themes in Argumentation Theory: Twenty Exploratory Studies (pp. 223-238). Amsterdam: Springer Publications.
Dove, I. (2013). Visual arguments and meta-arguments. In Mohammed, D. & Lewiński, M. (Eds.). Virtues of Argumentation. *Proceedings of the 10th International Conference of the Ontario Society for the Study of Argumentation (OSSA)*, 22-26 May 2013 (pp. 1-15). Windsor, ON: OSSA. http://scholar.uwindsor.ca/ossaarchive/OSSA10/papersandcommentaries/38
Dove, I. (2016). Visual scheming: Assessing visual arguments. *Argumentation and Advocacy* 52(4), 254-264.
Fleming, D. (1996). Can pictures be arguments? *Argumentation and Advocacy* 33(1), 11-22.

Godden, D. (2013). On the norms of visual argument. In Mohammed, D., & Lewiński, M. (Eds.). *Virtues of Argumentation. Proceedings of the 10th International Conference of the Ontario Society for the Study of Argumentation (OSSA), 22-26 May 2013* (pp. 1-13). Windsor, ON: OSSA. https://scholar.uwindsor.ca/ossaarchive/OSSA10/papersandcommentaries/54

Godden, D. (2017). On the norms of visual argument: A case for normative non-revisionism. *Argumentation* 31(2), 395-431.

Groarke, L. (1996). Logic, art and argument. *Informal Logic* 18(2), 105-129.

Groarke, L. (2003). Why do argumentation theorists find it so difficult to recognize visual arguments?

Groarke, L. (2003). Reply to R. Johnson's "Why 'visual arguments' aren't arguments." In J.A. Blair et al. (Eds.) *Informal Logic at 25: Proceedings of the Windsor Conference* (OSSA) (pp. 1-4). Windsor, ON: OSSA. http://scholar.uwindsor.ca/ossaarchive/OSSAS/papersandcommentaries/50/

Groarke, L. (2015). Going multimodal: What is a mode of arguing and why does it matter? *Argumentation* 29(2), 133-155.

Groarke, L. (2017). Informal logic. *The Stanford Encyclopedia of Philosophy* (Spring 2017 Edition), ed. E. N. Zalta, https://plato.stanford.edu/archives/spr2017/entries/logic-informal/

Groarke L., Palczewski, C. H. & Godden D. (2016). Navigating the visual turn in argument. Argumentation and Advocacy, 52, 217-235.

Hammer, E. (1995). *Logic and visual information.* Stanford: CSLI Publications.

Hitchcock, D. (2007). Informal logic and the concept of argument. In D. Jacquette, *Philosophy of logic*, pp. 101-129. Amsterdam: Elsevier.

Johnson, R. H. (2003). Why "visual arguments" aren't arguments. In A. J. Blair, D. Farr, H. V. Hansen, R. H. Jonson & C. W. Tindale (Eds.). Informal logic 25: Proceedings of the Winsor Conference (pp. 1-13). Ontario: OSSA

Johnson, R. H. (2010). On the evaluation of visual arguments: Roque and the autonomy thesis. [Unpublished conference paper, presented to] *Persuasion et argumentation: Colloque international* organisé par le CRAL à l'Ecole des Hautes Etudes en Sciences Sociales, 7-9 Septembre 2010.

Lake, R. A. & Pickering, B. A. (1998). Argumentation, the visual, and the possibility of refutation: An exploration. *Argumentation* 12(1), 79-93.

Peirce, C. S. *Collected Papers of Charles Sanders Peirce*, 8 Vols. Charles Hartshorne, Paul. 1931–58. [Cited as CP followed by volume and abstract number].

Peirce, C. S. *Manuscripts in the Houghton Library of Harvard University*, as identified by Richard Robin, 1967. // Annotated Catalogue of the Papers of Charles S. Peirce. Amherst. [Cited as MS followed by manuscript number].

Pietarinen, A.-V. (2015) Two Papers on Existential Graphs by Charles Peirce. *Synthese*, 192(4), 881-922.

Roberts, D. *The Existential Graphs of Charles S. Peirce*. Mouton. 1973.

Roque, G. (2009). What is Visual in Visual Argumentation? In: J. Ritola (Ed.), *Argument Cultures: Proceedings of OSSA 09* (pp. 1-9). Windsor, ON: OSSA.

https://scholar.uwindsor.ca/ossaarchive/OSSA8/papersandcommentaries/137

Tseronis, A. (2013). Argumentative functions of visuals: Beyond claiming and justifying. In Mohammed, D. & Lewiński, M. (Eds.). *Virtues of Argumentation. Proceedings of the 10th International Conference of the Ontario Society for the Study of Argumentation (OSSA)*, 22-26 May 2013 (pp. 1-17). Windsor, ON: OSSA. Windsor, ON. http://scholar.uwindsor.ca/ossaarchive/OSSA10/papersandcommentaries/163

Stjernfelt, F. (2011). On Operational and Optimal Iconicity in Peirce's Diagrammatology, *Semiotica,* 186(1/4), 395-419.

Zeman, J. (1964). *The Graphical Logic of C.S. Peirce*, dissertation, University of Chicago. (Online edition: 2002, http://web.clas.ufl.edu/users/jzeman/)

Issues leading to argumentative discussions during family mealtime conversations.

ANTONIO BOVA
Università Cattolica del Sacro Cuore, Milan, Italy
antonio.bova@unicatt.it

This paper sets out to investigate the issues leading parents and children aged 3-7 years to argumentative discussions during mealtimes. The research design implies a corpus of 30 video-recorded separate meals of 10 middle to upper-middle-class Swiss and Italian families. The findings of this study indicate that the argumentative discussions between parents and children unfold around issues that are generated both by parental directives and children's requests. The results of this study suggest that argumentative discussions during family mealtime are not mere conflictual episodes that must be avoided, but they have a crucial educational function.

KEYWORDS: Family; Argumentation, Parent-child interaction, Issues, Pragma-dialectical approach, Argumentum Model of Topics

1. INTRODUCTION

During family mealtime it is possible to observe how behaviors and opinions of family members are frequently put into doubt and negotiated (Bova, 2019; Bova & Arcidiacono, 2018; Bova et al., 2017; Fiese et al., 2006). The parents could easily avoid engaging in a discussion by advancing arguments in support of their standpoint, and yet resolve the difference of opinion in their favor, forcing children to accept, perhaps unwillingly, their standpoint (Bova & Arcidiacono, 2014, 2015). The difference in age, role, and skills with their children would allow them to do so (Arcidiacono & Bova, 2015; Blum-Kulka, 1997). Now it is evident that this happens frequently. However, equally frequently during mealtime, we can observe argumentative discussions, in which parents and children put forward arguments to convince the other party that their standpoint is more valid, and therefore deserves to be accepted (Bova, 2015; Bova & Arcidiacono, 2013a, 2013b). In this study, we shall try to understand when this happens. In particular, the present paper sets out to investigate the issues that lead parents to start an argumentative discussion with their children aged 3-7 years during mealtimes. The research question that I aim to answer is the following: "Which types of issues lead parents to start an argumentative discussion with their children during mealtimes?" This research question will be answered through a qualitative analysis of argumentative discussions between parents and children. The analytical approach is based on the pragma-dialectical ideal model of a critical discussion (van Eemeren & Grootendorst, 2004) that proposes a definition of argumentation according to the standard of reasonableness: an argumentative discussion starts when the speaker advances his/her standpoint, and the listener casts doubts upon it or directly attacks the standpoint. Accordingly, confrontation, in which disagreement regarding a standpoint is externalized in a discursive exchange or anticipated by the speaker, is a necessary condition for an argumentative discussion to occur. This model particularly fits this study, and, more generally, the study of argumentative interactions occurring in ordinary, not institutionalized, contexts such as family mealtime conversations, because it describes how argumentative discourse would be structured when aimed at resolving differences of opinion.

The paper is structured as follows: in the first part, a concise review of the most relevant literature on family argumentative discussions is presented and critically discussed. Afterward, I will describe the methodology that our study is based on. In the last part of the paper, I will present and discuss the results obtained from the analysis.

2. METHODOLOGY

2.1. Data corpus

The data corpus is composed of 30 video-recorded separate family meals (constituting about 20 hours of video data), constructed from two different sets of data, named sub-corpus 1 and sub-corpus 2. All participants are Italian-speaking and did not receive any financial support to take part in the study. The length of the recordings varies from 20 to 40 min. Sub-corpus 1 consists of 15 video-recordings of mealtime conversations of five Italian families living in Rome. For the selection of the Italian families, we recruited families including both parents and at least two children, with the younger of preschool age (three to six years old) and the other of primary school age. Based on the parental answers to questionnaires about socio-economic status (SES) and personal details that family members filled before the video-recordings, it was established that participants were middle to upper-middle-class families. Most parents at the time of data collection were in their late 30s (M = 37.40; SD = 3.06). All families in sub-corpus 1 had two children. Sub-corpus 2 consists of 15 video-recorded meals in 5 middle to upper-middle-class Swiss families with high socioeconomic status, all residents in the Lugano area. The criteria adopted in the selection of the Swiss families mirror those adopted in the creation of sub-corpus 1. At the time of data collection, most parents were in their mid-30s (M = 35.90; SD = 1.91). Sub-corpus 2 families had two or three children.

3. ANALYTICAL APPROACH

The pragma-dialectical ideal model of a critical discussion is used, in the present study, as a grid for the analysis, since it provides the criteria for the selection of the argumentative discussions as well as for the identification of the types of issues, which lead parents and children to engage in them. The pragma-dialectical ideal model of a critical discussion spells out four stages that are necessary for a dialectical resolution of differences of opinion between a protagonist that advances and sustains a standpoint, and an antagonist that assesses it critically: at the confrontation stage is established that there is a dispute. A standpoint is advanced and questioned; at the opening stage, the decision is made to attempt to resolve the dispute through a regulated argumentative discussion. One party takes the role of protagonist, and the other party takes the role of antagonist; at the argumentation stage, the protagonist defends his/her standpoint, and the antagonist elicits further argumentation from him/her if he/she has further doubts; at the concluding stage, it is established whether the dispute has been resolved on account of the standpoint or the doubt concerning the standpoint having been retracted. For the present study, the analysis will be focused on the first stage of the model of critical discussion, i.e., the confrontation stage, to identify the issues leading parents to start an argumentative discussion with their children. In this stage, the interlocutors establish that they hold different opinions about a certain issue: "the dialectical objective of the parties is to achieve clarity concerning the specific issues that are at stake in the difference of opinion" (van Eemeren & Grootendorst, 1992, p.138). The discussions between parents and children will be considered as argumentative if the following criteria are satisfied: a) a difference of opinion between parents and children arises around a certain issue; b) at least one standpoint advanced by one of the two parents is questioned by one or more children or vice versa; c) at least one family member (parent or child) puts forward at least one argument either in favor of or against the standpoint being questioned.

4. RESULTS

Within the 30 video-recorded meals constituting the general corpus of this research, I selected 127 argumentative discussions among family members. The argumentative discussions between parents and children represent a large part of the corpus of argumentative discussions (N=107; 84%). In particular, what emerges from the analysis of the 107 selected sequences is that the argumentative discussions unfold around two different types of issues that can be described through one of the following two questions: "Should child X do Y?" "May child X do Y?" The first question allows consideration of all issues generated by an initial request by one of the children with which (at least) one of the two parents showed to disagree. The second question, instead, allows one to consider all issues generated by an initial directive by one of the two parents with which (at least) one of the children showed to disagree. In most cases, the issues leading parents and children to start an argumentative discussion were generated by parental directives (N=76; 71%) and were related to the following categories (cf. Figure 1): feeding practices, the teaching of correct table manners, and social behavior of children outside the family context.

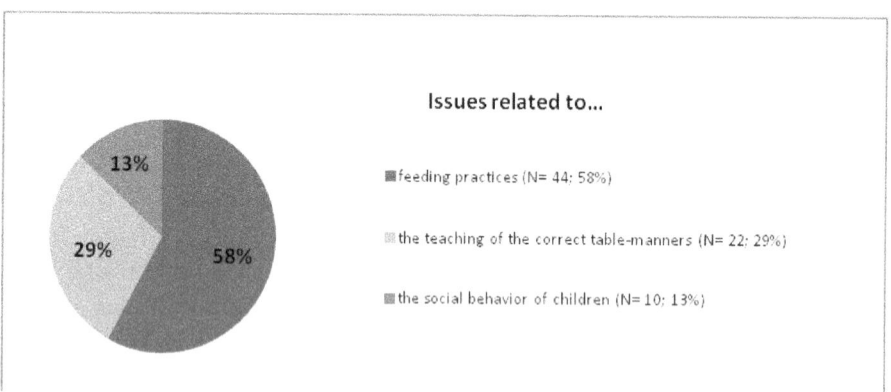

Figure 1: Types of issues generated by parental directives

Children's requests generated almost one-third (N=31; 29%) of the issues leading parents and children to start an argumentative discussion during mealtimes. I observed that these issues related to the following categories (cf. Figure 2): eating behaviors, the teaching of correct table manners, and behavior of children and parents both outside and within the family context. The categories are somewhat like the issues used by parents in their directives to children.

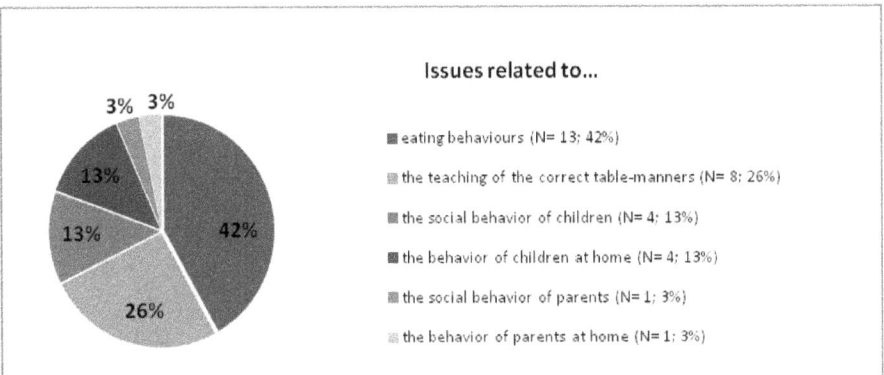

Figure 2: Types of issues generated by children's requests

Overall, the findings of this study indicate that the argumentative discussions unfold around issues that are generated most frequently by parental directives and less frequently by children's requests. These findings are like the results of other studies on family mealtime conversations. For example, Kendall (2008) has shown that the discursive positions fathers and mothers take up are oriented to negotiating authority and favoring connection with children. When mothers perform more meal-related and sociable functions, fathers support them discursively. Further evidence of these aspects is offered by Arcidiacono and Pontecorvo (2010) on parents' discursive positions and by Aronsson and Gottzén (2011) concerning how people shift between distinct intergenerational positions during family interactions at dinnertime.

In discussing the results, I will present a selection of some excerpts representative (regarding modality and frequency) of the results obtained from the entire corpus of data to offer a view of the types of issues leading parents and children to start an argumentative discussion during mealtimes.

4.1. Issues Generated by Parental Directives

Even though the issues generated by parental directives also include the social behavior of children outside the family context, the argumentative discussions related to issues generated by parental directives are in large part strictly bound to the specific situational activity children are involved in, i.e., the activity of mealtimes. These issues frequently concern feeding practices. The following discussion between a father and his 7-year-old son, Samuele, offers an illustration of how a parental directive related to feeding practices can trigger the beginning of an argumentative discussion.

Excerpt 1
Italian family. Participants: father (DAD, 38 years), mother (MOM, 34 years), Samuele (SAM, 7 years and 2 months), Daniele (DAN, 5 years and 4 months).

		%sit:	SAM is drinking a soft drink
1.		*DAD:	stop drinking XXX ((name of the brand of the soft drink)) Samuele!
→		*DAD:	now I'll give you some rice.
2.		*SAM:	no, I don't want anything else: ((sitting on the chair))
→		*SAM:	please, no more. [:! shaking his head in refusal]
3.		*DAD:	no:: you haven't eaten enough.
4.		*SAM:	no:::
→		*SAM:	no:: I'm full:
		%act:	SAM looks towards DAD and starts drinking the soft drink again
5.		*DAD:	I told you:: Samuele stop drinking this stuff ((the soft drink))
		%act:	DAD takes SAM's glass and takes it to the kitchen

The excerpt is opened by a father's directive (line 1) that can be interpreted as implicitly condensing a standpoint and a justification: in the analytical reconstruction of argumentation, the father's claim concerns an invitation to the child ("you should eat some food"), followed by a justification ("because you are drinking too much"). It is in line 2 when a difference of opinion between Samuele and his father arises. The child's intervention constitutes the beginning of the argumentative discussion, as the child replies to the father that he does not want to eat anything else. What is interesting in argumentative terms is the fact that Samuele does not consider that he must stop drinking, but immediately focuses on the central claim of the parent, namely to convince the child to eat the rice. Samuele's choice, i.e., a refusal to the father's proposal through a counter-position on his argument, determines the orientation of the discussion exclusively around the food. The father ratifies this specific direction of the argumentative discussion in line 3, as he puts forward an argument based on the quantity of food. However, as we can observe from Samuele's answer in line 4, this argument is not effective enough to convince the child to accept the father's standpoint. The opposition of Samuele ("no::: no:: I'm full") determines a change of strategy in the father's position. The adult turns back to the first directive (to stop drinking) to make explicit the fact that Samuele cannot discuss the parental issue anymore. The father's directive is advanced again using the expression "I told you:: Samuele" (line 5) and through the action of taking the soft drink away from Samuele. From the father's perspective, this last intervention is a way to re-conduct the discussion to the first level, giving as an argument the inappropriate conduct of Samuele who is drinking instead of eating. Gaining further insights into the context of family conversations can provide a richer perspective on the goals dominating the participants' argumentation: what could be interpreted as an imposition of the order could turn out to be a constructive move aiming at teaching the value of argumentation as a rational way to solve differences of opinion.

Parental directives did not pertain exclusively to feeding practices, but also the teaching of correct table manners. The following example shows how other types of parental directives can generate issues that lead parents to start an argumentative discussion with their children.

Excerpt 2
Italian family. Participants: father (DAD, 37 years), mother (MOM, 37 years), Gabriele (GAB, 7 years and 5 months), Adriana (ADR, 4 years and 4 months).

	%act:	GAB gets down from the table, and he is about to go and sit on the couch
1.	*MOM:	Gabriele, you can't go to watch TV on the couch
	%act:	GAB comes back to sit at the table
2.	*GAB:	but I want to watch TV on the couch!
3.	*MOM:	Gabriele, during mealtimes you cannot get down from the table
4.	*GAB:	why not?
5.	*MOM:	because it is ill-mannered to do it
6.	*GAB:	mmm
	%act:	GAB remains seated at the table and continues to eat.

In this sequence, the difference of opinion is between the mother and her 7-year-old son, Gabriele. The child leaves the table and is about to go and sit on the couch to watch TV. The mother disagrees with her son's behavior and makes her standpoint explicit in line 1. However, the adult's directive, in its actual form, does not provide any reasons. Gabriele interprets the fact that he is not allowed to watch TV as a directive against his wish. In fact, in line 2 the child, who came back to sit at the meal-table, disagrees with his mother and advances his standpoint using the adversative conjunction "but" to mark the different position concerning the adult statement. In this phase of the discussion, the issue leading the mother to start an argumentative discussion with her son is related to the teaching of correct table manners.

To understand the issue discussed in the presented sequence, the circumstances in which the argumentation takes place must be considered. In the present case, the possibility of watching TV is not a topic of discussion per se, but it is the fact that family rules, at least for this family, imply finishing dinner before going engaging in other activities (including watching TV on the couch). Accordingly, in this case, the mother-child argumentative discussion is evidence for semiotic regulation of new behavior acquisition because it has not only the purpose of teaching a new, good behavior, but also regulating their action (or, in this case, not acting). The discursive interventions of Gabriele have played a crucial role since his mother has been challenged to defend her standpoint. Based on this issue, the mother has been forced to specify the reasons for her directive, and to justify why he was not allowed to leave the table at that point. The implicit accusation made by Gabriele (the impossibility of going to watch TV despite his wish, "I want...") requires the parent to give a justification. The question is whether and how the participants use the potential of dissent to handle the critical question argumentatively. Finally, after the unilateral directive, the mother offers a strong disagreement preventing the possibility of continuing the debate.

4.2. Issues Generated by Children's Requests

Mostly, the issues generated by children's requests concern activities strictly related to mealtimes, such as eating behavior and teaching of correct social norms and behavior by parents. These findings are in line with the frequencies we found in the analysis of parental directives. The following dialogue between a father and his 7-year-old daughter, Manuela, is an example of how a child's request to the adult-related to having to eat a particular food can trigger the beginning of an argumentative discussion.

Excerpt 3
Swiss family. Participants: father (DAD, 39 years), mother (MOM, 34 years), Manuela (MAN, 7 years and 4 months), Filippo (FIL, 5 years and 1 month), Carlo (CAR, 3 years and 1 month).

1. MAN: can I leave this little bit of pasta? ((slightly raising the plate to show the contents to her father))
2. DAD: no, you can't
3. MAN: why, dad?
4. DAD: you haven't eaten anything, Manuela

In this exchange, the difference of opinion between the child, Manuela, and her father concerns the amount of pasta to be eaten: Manuela wants to leave a little bit of pasta that is still on her plate, but the father disagrees with her (line 2, "no, you can't"). While the child's expression, "this little bit," aims to obtain a concession, the father, on the contrary, replies with a prohibition. The adult's contribution opens the ground for an argumentative discussion because the participants express two opposite standpoints. By asking a why-question (line 3), Manuela is challenging the parental prohibition and shows her willingness to know the reasons on which the father's prohibition is based. This position is argumentatively strategic because it obliges the father to put forward an argument in support of his standpoint (line 4, "you haven't eaten anything, Manuela"), refuting the daughter's argument based on this little bit. However, the argument related to the presumed quantity of food that must be eaten closes the child's possibility to extend the argumentative exchange. The reason for the directive is connected to the non-consistent behavior of Manuela during the dinner and, for this reason, there is no further space for debating about the reasons why the child cannot leave some food.

The issues leading parents to start an argumentative discussion with their children were also generated by children's requests about the possibility of teaching the children how to behave correctly, both in social interactions within and outside the family context, especially at school. For instance, the following example illustrates how a request by the 5-year-old son, Alessandro, who wants to take a pill from the medicine container, can lead to an argumentative discussion with his mother.

Excerpt 4
Swiss family. Participants: father (DAD, 36 years), mother (MOM, 34 years), Stefano (STE, 8 years 5 and months), Alessandro (ALE, 5 years and 6 months).

	%sit:	ALE touches and looks at the container with the pills
1.	*ALE:	I'm: going to take one of these.
→	*ALE:	yes!
2.	*MOM:	you can't, Alessandro!
3.	*ALE:	what?
4.	*MOM:	you can't. ((shakes her head))
5.	*ALE:	why not?
6.	*MOM:	because children, have to take special medicine
→	*MOM:	they can't take medicine for adults
→	*MOM:	otherwise, they will get sick.

7.	*ALE: and before did you also feel sick?
8.	*MOM: no, because I'm an adult
9.	*ALE: and me?
10.	*MOM: you are still: a child
	%pau: 1.0 sec
%sit:	Alessandro bangs the medicine container on the table. MOM reaches toward him to try to make him eat a piece of fruit. ALE turns his head away quickly and slowly leaves the kitchen to go toward DAD and STE

In this exchange, the issue leading the mother to start an argumentative discussion with Alessandro is related to teaching the child proper behavior at home. The sequence begins when the child tells the mother of his intention to take a pill from the container. The argumentative discussion is opened by the mother, in lines 2 and 4, when she disagrees with the child's behavior, twice repeating, "You can't.". In this phase, we can observe that the child's standpoint (I want to take a pill from the container) meets with the mother's refusal (You can't, Alessandro). Interestingly, in the corpus, disagreements between parents and children are not only related to the generation of reasons regarding the truth-value of an assertion advanced by children, but also to the control of desired/undesired behavior by parents, e.g., on how to behave appropriately, both within and without the family context. As in previous cases, the argumentative strategy used by the child is the why-question to the adult to challenge the mother to defend her standpoint. In doing so, Alessandro makes no effort to defend his position by advancing arguments on his behalf; instead, he assumes a waiting position before accepting or putting into doubt the parental directive. The mother does not avoid justifying her prohibition, putting forward her argument and evoking a general rule – children have to – to which Alessandro is also subject.

Interestingly, in the corpus, I also observed one case where the issue leading to an argumentative discussion was related to the behavior of parents outside the family context, and one case to the parental conduct within the family context. These two issues, both generated by children's requests, can be described as follows: May Mom go to the sports hall to pick up Paolo? May Mom prepare breakfast for Dad every day?

6. DISCUSSION AND CONCLUSION

The results of this study indicate that argumentative discussions are not primarily aimed at resolving verbal conflicts among family members, but they mainly appear to be an instrument that enables parents to transmit, and children to learn, values and models about how to behave in a culturally appropriate way. Mealtimes appear as activity settings and opportunity spaces where family members intentionally and unintentionally express their feelings and expectations. Of course, just because opportunities exist does not mean they are taken. In our case, we observed that the argumentative discussions unfold around issues that are generated both by parental directives and by children's requests. The parental directives mostly concern context-bound activities such as having to eat a particular food or teaching correct table manners. The issues triggered by children's requests refer to a wide range of activities, mainly context-bound, but also in some cases context-unbound, such as the children's behavior outside and within the family context.

The observed dynamics characterizing family discussions reveal that argumentation is a co-constructed activity in which children play a role that is equally fundamental to that of their parents. Using a qualitative approach of analysis, i.e., the pragma-dialectical approach to argumentation, we have observed that by engaging in argumentative discussions, parents accept (assume) the commitment to transmit rules, values, and correct behaviors to their children. By participating in argumentative discussions with their parents, children can become more aware of their active role within the family context. The analytical reconstruction of how family members dialectically solve differences of opinion is thus a useful way to highlight choices, forms, and dynamics adopted by adults and children at mealtimes.

Furthermore, the argumentative exchanges we have observed in our data appear as areas of socialization in which accusations are used as declarative statements where a family member explicitly mentions the activity or the attitude that constitutes a violation, attributing a negative quality to it. Argumentative discourse in the family context enables behavioral change and behavioral control. The argumentative interactions between parents and children generate not only a cognitive effect but, also, a behavioral one. For example, parents start an argumentative discussion with their children to teach them how to behave appropriately not only at the meal-table but also in all situations in which children can be in contact with other people outside the family context. The school context and the children's behavior with their peers, e.g., schoolmates, represent the issues parents are most concerned about. A possible implication of this aspect concerns how the children's capacity to start an argumentative discussion with their parents could influence their future ability to be actively involved in exchanges with other adults.

Within the family context, argumentation can contribute to improving the conversational, social, and cognitive skills of adults and children. Argumentation is constitutive, not just disruptive, of social life, leading adults and children to continually renegotiate the norms of interaction and contributing to the construction of the family borders from a social and linguistic point of view. For this reason, a focus on the moves of people during daily argumentative discussions is, in our opinion, a way to understand how parents and children recognize continuously what they are doing and what they have to do with the interlocutors. As people can deal with disagreements through reasonable argumentative exchanges, this capacity is considered as a resource within the family context.

From an argumentative perspective, I believe that the role of children is not less important than the role of their parents. Through the analysis of argumentative sequences, we have seen that their presence and involvement in family conversations favor the beginning of argumentative discussions and represents a stimulus factor, inducing parents to reason with their children. Through their continuous questioning, children show their desire to find out the – often implicit – reasons on which their parents' standpoints are based. Children's questions reflect the children's desire to know and find out what is, until that point, unknown to them (Bova & Arcidiacono, 2013b). The questions asked by children to their parents and caregivers, in general, represent a great educational opportunity; because of the children's questions, the parents need to advance arguments in support of their standpoint. It is a responsibility of parents and caregivers, in general, to take advantage of the opportunity offered by children's questions, providing the educational responses that children need. This feature is connected to the value of family conversations as spaces in which the dynamics of generational positions can be developed as part of language socialization and interactional events.

REFERENCES

Arcidiacono, F., & Bova, A. (2015). Activity-bound and activity-unbound arguments in response to parental eat-directives at mealtimes: Differences and similarities in children of 3-5 and 6-9 years old. *Learning, Culture and Social Interaction, 6*, 40-55.

Arcidiacono, F., & Pontecorvo, C. (2010). The discursive construction of the fathers' positioning within family participation frameworks. *European Journal of Psychology of Education, 25*(4), 449-472.

Aronsson, K., & Gottzén, L. (2011). Generational Positions at a Family Dinner: Food Morality and Social Order. *Language in Society, 40*(4), 405-426.

Blum-Kulka, S. (1997). *Dinner talk: Cultural patterns of sociability and socialization in family discourse.* Mahwah, NJ: Erlbaum.

Bova, A. (2015). Adult as a source of expert opinion in child's argumentation during family mealtime conversations. *Journal of Argumentation in Context, 4*(1): 4-20.

Bova, A. (2019). *The Functions of Parent-Child Argumentation*. Cham: Palgrave Macmillan.

Bova, A., & Arcidiacono, F. (2013a). Invoking the authority of feelings as a strategic maneuver in family mealtime conversations. *Journal of Community and Applied Social Psychology, 23*(3), 206-224.

Bova, A., & Arcidiacono, F. (2013b). Investigating children's Why-questions. A study comparing argumentative and explanatory function. *Discourse Studies, 15*(6), 713-734.

Bova, A., & Arcidiacono, F. (2014). "You must eat the salad because it is nutritious". Argumentative strategies adopted by parents and children in food-related discussions at mealtimes. *Appetite, 73*, 81-94.

Bova, A., & Arcidiacono, F. (2015). Beyond conflicts. Origin and types of issues leading to argumentative discussions during family mealtimes. *Journal of Language Aggression and Conflict, 3*(2), 263-288.

Bova, A., & Arcidiacono, F. (2018). Interplay between parental argumentative strategies, children's reactions, and topics of disagreement during mealtime conversations. Learning, Culture and Social Interaction. *Learning, Culture and Social Interaction, 19*, 124-133.

Bova, A., Arcidiacono, F., & Clément F. (2017). The transmission of what is taken for granted in children's socialization: The role of argumentation in family interactions. In C. Ilie & G. Garzone (Eds.), *Argumentation across communities of practice: Multi-disciplinary perspectives* (pp. 259-288). Amsterdam: John Benjamins.

van Eemeren F. H., & Grootendorst, R. (1992). *Argumentation, communication, and fallacies. A pragma-dialectical perspective*. Hillsdale, NJ: Erlbaum.

van Eemeren, F. H., & Grootendorst, R. (2004). *A systematic theory of argumentation: The pragma-dialectical approach*. Cambridge: Cambridge University Press.

Kendall, S. (2008). The Balancing Act: Framing Gendered Parental Identities at Dinnertime. *Language in Society, 37*(4), 539-568.

Value-based argumentation and the Transition to Low Carbon Economy in Turkey and Portugal: Values, Uncertainty and Actions

HUTHAIFAH BUSUULWA
Ibn Haldun University TR
Huthaifah.busuulwa@ibnhaldun.edu.tr

This paper goes deeper into the values debate, according greater importance to the role of values in practical reasoning and argumentation. I argue that action on uncertain events, depend profoundly on values of the agents or the values that are incited by circumstances and the desire to achieve a desired goal. In such situations even if value premises are not always made explicit, they are nevertheless present in the decision-making process.

KEYWORDS: Circumstantial values, Endogenous motivations, Environmental argumentation, Exogenous motivations, Low carbon economy, Normative values, Practical reasoning.

1. INTRODUCTION

Argumentation analysis as a distinct part of discourse analysis has become one of the most suitable tools to discuss environmental issues. (Rodrigues, Lewinski & Üzelgün, 2019; Dryzek, 2013) Argumentation theory is better poised to be espoused to understand environmental issues because of its precise methods of analysis and evaluation of naturally-occurring argumentation (Lewinski & Üzelgün, 2019; Van Eemeren et al, 2014; Lewinski & Mohammed, 2016). Environmental argumentation has existed over the last century with various arguments and positions on environmental concerns (Lewinski & Üzelgün, 2019, p.1). Most of the time reasoning about the environment is **Practical reasoning**, people who are engaged in mitigation efforts deal with the question of what to do. It arises from problems that we face as agents in the world (Fairclough & Fairclough, 2011, p.6) it *"starts from an action-question: What shall we do?"* (Rodrigues, Lewinski & Üzelgün, 2019, p. 23) and involves arguing in favour of a conclusion, using selected means to reach some desirable goal. The World is in a period of transition towards the desired goal of a low carbon economy where the sustainability of the world's resources and economic activities go together hand in hand. A transition is difficult to achieve and it is often a messy process. It is getting widely recognized that the challenge of the 21st century is to transition towards a more sustainable energy system efficient and characterized by low carbon sources. This transition effort mirrors an earlier version from the 19th century that emphasized a transition from wood to water and from coal to oil in the twentieth. As (Dryzek, 2013) noted, a lot has happened in the field of environmental affairs.

Differing views are clearly manifested when it comes to the **practical reasoning**, which include debates about climate change. Like other political processes, actions are based on a deliberative process where decisions are arrived at cooperatively (Fairclough & Fairclough, 2012). Some believe that protecting the environment makes good economic sense (Balmford et al. 2002; Turner et al. 2003). Others disagree, arguing that protecting the environment because of an expected economic gain is not a right thing to do, simply because the environment ought to be valued intrinsically, not just for its instrumental value. They aver that it is because the environment exists that we can even take a step to think about economics, a step which in other words diminishes the value we ought to attach to the platform which makes our human existence possible in the first place. They therefore find it appalling that the environment is not prioritised by the economic-oriented people. Yet others take a middle position arguing that whether the environment is protected because of its intrinsic or its instrumental value, the end is the same; that at least efforts would be put in place to protect it. Whether the purpose is achieved through proper, morally thoughtful methods or without regard to moral values at all, if the end result is an action that protects the environment no one should complain. The best thing to do, they say, is to be sensitive to all environmental values and develop institutions that enable broad participation in the making of difficult environmental decisions.

Disagreements on difficult issues like climate change may not be easily allayed and the best outcome may not be winning the debate (for this could be hardly attained, especially where uncertainty reigns) but instituting processes that more people can agree upon to guide all through an outcome that could be welcomed by the majority. This is a challenge, but not impossible to achieve. Studying people's Values is a necessary condition for efforts towards compromise to bear fruit. Some people favour radical solutions now for fear that postponing action may lead to devastating and irreversible impacts. On the other hand, taking action immediately without getting to know more about the impacts risks potentially irreversible investments that would have been to greater benefit. This is also the case with the debate on lowering carbon emissions. There are no clear-cut criteria to know the consequences of either action, as a result, the actors turn to their value system for a verdict at that particular point in time. Uncertainty, with or without the ability of learning is sure to elicit divergent views which further engenders differing actions.

2. VALUES

It is generally accepted that values are an important element in every argument and practical reasoning (Allport, 1961; Perelman & Olbrechts-Tyteca, 1969; Audi, 2004; Brandom, 1998, Fairclough & Fairclough, 2011; 2012). Recently interest in the subject of values has resurfaced under the novel field of argumentation research. We need to look at the various ways that values have been utilized in literature and in natural language. This will help us be in position to identify values whenever they are used in a deliberation or argumentation.

2.1 Values as a tool for evaluation

Early usages of the word denote the act of appraising something's worth. It was used to "refer to the fairness and equivalence of the amount of a commodity in an exchange" (Rohan, 2000, p. 256). This understanding of values continues to be most prevalent in the day to day usage. When someone says she values (x) it often means that she attaches a deserving degree of worth to it. Early value theorists faced a dilemma of determining whether values should be investigated from the perspective of the entity being investigated or from the perspective of the person doing the evaluation? (Feather, 1975, p.3; Rohan, 2000, p. 256). According to Rohan, the dilemma has since been settled as contemporary value theorists investigate values from the perspective of the person, their priorities and the stimuli in environment in which they develop those motivations (Rohan, 2000, p. 256). Values now assumed broader meaning, not just evaluating something's worth, but also seen as a guide for action, motivation and adherence to beliefs (Audi, 2004, Fairclough & Fairclough, 2011;2012).

2.2 Values as Conformity, adherence (to beliefs and norms)

Another prominent usage of the term values is used to denote a principle for conformity to some standard or forms of behaviour such as obligation, fidelity, loyalty, solidarity and discipline (Perelman & Olbrechts-Tyteca, 1969, 1969, p.77). People use the term to require themselves to live up to what they believe in as persons and as a collection of persons. As a result, several scholars have distinguished two kinds of values; personal and societal values (Rohan, 2000; Williams, 1979). There is consensus that values are important predictors of behaviour and attitudes and are cherished across cultures. (Pakizeh et al., 2007; O'Brien, 2009, p. 166). Some values are universal while others are particular to an individual or group. Confucius's five universally binding obligations- between rulers and the ruled, father and son, husband and wife, older brother and younger brother, friend and friend can apply in most societies and therefore can be said to be universal values. Here the values are norms, forming a part of the social/ cultural background (Durnova p. 721; Blakeley & Evans, 2006, p.29). Schwartz (1994) for instance identified ten types of universal values found in all societies; security, tradition, conformity, power, achievement, hedonism, stimulation, self-direction, universalism and benevolence (see also O'Brien, 2009, p. 167). Before Schwartz, Rokeach (1958) identified thirty-six values which he believed existed within all societies, he classified them into two, eighteen 'terminal' and eighteen 'instrumental' values. Generally, here values are seen as conceptions of the desirable, and often times people are eager to follow or to see them followed. This is because values are emotionally charged and people are mobilized to follow those with whom they share similar values creating identities and clashes (Durnova, 2018, p. 722; Barnes, 2008; Hochschild, 1990; Stavrakakis, 2008; Honneth, 1996). These are reflected in differences of opinion, policy and planning controversies (Durnova, 2018, p. 721; Griggs & Howarth, 2004, Gualini & Majoor, 2007; Huxley, 2010; Schön &Rein, 1994).

2.2 Values as motivation, Principles for guiding actions

The third broad conceptualisation of values denotes values as motivation, guiding principles for action. According to Rokeach, (2000, p.2) they serve as standards or criteria to guide judgment and action. In this case, concepts such as choice, attitude, evaluation, rationalization, attribution of causality among others are espoused (O'Brien, 2009, p. 166). As rational beings, people's actions are formed as a result of some kind of deliberation or thinking, and the resultant actions then deserve to be called rational judgements. There is always an element of beliefs in actions, whether they are due to ethical reasoning or practical reasoning and argumentation or not. According to Robert Audi (2004, p.123) beliefs are needed to guide action. Audi suggested that motivational reasons are one of the main kinds of reasons for action others being normative and explanatory reasons (Audi, 2004 p 120). For Schwartz (1987) motivational goals underlie value priorities. People are always motivated to engage in situations or actions they deem to produce positive affect (Schwartz, 1987; Rohan, 2000) In this case, values are defined as motivating factors, the principles upon which a person is moved to act, evaluate action or inaction and potential consequences. They are what we care about, and they portray our internal and external inclinations. (Fairclough, 2011, p.8, Keeney,1992, pp 3-6).

2.3.1 Fairclough's method

In Fairclough and Fairclough (2012), values as premises are a part of their proposed structure of practical argumentation scheme. This structure includes a circumstantial premise, or the current state of affairs which is seen as a problem and ought to be solved, a goals premise which depicts the desired action that should be achieved in the future once action is taken, a claim for action premise which discuss what needs to be done considering the current state of affairs, a means-goals premise which shows how the desired outcome can be achieved and a values premise which supports the other values. Values and emotions are necessary premises in a practical argumentation because without them nothing would matter (Fairclough & Fairclough, 2011, p.5; Blackburn,1998 cited in Fairclough). He emphasizes that all arguments have a motivational component and the results portray what one wants and values (Fairclough & Fairclough, 2011, p.5). Values justify the goal that an individual or party chooses to adopt, points out the problematic circumstances that calls for action and what ought to be done to ameliorate them in the best possible way.

When a string of positions is studied, it becomes easier to discern the value system with or without prior background knowledge about the subject under consideration. For instance, a college educated pro-choice, pro-marriage equality, pro-cannabis, climate change activist American even without describing her political inclinations would most likely identify or be identified as a progressive liberal acting under the influence of progressive liberal values. A person who has a strong conservative value system would most likely be against allowing same sex marriage. Values can therefore be empirically investigated.

3 DISAGREEMENTS, VALUES AND ACTION: POPULATION PRESSURE AND THE GOAL OF LOW-CARBON ECONOMY IN PORTUGAL AND TURKEY

3.1 On Normative and Circumstantial Values

I classify Fairclough's values as Normative values. The key concern they address is whether or not the agent acts on what he/she is concerned about or ought to be concerned about. Disagreements on the necessary course of action engender actions that can be said to follow normative rules and those that do not. Here agents act consciously or otherwise with cognition of what they ought to do, or what they ought to be concerned about in the course of their actions.

In addition to this normative understanding of values, I find it necessary to distinguish such values that are shaped primarily by the context, the conditions without which a different action would have been undertaken. These I call Circumstantial values.

Disagreements emanating from this category often stem from the means-goals, the strategies to achieve a desired future state of affairs. In this case, there is often a general agreement on the circumstances leading to the dilemma. What causes disagreement is how to achieve the goal.

Circumstantial values accrue from the background informing an action. They can be constrained by normative obligations, but deciding on whether to act in a certain way depends on the weight of the circumstances faced. Both normative and circumstantial values are mutually constraining.

From our research it was noticed for instance that most respondents agree on the desirability of achieving a low carbon economy. However, there is no consensus on how to address the present challenges. At this stage, values come in. What motivates some to believe in one course of action and others to be opposed to it?

To answer this question, I analysed the discourse on population and found out that respondents implicitly espouse both normative and circumstantial values.

Many respondents pointed out that overpopulation is a huge challenge that threatens to derail achieving the desired state of low carbon economy. Here I notice a general acceptance of the goal and the circumstance premises as exemplified below;

> "I think that the largest problem that we have is over-population and that's driving everything. Because, while we know that ten people can live in this planet, because we have enough resources, right? But the question is, can ten billion people live, even if it's 100% efficient, even if you're absolutely efficient in the way you use your resources, can ten billion people live on this planet? Maybe. Can 20 billion people live on this planet, probably not. Can 100 billion people live in this planet, absolutely not" (Interviewee P.1)

> "Not one of the problems is population. Major problem is population. You know for example in 2014 there was the election for mayor here in İstanbul, and everybody wanted to be a mayor here in İstanbul, you know, for money, for power, for everything. So, they asked me they actually, one of the

candidates let's say had a meeting with the scientists and asked "what is the major problem of İstanbul?" Everybody said that "traffic etc." I said population..." (Interviewee T22)

Argument reconstruction

1. The largest problem we have is over-population
1.1. We are faced with a circumstantial problem of over-population
1.1.1. Over-population takes precedence above all other problems

We notice that both respondents agree on the circumstance and the goal. They both agree that population is a challenge that needs to be ameliorated. From here disagreement on the course of action begins to surface.

> "So, I think that the real, big question, is how do you limit the growth of the population. Because, the problems that arise from climate change, from everything, are much bigger because of the extensively large amount of people that live on this planet" (Interviewee P.1)

> "Yes. I mean, we are growing steadily - except the big epidemic in the middle ages...but apart from that, there has to be a correction. Either we have, we will have to correct our population by ourselves. By saying that no more than one child because there is no food for us. Or the government will say, like in China, "No more than one child." Or, we will be 15 billion people, and suddenly there will be a famine, and we will die" (Interviewee T2)

Despite an agreement on the goal, circumstances may require different means- goals because of differences in values and circumstances at play. In China, for instance, the government could easily come in and put a limit on the number of kids one could have. Yet, as one respondent noted, the same policy would not be necessary in a country like Portugal. The difference in circumstances between the two countries play a fundamental role in having different course of action as observed below;

> "And this comes to a sociological problem, that the Chinese have kind of solved, because in China, which is you can only have one child, right? But you cannot do this in Portugal. You cannot say "hey, you can only have one child", actually because we should have more here. So the problem that the Chinese have is the opposite of the problem that we have, but overall the population is growing and growing faster every year, so,

right now, there are more living people, than people that ever lived in the face of the Earth, from the beginning of time" (Interviewee P1)

Argument Reconstruction

1.1 Whereas there is a general problem of population,
1.2. And the goal is the same,
1.3. The route that was taken in China should not be taken in Portugal,
1.4. Because the circumstances are different

Faced with such a dilemma the normative and circumstantial values are weighed and the actor may select from either depending on what he or she prioritises. This is done either implicitly or explicitly. Let us take for instance these two submissions below;

> "And the problem is that, from time to time there were like these huge wars that killed twenty, thirty, forty million people. Right now, even if one of those wars occurred, forty million people it's absolutely nothing. And I'm not saying that I want a war, of course (smiles), I'm just saying it's historical. You know, there has been recycling, the plagues have recycled, the wars have recycled people, so less people were there. Right now, there is nothing beside a huge cataclysm that can recycle what's happening... So ... I don't know, I don't have an answer for that" (Interviewee P1)

Argument Reconstruction
1.1. There is a problem of over-population
1.2 a. It can only be solved by a huge catalyst to solve that problem
1.2 b. I think we need that catalyst
1.3. I cannot say it explicitly, it does not sound right to want a war

> "So, they asked me they actually, one of the candidates let's say had a meeting with the scientists and asked "what is the major problem of İstanbul?" Everybody said that "traffic etc." I said population. Everything depends on population. The water, because we are 17 million, the traffic and the pollution, everything comes from the population. We have to kill some, like Hitler." (Interviewee T22)

Argument Reconstruction
1.1. To solve Istanbul's problems
1.2 a. There are too many people in Istanbul
1.2 b. We have to be concerned about the over-population
1.3. We have to solve the problem of population

1.4. The way to do it is by killing some people.
Respondent T 22, explicitly gives a solution that respondent P1 implied but was hesitant to make explicit, perhaps because of normative reasons. Respondent T22 appears to have weighed that the circumstantial values needed to take precedence over normative values, and the value priority in this case was in favour of circumstances rather than the moral implications of killing people and thus did not find it impulsive to prescribe the remedy of death to thousands or even millions of human beings.

Those that are explicit in most cases do so when attaching a worth to something which acts as stimuli. Here it is easy to identify that a value has been invoked, because they are often overt. However, when it comes to other ways of espousing values, more detailed investigation is necessary to detect that indeed values are present. We can only do so using inference. Conclusions in such cases can only be plausible, there is room for error. Yet, like Brandon (1998) noted, we licence others to infer our beliefs from our explicit claims and our implicit or overt actions (Brandom, 1998: 129). Two ways can be identified as a starting point to highlight values from texts, endogenous and exogenous motivations.

3.2 *Endogenous motivations*

Some people are motivated to act on climate change because of reasons that can best be described as endogenous. They are inspired to act because of reasons such as personal interest in the field, fear for what the future could be, love for their family among others. These are often independent motivations forged not from the out but from within the actor's interests. Apportioning blame for climate change and its negative consequences is a major source of differences. A sizeable number think that it is humans who are responsible for it, the Anthropocene are contrasted by others who believe that the role of humans is very limited or negligible in the broader sense. The Anthropocene fear that the recent actions of man have worsened the environment and fear that if things continue, the future will be bleak.

Fear is one of the main drivers that's inspiring action on Climate Change in both Turkey and Portugal. The general fear is that without proper action, the following generations might suffer irreparable consequences if action on ensuring a low carbon economy is not implemented. This fear is often exacerbated by events such as changing weather patterns, disasters among others which upon being attributed to climate change leads to fear that trouble is coming even sooner than expected.

Those with families and children take fear even further, imagining a bleak future for their children. One interviewee from Portugal had this to say;

> "Unfortunately, I'm very sceptic about the future, of the environment and the world, I have four kids, four daughters, and I'm really worried about their future, because the climate change and the warming of the world, cannot be addressed with the measures that we have now. We have to invest a lot more than we have invested until now" (Interviewee P8)

For that person all other considerations including money are not as important as the future of her children, which in this case is the priority value. This can be seen from the argument below;

> "There is this economic reasoning beside the company, that I have of course to be concerned, but I think you can be balanced, the sense and the concern that I have in my guts, talking like this, about the environment, about the climate, about the future of my own children, I don't see it shared by all the directors of this company or other companies worldwide" (Interviewee P8)

We can see several values from the statement, economics which the respondent acknowledges is important. But even more important to her is concern about the environment and the future of her children.

3.3 *Exogenous motivations*

The source of motivation in this category is external to the actor or her immediate others like children and relatives. It does not mean however that the strength of motivation is questioned, far from it. An externally generated motivation can still be the key reason for action surpassing all others in value attached.

> "we have to deal with the idea that we have a target, we have a deadline.... and that we still don't know how to deal with that deadline, but we have to do it.... And we have to keep investigating and trying new things and... fast tracking everything that we try" (Interviewee P1)

External conditions may present the urgency requiring to act immediately rather than later. A growing number of actors have warned of a looming danger of catastrophes, and deadlines have been proposed in order to avert tragic occurrences. This has inspired many to want to act as soon as possible.

4. CONCLUSION

In a nutshell, values are the reason to dissent. It is because of one's values that they are motivated to act in one way or another. In environmental argumentation, values have featured prominently as the supporting premise to the goal, means to goal, circumstantial premises. In this paper we have endeavoured to portray the extent of the relevance of values in argumentation. To this end, we have distinguished two main separate but complimentary categories of values; the normative values and the circumstantial values. The normative values are the ones dealing with what one is concerned about or ought to be concerned about. On the other hand, Circumstantial values accrue when circumstances, rather than the deontic influence is prioritised before an action is taken. Motivation being a key component of the value structure is engendered by either endogenous or exogenous influences. From our research, we noticed that the actor's determination to do to not to do something was influenced either by factors that could be categorised as personal like the desire to earn more, to protect their family and children and fear. Others were influenced more by external considerations such as the policy of the European Union or market considerations.

Trying to solve differences necessitates understanding the value considerations that are fronted or that are driving a person or group. To Resolve disputes and conflicts generally requires careful deliberation targeting the opposite side's values and motivations. It helps to see the person's point of view and makes coming to a middle ground more achievable.

AKNOWLEDGEMENT

This research was made possible because of the grant by TUBITAK, 115K370, ARDEB, SOBAG.

REFERENCES

Audi, Robert (2004) Reasons, Practical Reason, and Practical Reasoning, *Ratio (new series) XVII,* Blackwell Publishing Ltd. pp.119-149

Allport, G. W. (1961). *Pattern and growth in personality.* New York: Holt, Rinehart & Winston.

Barnes, M. (2008). Passionate participation: Emotional experiences and expressions in deliberative forums. *Critical Social Policy,* 28 (4), 461–481.

Blackburn, S. (1998) *Ruling Passions. A Theory of Practical Reason.* Oxford: Clarendon Press.

Blakeley, G., & Evans, B. (2009). Who participates, how and why in urban regeneration projects? The case of the new 'city' of East Manchester. Social Policy & Administration, 43(1), 15–32.

Brandom, Robert (1998) Action, Norms, and Practical Reasoning, *Language, Mind and Ontology, Nous,* Vol.32, Supplement: *Philosophical Perspectives*, 12,127-139.

Dryzek, J. S. (2013). *The politics of the Earth: Environmental discourses.* 3rd edition. Oxford: Oxford University Press.

Durnova, A. (2018) A tale of 'fat cats' and 'stupid activists': contested values, governance and reflexivity in the brno railway station controversy. *Journal of Environmental Policy & Planning*, 20:6, 720-733

Eemeren, F. H. van, Garssen, B., Krabbe, E. C. W., Snoeck Henkemans, A. F., Verheij, B., & Wagemans, J. H. M. (2014). *Handbook of argumentation theory.* Dordrecht, Netherlands: Springer.

Fairclough, I., & Fairclough, N. (2012). *Political discourse analysis: A method for advanced students.* London: Routledge.

Fairclough, Isabela and Fairclough, Norman (2011) Practical reasoning in political discourse: The UK government's response to the economic crisis in the 2008 Pre-Budget Report. *Discourse & Society,* 22 (3). pp. 243-268.

Feather, N. T. (1975). *Values in education and society.* New York: Free Press.

Griggs, S., & Howarth, D. (2004). A transformative political campaign? The new rhetoric of protest against airport expansion in the UK. *Journal of Political Ideologies*, 9(2), 181–201.

Gualini, E., & Majoor, S. (2007). Innovative practices in large urban development projects: Conflicting frames in the quest for 'new urbanity'. *Planning Theory and Practice*, 8(3), 297–318.

Hochschild, A. R. (1990). Ideology and emotion management: A perspective and path for future research. In T. D. Kemper (Ed.), *Research Agendas in the Sociology of Emotions* (117–179). New York: State University of New York Press.

Honneth, A. (1996). The struggle for recognition the moral grammar of social conflicts. Cambridge, MA: MIT Press.

Huxley, M. (2010). Problematising planning: Critical and effective genealogies. In P. Healey & J. Hillier (Eds.), *The Ashgate research companion to planning theory: Conceptual challenges for spatial planning* (135–158). Farnham: Ashgate.

Lewiński, M., & Üzelgün, M. A. (2019). Environmental argumentation: Introduction. *Journal of Argumentation in Context*, 8(1), pp. 1–11.

O'Brien L. K. (2009). Do values subjectively define the limits to climate change adaptation? In Adger W. N, Lorenzoni I., O'Brien L. K. (ed.) *Adapting to Climate Change: Thresholds, Values, Governance.* (pp 164-181), Cambridge University Press.

Pakizeh, A., Gebauer, J. E. & Maio, G. R. (2007). Basic human values: inter-value structure in memory. *Journal of Experimental Social Psychology* 43: 4 58-4 65.

Perelman Ch. & Olbrechts-Tyteca L. [1958] = (1958). *The New Rhetoric. A Treatise on Argumentaion*. Trans. Wilkinson, J. & Weaver, P. Notre Dame: University of Notre Dame Press.
Rodrigues, S., Lewiński, M., & Üzelgün, M. A. (2019). Environmental manifestoes: Argumentative strategies in the *Ecomodernist Manifesto*. *Journal of Argumentation in Context*, 8(1), 12–39.
Rohan, Meg. J. (2000). A rose by any name? The values construct. *Personality and Social Psychology Review, 4*, 255-277
Rokeach, M. (1968). *Beliefs, attitudes, and values*. San Francisco: Jossey-Bass.
Rokeach, M. (1973). *The Nature of Human Values*. New York: Free Press.
Rokeach, M. (ed.) (1979). Understanding Human Values: Individual and Societal. New York: Free Press.
Rokeach, M. (2000). *Understanding Human Values*, 2nd edn. New York: Simon and Schuster.
Schön, D. A., & Rein, M. (1994). Frame reflection toward the resolution of intractable policy controversies. New York: Basic Books.
Schwartz, S. (1994). Are there universal aspects in the structure and contents of human values? *Journal of Social Issues, 50*, 19–45.
Schwartz, S. H. 1996. 'Value priorities and behaviour', in Seligman, C., Olson, J. M. and Zanna, M.P. (eds.) *The Psychology of Values, Ontario Symposium, vol. 8*. Mahwah: Lawrence Erlbaum, pp. 1–24.
Silverstein, M. (2017). Ethics and Practical Reasoning, *Ethics (January 2017)* The University of Chicago. 353-382.
Stavrakakis, Y. (2008). Subjectivity and the organized other: Between symbolic authority and fantasmatic enjoyment. *Organization Studies*, 29(7), 1037–1059.

Dissent: Considering Culture and Personality

LINDA CAROZZA
York University, Toronto
lcarozza@yorku.ca

If "argument" and "arguing" (O'Keefe 1977) are both mired in culture and the personal, then how we conceive of argument-making and argument-having can both be influenced by where we come from and who we are. This paper functions as a discussion based on investigations into the factors of culture and personality within critical reasoning classrooms. Contrary to what we may believe or want, who you are and where you come from might matter more than we want them to.

KEYWORDS: argument, critical reasoning, critical thinking, culture, general education, temperament

1. INTRODUCTION

My approach to arguments is open-minded, valuing theories and methods that acknowledge and incorporate different modes of argument (Gilbert 1994). A pivotal attribute of this approach is that it acknowledges that the field continuously grows and changes, so that it is not a stagnant outlook of argumentation. People and their means of communication change over time depending on culture, subculture, age, status, gender, relationships, and contexts that they find themselves in; it follows from this that the theories that describe and address the communication of arguments should be amenable to such changes as well. I share this as an introduction to provide a glimpse into the spirit behind current research, empirical in nature.

If argument and arguing, argument$_1$ and argument$_2$ (O'Keefe 1977), are both enmeshed in culture and the personal, then how we conceive of argument-making and argument-having can both be influenced by where we come from and who we are. This paper functions as a discussion based on investigations into the factors of culture and personality within multiple deliveries of a general education course that has a specific focus on informal logic. I share empirical results of two ongoing studies within the Scholarship of Teaching and Learning. The first study that I discuss reviews reflective and critical practices of English language learners (ELL) in a general education course titled *Reasoning in Everyday Language* (REL). The second study that I discuss relates to temperament and student success in a course titled *Techniques of Persuasion* (ToP). General education courses function as courses that are interdisciplinary in nature and offer breadth to a degree. Being a general education course at a large metropolitan Canadian university, the students who enrol in such courses are heterogeneous in nature. Students could be from any faculty, studying any major, and they can be taking one of the first courses within their degrees or their last. So, other than being students, there's little else that is similar across all enrolled students.

2. CULTURE

REL is a new general education course, developed and delivered inaugurally in 2016-2017. It is open to students who identify as ELL, as there is an emphasis on language skills. Students were given participation assignments throughout the course. 10% of the participation grade involved completion of text exercises in a critical thinking textbook, and 10% of the participation grade involved writing a reflection. The reflection responded to questions from Brookfield's Critical Incident Questionnaire (2011). The questionnaire encourages students towards reflexivity by creating, "a habit of looking back at learning" (Hessler & Rupiper Taggart, 2011). The questions include: At what moment in class this week did you feel most engaged with what was happening? At what moment in class this week were you most distanced from what was happening? What action that anyone (teacher or student) took in class this week did you find most affirming and helpful? What action that anyone (teacher or student) took in class this week did you find most puzzling or confusing? What about the class this week surprised you the most? (This could be something about your own reactions to what went on, or something that someone did, or anything else that occurs to you). These are open-ended questions, and students were encouraged in the instructions to justify and expand their thoughts.

In analyzing reflections from the 2016-2017 course only, Ryan's scale for reflective practice was used (2013). The four levels in this scale include Level 1: Reporting and Responding (What was the incident/challenge/opportunity?); Level 2: Relating (Have you seen this before? What skills can you apply?); Level 3: Reasoning (Add a critical perspective to deepen reflection; use sources to support your ideas); Level 4: Reconstructing (Reconstruct future practice, or offer advice or recommendations, given your reflection). In particular the third and fourth levels of reflective ability encompass the skill of argument-making.

A summary of results follow. 80% of students completed all textbook activities; 60% of students completed all reflective activities.

Thus, the end-of chapter questions in a textbook had more student engagement. Only 10% of students took advantage of extra reflection opportunities. 40% of students had the hypothesized results: to continually perform better on reflective activities. Furthermore, 100% of learners had a higher-grade average in textbook exercises than reflections. Overall, students could discuss what happened in a particular class or unit of instruction, by summarizing relevant events, but they could not make connections with their experiences, their analytical thoughts, or think of paths moving forward given their observations. It follows from this study that all students were firmly rooted in Level 1 reflective ability, which involves mainly reporting and summarizing some thought or event.

Some of the suggested pedagogical outcomes from this inaugural study into ELL and reflective practice include: make no assumptions regarding students reflective backgrounds and experiences; reflective practice may need to be modelled to students; students need practice to hone the skill of critical reflection (Coulson and Harvey 2013). However some of the qualitative feedback from student reflections revealed that students had no familiarity with argument-making as they were being asking to demonstrate (i.e. Level 3 and/or 4 reflective writing). While ELL could evaluate arguments when given informal logic tools, expecting them to be able to employ the skill of argumente-making, in order to craft their own strong views, was presumptuous on the part of the instructor.

Students often expressed that they preferred that the instructor, "just tell them what to write because they want to do well, but they don't know how to do well." Exploration around this reveals that argument-making, in an informal logic sense, was foreign to many ELL because their educational background taught them that the term "argument" is associated with formal logic. Providing a claim or thesis, that is opinionated in nature, with supporting rationale was foreign to many students. There is scholarly dialogue about this cultural gap in argument-making and thinking critically. O'Sullivan and Guo (2011) and Guo and O'Sullivan (2012) engage in discussions of cultural gaps in Chinese ELL studying in graduate school in Ontario Canada, where thinking critically was conflated with formal logic.

What this shows, from a pedagogical perspective, is that being cognizant of learners' cultural backgrounds is important. While this study did not have a large enough sample size to make any firm conclusions, there is a general lack of critical thinking and reasoning skills with most ELL. Whether or not this relates to culture is impossible to determine. More data has been collected since this presentation was delivered though, and it seems to confirm the results rendered. There may be cultural differences with reflective practices.

Western culture values reflection levels 1 to 4 hierarchically, where level 1 (reporting) is inferior to the other levels (Ryan 2013). From an argumentation perspective, this study confirms that arguments that fall within the model of informal logic are not universal. Requiring students to develop a reflection that makes a claim and supports it with evidence should not be assumed as the "norm" for post-secondary students. It also prompts the question of whether introducing students to certain argumentation models, in a general education course where the instructor can make such discretions, is a result of the instructor's cultural bias. It is clear from this study that education outside of western culture can influence how students engage in argument-making. It seems, though, that culture can mold even instructors' discretions, not just learners' strengths and weaknesses, when we set aside expectations surrounding argument-making.

3. TEMPERAMENT

In the 1990s there was a resurgence of interest and progress in personality research (Rothbart et. al., 2000, p. 122). Temperament is an innate system of how a human is organized, which is revealed through particular behaviours, talents, values, and needs (Keirsey, 1998). A clinical distinction between two aspects of human personality are temperament and character (Cloninger, 1994, p. 266), and so understanding our temperament helps us understand aspects of our personality. The difference between the two is that while character is dispositional and addresses the configuration of our habits, temperament is pre-dispositional and addresses emotion-based habits and skills, our inclinations (Keirsey, 1998; Cloninger, 1994, p. 268).

There are four established temperaments according to neurobehavioral studies (Cloninger, 1994, pp. 267, 271); however, there has been a pattern of different researchers renaming temperament variables, even when the content of the previous and renamed constructs is similar. This yields an apparent lack of agreement about the subject matter of temperament that may not be merited (Rothbart, 1999). For the purposes of this presentation the names used are those associated with the commercialized test for temperament that was used in the study. The four temperaments are *gold, green, blue*, and *orange*.

For those whose primary temperament is gold, it is important to feel a sense of belonging or affiliation with others or groups, as well as a sense of responsibility (McKim, 2003, p. 33; Berens, 2006, pp. 12, 25). Responsible in nature, they seek to be dutiful, to protect, and preserve (Berens, 2006, p. 12). They tend to orient concretely, that is, in the present and tangible. As Golds tend to be task-oriented, and hard-working at that, they become anxious, or destabilized, when disorganization or conflict arise (McKim, 2003, p. 33). Given the need to be responsible, to meet deadlines and expectations, a Gold will excel in a classroom context, even if the material is not his/her main interest, because they are naturally taskmasters.

For those whose primary temperament is green, the acquisition of knowledge, being competent, and achieving mastery are primary needs (McKim, 2003, p. 28; Berens, 2006, pp. 14, 24). Greens are natural thinkers and theorists who seek to explore phenomena, have tendencies of skepticism, and expect rationale for everything (Berens, 2006, p. 14). They also tend to orient abstractly, as opposed to *in the here and now* (McKim, 2003, p. 29). Generally, most typical assignments in academia satisfy the needs, values, and talents of the Greens. Tests, essays, and presentations, for examples, assess mastery of knowledge in a manner that is abstract (e.g. hypothetical situations, assessment of world events using concepts and theories, summarizing knowledge learned). Characterized as *life learners,* Greens could thus thrive in an academic environment.

For those whose primary temperament is blue, they tend to seek identity, meaning, and significance in life (McKim, 2003, p. 37; Berens, 2006, pp. 16, 24). Blues tend to be relationship oriented, gravitating towards harmonious and cooperative social contexts, and they avoid conflict at all costs (McKim, 2003, p. 37). They, like Greens, orient more abstractly than concretely. A blue is typically more interested in having peaceful relationships, which lends to a deep concern for others and being empathic and supportive naturally. They tend to orient more emotionally (McKim, 2003, p. 37), in contrast with Greens who trust their heads before their hearts. Since Blues tend to be relationship-oriented and are satisfied, or content, when they feel connected to people, typical academic assignments do not meet the needs of their temperament. Essay and test writing do not require one to be mindful of other people. They also are not paths to finding life's meanings, at least not essay assignments that have strict rubric expectations. Blues are in optimal learning environments when the development of self and relationships are incorporated.

For those whose primary temperament is orange, core needs include the freedom to be oneself, to choose, and to act. It is important for an Artisan to make an impact, typically by achieving the intended results of his/her actions (McKim, 2003, p. 41; Berens, 2006, pp. 10, 25). They are improvisers and tend to be absorbed in the action of the moment, focusing on the present and concrete (Berens, 2006, p. 10). Typically Oranges are speedy in what they do, quick to make decisions, and are comfortable assuming various tasks, becoming bored more easily than other temperament types (McKim, 2003, p. 42). They can be more adventurous than others, but Oranges are practical and task-oriented (McKim, 2003, p. 42) and tend to bore easily of abstract ideas. They would tend to prefer action-oriented tasks, perhaps *moving* their bodies, as they are tactical in nature. Since many assignments are driven by written discourse, requiring abstract thinking and not our knowledge of the world, an Orange's needs typically are not met in an academic classroom.

For the most part, the vast research of temperament is in developmental psychology and focuses on infants and children. In the education sector, you can find plenty of studies on temperament; however, the literature tends to focus on children. Our temperament does not disappear as we age though; it influences our adult lives too. Rothbart et al. (2000) write that, "temperament arises from our genetic endowment. It influences and is influenced by the experience of each individual, and one of its outcomes is the adult personality" (p. 122) – which is of particular interest in the post-secondary, diverse, general education classroom.

3.1 Temperament Study in Post-Secondary Education

ToP is a full year general education course that has a significant emphasis on informal logic. The research conducted on this course intersects the areas of informal logic, general education/first year experience, the scholarship of teaching and learning, e-learning, personality/temperament. The overarching research question asks whether or not there are any connections between a student's temperament and his or her informal logic and/or critical thinking skills. While the project has a mixed methods approach, it is only the quantitative data that are included here. The quantitative data includes a collection of students' temperament preferences (based on an assessment tool) as well as their grades in all assignments, plus their final grade in the course. Data spans from January 2016 to 2018, from both blended and online deliveries of this particular general education course.

The extant literature on the relation between temperament and academic achievement consist of few studies and the corresponding findings are mixed. For example, whereas one study found that temperament is independent from critical thinking (Kreber, 1998), another study (Sefcik et al., 2009) found a significant relation between temperament and scores on a cognitive test. Furthermore, their results indicated that individuals who have blue as their primary temperament scored lower on these cognitive tests than individuals with other temperaments. This is not an identical trend to the current findings, but there is some overlap with results in the current study.

The initial hypothesis of this study was concerned with whether individuals who prefer blue and orange temperaments, given their characteristics, may not typically be engaged or satisfied in a traditional class format and setting. The assignments in the courses that were studied included: a test, a multiple choice quiz, a Rogerian argument style letter, and a critical essay. We have run different analyses (a regression analysis, ANOVA), and the results that are the most conservative indicate statistically significant results with respect to the Rogerian letter assignment only. Table 1 summarizes some of the data.

Primary Temperament	Sample size	Letter Mean (std. deviation)	Quiz Mean	Test Mean	Essay Mean	Final Grade
Golds	80	70.01 (16.5)	63.57	67.89	70.85	76.76
Blues	44	**73.9 (6.6)**	51.74	64.48	70.13	73.77
Oranges	27	61.89 (22.4)	49.45	62.68	65.63	70.52
Greens	23	70.96 (17.0)	63.51	69.6	67.24	74.13
	174	69.86	58.37	66.45	69.38	74.69

Table 1 – Summary of Temperament and Mean across all ToP

The argumentative letter assignment employs a Rogerian style of argument; it's between 750 and 900 words. In this assignment students were expected to choose an issue from a short list provided (e.g. Should we condone self-driving cars? Should women breast-feed in public? Is global warming real?), find a specific person/group's stance on the issue (e.g. Tesla Founder and CEO, Elon Musk) and aim to convince this person/group of an alternative perspective (e.g. it is dangerous to implement self-driving cars in society). Learning objectives relevant to this writing assignment included researching and summarizing arguments succinctly, comparing/contrasting arguments, justifying viewpoints, and most importantly addressing a hostile/resistant audience in an empathetic and collaborative manner as a technique of persuasion. The strategy inherent in a Rogerian letter is to delay one's thesis until rapport is built through commonalities between writer and resistance audience.

Prior to the study it was hypothesized that there would be a trend demonstrating that Greens and Golds would have an advantage in the success of academic assignments. The data does not show this though. It does show that Blues excelled at the letter writing assignment. From a pedagogical perspective: the argumentative letter assignment was completed mid-term, and it required demonstration of mostly *analyzing* according to Bloom's taxonomy of cognitive skills, but also *creating* (Bloom et al., 1956). Students researched and reviewed different positions of a given, timely, issue. Students asked questions about the implications of different views on the issue, and interpreted the best standpoint given their research and thinking. Developing a new, amalgamated position, so that both writer and audience could be satisfied was a strong outcome, whereas a weaker letter failed to negotiate wants and needs of both arguer and audience. This expectation of synthesizing competing views to develop a palatable conclusion by writer and audience is a more creative activity, arguably. In other assignments, the main cognitive skills tested were *applying* and *evaluating*. This assignment prompted learners to engage in a strategy that fosters collaboration and relationship building in contexts of dissensus. It does not seem coincidental that Blues tend to be relationship-driven if any temperament is.

So, while most learners may adapt to typical expected academic evaluations with training (which begins long before post-secondary education), and every type of person can succeed, overall not all learners did as well, in relation to Blues, with this *alternative* assignment. On the one hand this is informative for pedagogical reasons, but on the other it confirms what some may intuitively acknowledge within argumentation: that there are many successful ways to come to agreement in dissensus. Typical normative models (informal logic, formal logic, pragma-dialectics) may not always apply, or they may not be pragmatic. In reference to Walton's dialogues (1998), Rogerian arguing style does not fit precisely into any of the dialogues – it has some of the goals of persuasion but accurate information (truth) is important, and it certainly does not aim to result in a win-lose outcome. There seems to be a connection between these empirical observations of argument-making and argument-having with a gap under the broader umbrella of Argumentation Theory. This particular dissensus methodology aimed at changing one's mind by being persuasive, dialectical, truthful, empathic – while the rigours of informal logic are expected - goes well beyond just strong argument-making. The implication of the data may indicate that we do not all excel at being versatile in handling dissensus.

4. CONCLUSION

We can excel at argumentation methodologies or modes (e.g. Gilbert) based on cultural advantages or temperament preferences. It does not follow from this that we cannot learn the theory and methods of different argument modes though. Cultural restrictions on education can affect one's critical thinking and reasoning skills, but explicit instruction of these skills (demonstarted in a study of REL subsequent to the one relayed in this presentation) can close the gap between students who can and cannot develop strong critical reflective ability.

Contrary to initial hypotheses temperament does not predict student academic success. However, interlocutors may argue better and gravitate towards argumentation models that appease their personality. The implications as I see them for educators in general education (or Critical Thinking at large) is that if we tend to look at informal logic, or similar *standard* models of argument, as a means for good argumente-making, then this is narrow in nature. Contrary to what we may believe, know, or want, who one is and where one comes from might matter in contexts of argument-making, but more especially in argument-having. While students may be able to learn the tricks to excel in a general education course, it does not follow that students can apply these skills to arguing in real-world contexts such as the Rogerian letter.

REFERENCES

Berens, L. V. (2006). *Understanding Yourself and Others: An Introduction to the 4 Temperaments.* Huntington Beach, CA: Telos Publications.

Bloom, B. S., Engelhart, M. D., Furst, E. J., Hill, W. H., & Krathwohl, D. R. (1956). *Taxonomy of educational objectives, handbook I: The cognitive domain* (Vol. 19). New York: David McKay Co Inc.

Brookfield, S. D. (2011). Teaching for critical thinking: Tools and techniques to help students question their assumptions. John Wiley & Sons.

Cloninger, C. R. (1994). Temperament and personality. *Current opinion in neurobiology, 4*(2), 266-273.

Coulson, D., & Harvey, M. (2013). Scaffolding student reflection for experience-based learning: A framework. *Teaching in Higher Education, 18*(4), 401-413.

Gilbert, Michael A. 1994. Multi-modal argumentation. *Philosophy of the Social Sciences, 24*(2): 159–177.

Guo, L., & O'Sullivan, M. (2012). From Laoshi to Partners in Learning: Pedagogic Conversations Across Cultures in an International Classroom. *Canadian Journal of Education, 35*(3).

Hessler, H. B., & Taggart, A. R. (2011). What's Stalling Learning? Using a Formative Assessment Tool to Address Critical Incidents in Class. *International Journal for the Scholarship of Teaching and Learning, 5(1)*, 1-18.

Keirsey, D. (1998). *Please understand me II: Temperament, character, intelligence.* Prometheus Nemesis Book Company.

Kreber, C. (1998). The relationships between self-directed learning, critical thinking, and psychological type, and some implications for teaching in higher education. *Studies in Higher Education, 23*(1), 71-86.

McKim, L. (2003). *Personality Dimensions.* Concord, ON: Career/LifeSkills Resources, Inc.

O'Keefe, D. J. (1977). Two concepts of argument. *The Journal of the American Forensic Association, 13*(3), 121-128.

O'Sullivan, M. W., & Guo, L. (2011). Critical thinking and Chinese international students: An East-West dialogue. *Journal of contemporary issues in education, 5*(2).

Rothbart, M. K. (1999). Temperament, fear, and shyness. In L. A. Schmidt & J. Schulkin (Eds.), *Extreme fear, shyness and social phobia: Origins, biological mechanisms, and clinical outcomes* (pp. 88-93). New York: Oxford University Press.

Rothbart, M. K., Ahadi, S. A., & Evans, D. E. (2000). Temperament and personality: origins and outcomes. *Journal of personality and social psychology, 78*(1), 122.-35

Ryan, M. (2013). The pedagogical balancing act: Teaching reflection in higher education. *Teaching in Higher Education, 18*(2), 144-155.

Sefcik, D. J., Prerost, F. J., & Arbet, S. E. (2009). Personality types and performance on aptitude and achievement tests: Implications for osteopathic medical education. *The Journal of the American Osteopathic Association, 109*(6), 296-301.

Walton, D. N. (1998). *The new dialectic: Conversational contexts of argument.* University of Toronto Press.

Heroic Argumentation:
On Heroes, Heroism, and Glory in Arguments

JOHN CASEY
Northeastern Illinois University
j-casey1@neiu.edu

DANIEL H. COHEN
Colby College
dhcohen@colby.edu

Despite objections, the argument-as-war metaphor remains conceptually useful for organizing our thoughts on argumentation into a coherent whole. More significantly, it continues to reveal unattended aspects of argumentation worthy of theorizing. One such aspect is whether it is possible to argue heroically, where difficulty or peril preclude any obligation to argue, but to do so would be meritorious if not indeed glorious.

KEYWORDS: argument-as-war, adversarial argument, disagreement, heroism metaphor, supererogation, virtue argumentation

1. INTRODUCTION

A broad coalition that includes argumentation theorists, feminists, and educators, among others, has waged war on the argument-is-war metaphor, warning us against the negative elements and consequences of the metaphor (e.g., Cohen 1995, Rooney 2010, Hundleby 2013). Nevertheless, it persists. And for good reason: the metaphor organizes what and how we think about arguments, thereby organizing our thoughts into the kind of coherent whole that transforms knowledge into understanding. That understanding, in turn, gives us new ways to look at arguments, which unearth new aspects of argumentation – e.g., by revealing argumentative counterparts to such martial concepts as collateral damage, proportionality, or just war theory. That additional knowledge then becomes fodder for even greater understanding.

We could, in earlier times, sing of the glories of war without irony. That is no longer possible now that humanity's technological prowess has progressed to the point that our capacity for inhumanity – and our ability to witness and broadcast it – challenge our ability to ignore the horrors. Because those horrors are so manifest, we have added reason to shy away from identifying, even metaphorically, arguments and wars. The metaphor shines too bright a light on the destructive aspects of arguing. An unfortunate side-effect of abandoning the argument-is-war metaphor is that we no longer sing the praises of heroic arguers or the glories of arguing.

We are going to push back against that trend.

2. ON METAPHORS IN GENERAL AND THE ARGUMENT-IS-WAR METAPHOR IN PARTICULAR

There are nine basic points to note about the argument-is-war metaphor that frame the subsequent discussion:

First, the assertion that argument is war is indeed a metaphor. We can choose to argue instead of going to war so the claim is not literal. If a hard literal-metaphorical dichotomy is rejected (as it ought to be) in favor of a spectrum, then this would be closer to the metaphorical end.

Second and third, it is a viable metaphor and it is a natural metaphor. It is viable because there are enough points of similarity to make it work – a very low bar since clever readers can make almost any metaphor work – but it can be termed "natural" because the comparison comes so easily that no explanation would generally be needed.

Fourth and fifth, it is metaphor that works and it is a useful metaphor. It serves as a broad organizing scheme for much of our thinking about arguments, reflecting the broad contours of how we think about arguments; and it provides the material for articulating and extending those thoughts.

Sixth and seventh, it is an entrenched and even dominant metaphor. It is embedded in how we talk – we want impregnable defenses to go along with strategies for attacking using strong or even killer arguments that are right on target – and consequently it informs both how we think about arguing and how we go about arguing. It is a necessary or immutable part of our linguistic practice, but it would take some heavy lifting to uproot it.

Eighth (and now we're getting somewhere), despite its viability, naturalness, utility, and ubiquity, it is also an awful metaphor because it deforms as much as it informs argumentation. It elevates differences into disputes, thereby turning co-inquirers into competitors; instead of working things out, we fight it out and we altercast our fellow interlocutors into enemies (Stevens 2019). Worst of all, the Dominant Adversarial Model – the DAM account – for argumentation conceptually equates learning through argumentation with losing an argument: if you convince me of something, I'm the one who has made a cognitive advance – usually the only one – and yet I am the one who is described as the loser. Something is wrong with this picture (Cohen 1995).

Nevertheless, there is a ninth point to emphasize: this awful metaphor is also a great metaphor because it is conceptually so very fertile. It is a rich mine of meaning whose veins have not yet been depleted. By juxtaposing the concepts of war and argument, the metaphor effectively superimposes the whole constellation of concepts associated with wars on the whole constellation of concepts associated with arguments.[1] In order for the metaphor to work, the major components of war, such as allies and enemies, victory and defeat, and offensive tactics and defensive strategies, need to have more or less obvious counterparts among the major components of argument. And they do. Great metaphors do more. Like all metaphors, they invite comparisons of the conceptual clusters' secondary components, but the best ones successfully reveal new ways of thinking about the target concepts.

The literature on metaphors provides ample resources for elaborating the first five points. The sixth and seventh points have also been addressed at length to the point that they are now routinely part of the discussion. More recently, the negative aspects of the argument-is-war metaphor have come under increasing critical scrutiny from several directions. That has been all to the good, except insofar as it has crowded out the final point: there are still valuable lessons to be learned from thinking of arguments as wars. There is more to be said about how such war-related concepts as reparations, Just War Theory, appeasement, strategic alliances, exit strategies, collateral damage, proportionality, and post-war policies might fruitfully be applied to arguments. Every one of those is a rich topic worthy of further research.

In this paper, however, we will explore a less likely and more challenging set of possible corollaries to the argument-is-war metaphor, namely, whether the ideas that war brings out our heroic best and brings meaning to our lives have counterparts in the conceptual neighborhood of argumentation. This is not a merely academic game of connect the dots, however, because there is something important to be learned about argumentation from the exercise.

---

[1] Black 1954 offers this account in terms of background "associated commonplaces" for each concept that serve to filter, reveal, emphasize, and transform the concepts with which they are juxtaposed.

3. HEROISM AND ARGUMENTATION?

At first, it would seem to be a rather big stretch to pair arguments and wars by way of heroism and glory.[2] Bravery, danger, and self-sacrifice do not immediately spring to mind when the topic of argumentation comes up, and it is precisely the absence of a common association that gives Ralph Waldo Emerson's claim, "There is no true orator who is not a hero," its punch. However, we can cite precedent for bringing them together: we are reviving a classical association. The juxtaposition of entering the deliberations of the assembly and entering the action on the battlefield is a recurring trope in the Iliad, where fighting well in battle and speaking well in council are the two characteristic virtues of a hero. It starts in Book One when Achilles is described as going "neither to glory-bringing assembly nor to war" (1:490-491), and references to "fighting with words" and "striving with speeches" recur throughout.[3] Cicero, too speaks of the glory of orators. We appreciate and think it is no accident that the greatest hero in the first 100 years of American cinema, as determined by the American Film Institute,[4] was not Indiana Jones, not James Bond, and not even Ellen Ripley, but Atticus Finch, the small-town lawyer in *To Kill a Mockingbird*, whose very name recalls Greek and Roman rhetorical traditions[5] and whose heroic act was not to fight aliens, rescue artifacts and maidens, or save the world from nefarious conspiracies. His heroism was simply to stand up and argue.

What is it that makes a lawyer armed only with arguments such an exemplar of heroism?

Part of the answer, of course, is the context in which he argued which did require great personal bravery. The act of arguing was heroic. But that is only part of the answer because we also want to say that in a deeper and more important sense the content of his argument itself was also heroic.

In the next section, we address the meaning of heroism to answer the first part; the concluding section briefly addresses what makes heroism meaningful to answer the second. The meaning of heroism is not what makes heroism meaningful – and by the same token, the meaning of argumentation is not what makes argumentation

---

[2] We are using the classical, pre-Christian concept of "glory".
[3] Other passages from the *Iliad* that speak of the glory to be had from council or the assembly include 9:53-54; 9:440-441; and 15:282-284. In addition, there are several phrases that combine war and rhetoric, such as "striving with speeches" and "fighting with words."
[4] https://www.afi.com/100years/handv.aspx
[5] https://1.cdn.edl.io/OOlrFqLm9fdGBP1oNLPtVX13cRQ0CRrgCEGUVhEtAuoWlrlP.pdf

meaningful. What makes argumentation meaningful is also what can make it heroic.

4. HEROISM IN WAR – AND ARGUMENTATION

Can argument be heroic? War is the classical scene of heroism where there is danger, hardship, and sacrifice. Argument has none of those things. But war is also an occasion for great nobility noble because it is a common struggle, perhaps in pursuit of a higher purpose like the defense of the innocent or the liberation of the oppressed. We are at least getting closer to things that arguments can do.

War is not the only place to find heroes, of course. The sports pages of any newspaper routinely laud the heroes of last night's games, but those are heroes without heroism. For heroism outside of war, we can point to firefighters who rush into burning buildings to save people or doctors who work in quarantined epidemic areas, and no one will object. What about arguers?

As a starting point, we will follow J. O. Urmson (1958) in thinking that heroic acts, like saintly ones, are supererogatory – no one has to be a hero – and we will follow Joel Feinberg in characterizing a supererogatory act as "a meritorious, abnormally risky non-duty" (1962, p. 281) If these criteria are taken as individually necessary and jointly sufficient, it is possible to generate legalistic counterexamples. Great excellence in one area apparently can compensate for slight deficiencies in another, but the characterization generally accords with our pre-analytic sense of heroism. You don't get to be a hero for simply doing your job, unless, like firefighters, the job is both dangerous and worthy. Similarly, going far above and beyond what is called for in highly praiseworthy but relatively safe activities can also be thought of as heroic, like the quiet heroism of dedicated kindergarten teachers.

The controversial criterion is that heroic acts be meritorious for noble ends. No matter how risky and difficult rock-climbing may be, it is not heroic. But what are we to say of brave, self-sacrificing actions on behalf of ignoble causes? There is something about them that people find honorable. The American South is replete with statues to "heroes of the lost cause" – the euphemism of choice for defenders of slavery. Could their acts be heroic while they themselves are not heroes, or the other way around? We will return to this question later.[6]

---

[6] We will sidestep entirely the question of whether the concept of the supererogatory is incoherent and ought to be abandoned. See Pybus 1982 for a skeptical view.

Can argument be heroic by these criteria – (a) dangerous and risky, (b) noble or meritorious, and (c) above and beyond the call of duty?

4.1 Risky and dangerous arguments

It is easy to find cases of very difficult, challenging, or even scary arguments, but risk is something else. Let us distinguish two categories of risk here, those that come from the act of arguing and those that concern the content of the argument.

The most obvious examples are in the first category. Because arguing is a social activity, challenges come from the social context, especially the other participants – arguing with difficult, dangerous, or ignorant people in difficult, dangerous, or trying circumstances. A hero will be unafraid to take on difficult opponents, whether they are uninformed and irrational, hostile and uncivil, or dogmatic to the point of being pig-headed. The danger can be very real because it is hard for such vicious arguers to stay within the space of reasons. As with war, there is a fine line between brave and foolhardy, the rash and the heroic, so arguers need to weigh the pros and cons. It is with good reason that the first principle of on-line argumentation is, "Do not feed the trolls." There may be something romantically heroic about tilting at windmills, but there is no glory in taking a troll's bait.[7]

Atticus Finch did indeed risk his family, friends, career, and even physical safety. Something similar might be said about the mathematician Andrew Wiles for his work on "Fermat's Last Theorem". He has been explicitly described as a "hero" and hailed for his "glory" because of his decision to take on the Theorem, devoting years in secret, solitary efforts, risking his career, professional standing, and if not his physical safety, then perhaps his mental well-being (Leon 2016). There is an important difference between these cases. The kinds of dangers facing Atticus Finch may be less imminent and dire but they are still comparable to the dangers found in war. However, they are not specifically argumentative dangers. Wiles' risks are not comparable to a soldier's risks; they were cognitive. His risk was failure: a failure to reach the conclusion, i.e., to prove the theorem, but also the risk of a subsequent failure to persuade other mathematicians. The second risk concerns other arguers; the first one concerned the argument itself.

Thomas Aquinas, swimming in deep waters, raised a related problem that magnifies the difference: the possibility, after much time

---

[7]Although we appreciate the reasons for characterizing Socrates as perhaps philosophy's greatest argument troll, we are using the term "troll" in its more common pejorative sense. See Cohen 2017.

and effort, of failing to comprehend the truth (*Summa Contra Gentiles*, I 4.4). Is that enough of a risk to make the enterprise heroic? That's stretching things, but it serves to isolate specifically cognitive dangers. It takes a special kind of courage to follow an argument to an unexpected, unwelcome, and even unbearable conclusion. The risks may include all of the above, as well as one's epistemic well-being and mental health.

4.2 Noble and meritorious arguments

In addition to their intellectual bravery, Aquinas, Finch, and Wiles directed their respective efforts in metaphysics, social justice, mathematics to things of great value – indeed, Aristotle's transcendentals Truth, Goodness, and Beauty. Arguments on behalf of noble causes and the arguers who make them are common enough, but while this may be a necessary condition for heroic arguing, it might be observed more in the breach: we do not honor arguers who have argued on behalf of false or ignoble causes, no matter how heroic their arguing may otherwise be – unless we are determined to honor them. In that case, we can surely come up with something worthy. The State of Alabama annually celebrates a holiday honoring Jefferson Davis, but they make sure that everyone knows they are honoring the slave-owning President of the Confederacy for his bravery and dedication to a people's autonomy in the form of states' rights, not for being a champion of white supremacy. Conversely, Alabama resisted the federal holiday for Martin Luther King, Jr., insisting it was because of his anti-war efforts, not his civil rights work, before finally acceding – albeit by combining it with a holiday for the Confederate General Robert E. Lee. Hypocrisy, as de la Rochefoucauld said, is the homage vice to pays virtue.

Although most of the risks associated with arguing are contextual and external, there are some that are internal and specific to argumentation.

4.3 Arguing above and beyond the call of duty

Supererogation also involves going beyond one's duties. In order for that to be possible in argument there would have to be basic duties for arguers to fulfill and then to exceed. Argumentation theory would appear to have this covered since so much of the discipline is explicitly devoted to the normative dimensions of arguing. This is true no matter whether the approach is primarily logical – use all the available and relevant evidence, warrant all your premises, and reason well;[8] or dialectical – listen to your opponents, be open-minded, etc.; or rhetorical – consider ethos, pathos, and logos, take into account the audience, and so on.

However, a closer look reveals two gaps in our theorizing about arguing. First, a large part of the normative principles that have been articulated for argumentation are prohibitions and permissions rather than mandates and duties. Fallacy theory, for example, is explicitly concerned with what not to do in arguments. Critical thinking does a little better, but even there, the positive principles tend to be in the form of fairly general and vague strategies rather than specific obligations. The pragma-dialectical approach fares pretty well on this score: the original list of ten rules for critical discussions does include one positive duty: the second half of rule 10 is that participants must interpret others carefully and accurately. There are also three conditional obligations regarding retractions and replies to objections and (2, 3, and 9). The rest of the decalogue are prohibitions regarding what we must not do (1, first half of 10), what we may not do (5, 6, 7,) or what we may do only in certain circumstances (4, 8) (van Eemeren and Grootendoorst 1992).

Attention to the relative lack of mandates reveals the second gap: the striking absence of principles of when – and when not – to argue at all. That is, while we have normative principles for conduct within an argument, we have no real guidelines for arguing in the first place, i.e., for moving from the confrontation stage in a discussion to the opening and argumentation stages of a critical discussion. More generally, when should a difference be the occasion for an argument?

What makes this so relevant for the question of heroic argumentation is that one way to argue heroically is simply to argue at all. We can argue whether Atticus Finch argued beyond the call, but that is because we recognize that as an attorney, he did have some initial professional duty to argue. Is there a duty to argue for non-professional arguers? Except for philosophers, that is almost everybody else.

The problem is that most of argumentation theory is concerned with the thou-shalt-nots of argumentation rather than the thou-shalts. Either way, there has been little systematic work regarding what should be the First Commandment of Argument. Is it *Thou shalt argue* or *Thou shalt not argue*? If the default is to argue whenever there is a difference, we would need "defeaters" or "excusers" to prevent us from going down the someone-is-wrong-on-the-internet rabbit hole. If, in contrast, the general prohibition comes first, we would need "over-rides", reasons to argue, to avoid a world of disengaged monads without any of the

---

[8]This, roughly, paraphrases Govier's A-R-G criteria for good arguments. Govier 2010.

benefits of arguing. There are strong intuitions on behalf of each: Job was heroic in trying to take on God in an argument over the injustices done to him, but he was not obligated to do so. For contrast, consider the case of Jackie Robinson, the first African-American to play Major League Baseball after the so-called color line was drawn.[9] Part of his greatness – his heroism – was that he did not argue against the malicious way he was treated by opposing players, who we might suppose felt no special obligation to treat him fairly, and by umpires, who most definitely were obligated to treat him fairly. When the umpires violated their own code, Robinson clearly had a justifiable argument to make. His heroism, however, was restraining himself precisely because the weight of so many others was on his shoulders and the prospects for success in the long-term required extraordinary stoicism.

The decision to argue or not may provide the clearest examples of supererogatory arguers – Job and Jackie Robinson, respectively – but supererogation can occur within an argument, too. By way of an example, we will offer the parallel cases of the Noble Chess Master and the Noble Philosopher in debate who lose their respective endeavors because, beyond any duty, they offered strategic help to their opponents out of great love and respect for the game of chess and the institution of argumentation (Cohen 2105). Perhaps they are really tragic heroes.

Argumentation can be dangerous, it can serve a noble cause, and it can involve going above and beyond what duty calls for, so tragic or not, there is indeed room for argumentation so admirable to qualify as heroic by the three-part measure.

5. ARGUERS CAN BE HEROIC: SO WHAT?

It would be fair to ask if there is anything philosophically significant here, because the conclusion that the concepts of heroes and heroism have applications in the field of argumentation could be nothing more than an odd, hopefully interesting, observation. But are there lessons to be learned from pushing the argument-is-war metaphor in this direction? For example, could it alter our argumentative practice – as might result from greater attention to the phenomenon of collateral damage or the need for exit-strategies? Does it enhance our understanding of argumentation – the way that considering the effects of argumentation in the pre-, post-, and inter-argument periods does?

---

[9]Robinson was not the *first* African-American in the major leagues. That distinction belongs to Moses "Fleetwood" Walker, a name nearly lost to history. Walker played in the mid-to-late19th century *before* the prohibition on black ballplayers was put in place.

(Cohen 2018). Does it open up new areas for research – as retro-jecting Just War Theory back on to argumentation does? Or is it mostly just academic doodling?

Yes, to all of the above. We could leave it as an academic exercise in connecting the dots, but that would squander the value it holds for understanding arguments, for future theorizing about argumentation, and for the practice and pedagogy of arguing. The connection to follow is the idea that war is source of meaning – not semantic meaning but the meaningfulness that defines a life: "War is a force that gives us meaning," a thought echoed by Wittgenstein who claimed that only be facing death on the front lines in the Great War of 1914-1918 was he able to give life its meaning.[10]

We are not claiming that argumentation by itself can make a life meaningful and worthy of living, but we are suggesting that argumentation is a source of meaning. The reasons we offer for a standpoint define that standpoint: a position apart from its supporting argumentation is incomplete; a conclusion for which we have not argued is at best a work-in-progress. Argumentation provides the necessary background context against which a conclusion can have a well-defined meaning, but it is also the medium out of which meaning emerges. That is the glory of argumentation. It is also a topic for its own paper. Still, "There's glory for you!"[11]

ACKNOWLEDGEMENTS: Thanks to our many friends and admirers...

REFERENCES

Aikin, S. F. (2011). "A defense of war and sports metaphors in argument." Philosophy and Rhetoric, 44(3), 250-272.
Aquinas. (1955). Summa contra gentiles. Book I. Translated by Anton Pegis. University of Notre Dame Press. London.
Bailin, S. and Battersby, M. (2016). "DAMed if you do; DAMed if you don't: Cohen's "Missed Opportunities'." In P. Bondy and L. Benacquista (Eds.), Argumentation, Objectivity, and Bias: Proceedings of the 11th International Conference of the Ontario Society for the Study of Argumentation (OSSA), May 18-21, 2016. OSSA: Windsor, ON.

---

[10] See Hedges (2003), *War is a force that gives us meaning*, where he argues that an underappreciated attraction of war is that it gives life purpose and, paradoxically, builds community.

[11] The quotation, meaning, "There's a nice knock down argument for you" is used by both Donald Davidson and Keith Donnellan, following Humpty Dumpty's lead. See Carroll 1872 (ch. 6), Donnellan 1996, and Davidson 2005.

Black, M. (1954). "Metaphor." *Proceedings of the Aristotelian Society.* N.S. 55, 273-294.
Carroll, L. (1999, 1872). *Through the Looking Glass.* Dover Publications.
Cohen, D. H. (2017) "The Virtuous Troll: Argumentative Virtues in the Age of (Technologically Enhanced) Argumentative Pluralism," Philosophy and Technology, 30 (2), 179-189.
Cohen, D. H. (2015). "Missed opportunities in argument evaluation." In Garssen, B., Godden, D., Mitchell, G., and Snoeck-Henhemans, A. F. (Eds.) Proceedings of ISSA 2014: Eighth Conference of the International Society for the Study of Argumentation. SIC SAT: Amsterdam. Pp. 257-265.
Cohen, D. H. (2003) "Just and Unjust Wars – and Just and Unjust Arguments." In IL@25: Proceedings of the 2003 Meetings of the Ontario Society for the Study of Argument. Windsor, ON (CD-ROM).
Cohen. D. (1995). Argument is war. . . and war is hell: Philosophy, education, and metaphors for argument. Informal Logic, 17: 177-188.
Donaldson, D. (2005, 1986) "A Nice Derangement of Epitaphs" in *Truth, Language, and History.* Oxford University Press.
Donnellan, K. (1966). "Reference and Definite Descriptions," *Philosophical Review* LXXV, 281-304. Emerson, R. W. (1875). "Letters and Social Aims", p.94
Feinberg, J. (1961). Supererogation and Rules. Ethics, 71(4), 276-288. Retrieved from http://www.jstor.org.neiulibrary.idm.oclc.org/stable/2379643.
Hedges, C. (2003) War is a force that gives us meaning. Anchor Books.
Hundleby, C. 2013. "Aggression, politeness, and abstract adversaries." Informal Logic 33(2), 238-262.
Léon, Manuel D. (2016). "From the margins to Glory: the Story of Andrew Wiles."
https://www.bbvaopenmind.com/en/science/mathematics/from-the-margins-to-glory-the-story-of-andrew-wiles/
Mandelbaum, E. and Quilty-Dunn (2015). Believing without reason: Or, why liberals shouldn't watch FOX News. Harvard Review of Philosophy, 32. 42-52.
Pybus, E., 1982, "'Saints and Heroes'", Philosophy, 57: 193–199. [skeptic of SE]
Rooney, P. (2010). "Philosophy, Adversarial Argumentation, and Embattled Reason." Informal Logic 30(3), 203-234.
Stevens, K. on "altercasting" ECA 2018
Urmson, J., (1958), "Saints and Heroes", in Essays in Moral Philosophy, A. Melden (ed.), Seattle: University of Washington Press. [the ur-text]
Winter, Michael. (1997). "Aristotle, hos epi to polu relations, and a demonstrative science of ethics." Phronesis 42 (2), pp. 163-189.
XKCD: A Webcomic of Romance, Sarcasm, Math, and Language, http://xkcd.com/386

Relationships between narrative and argumentation. In defence of a functional account

GUILLERMO SIERRA CATALÁN
University of Granada
gsierra@correo.ugr.es

The objective of this investigation is to study the relationships between narrative and argumentation. These are apparently very different objects, but overlaps are frequent in literary works. The proposed classifications are based on the notion of speech-act, and are defined according to two different criteria: one is of a structural nature and generalizes some previous outlooks, while the second one is based on functional accounts. We defend our functional approach over the structural ones.

KEYWORDS: speech-act of arguing, argumentative text, narrative speech-act, narrative text, rhetoric, fiction, non-fiction.

1. INTRODUCTION

At first sight, it is easy to think that narratives and argumentations are communicative objects of a quite different nature. Through narratives it is possible to describe certain series of events as well as tell stories while, in argumentations, reasons are presented in order to justify certain points of view. In that way, we might tend to identify narrative with imagination, as well as argumentation with reason.

However, let's quote a fragment of the story "The Owl who wanted to save Humankind", by the Guatemalan writer Augusto Monterroso:

> Years later, he developed a great facility to classify, so that he knew exactly when the Lion was going to roar and when the Hyena was going to laugh, [...]
>
> So he concluded:
>
> "If the Lion did not what he does but what the Horse does, and the Horse did not what he does but what the Lion does [...] and so on until Infinity, Humankind would save itself because

> everyone would live in peace and war would be again as it was in the times when there wasn't any war."

In this text we can easily see an argumentation —inserted between quotation marks—, despite its evident narrative character.

In a similar way, there are narrative pieces of work that present argumentative parts—appearing without quotation marks—. That can be seen in the following fragment, extracted from Montaigne's Essays, chapter I, II ("Of the Inconstancy of Our Actions"):

> [...] for they [men] commonly so strangely contradict one another that it seems impossible they should proceed from one and the same person. We find the younger Marius one while a son of Mars and another a son of Venus. [...]; and who could believe it to be the same Nero, the perfect image of all cruelty, who, having the sentence of a condemned man brought to him to sign, as was the custom, cried out, "O that I had never been taught to write!" so much it went to his heart to condemn a man to death.

We can note that Montaigne represents certain facts—associated to Marius, Pope Boniface VIII and Nero—that act as reasons trying to prove that his point of view ("[men] commonly so strangely contradict one another that it seems impossible they should proceed from one and the same person") is correct or valid. Another example of a narration acting in an argumentative process is the fable "The fox and the grapes", attributed to Aesop:

> A Fox one day spied a beautiful bunch of ripe grapes hanging from a vine trained along the branches of a tree. The grapes seemed ready to burst with juice, and the Fox's mouth watered as he gazed longingly at them.
>
> The bunch hung from a high branch, and the Fox had to jump for it. The first time he jumped he missed it by a long way. So he walked off a short distance and took a running leap at it, only to fall short once more. Again and again he tried, but in vain.
>
> Now he sat down and looked at the grapes in disgust.
>
> "What a fool I am," he said. "Here I am wearing myself out to get a bunch of sour grapes that are not worth gaping for."
>
> And off he walked very, very scornfully.
>
> Moral: There are many who pretend to despise and belittle that which is beyond their reach.

In this fable, the narrated facts are presented as reasons for the conclusion stablished by the moral, so it is easy to notice that the narrative constitutes a part of an argumentative process between the author and the reader.

In this way, it is not strange to find situations in which the distinction between narrative and argumentation vanishes. Although it is important to point out that this fact can happen in different ways: it is clear that Monterroso's text is of a different nature that the ones by Montaigne or Aesop.

The main objective of this paper is —focusing on literary texts,— to study which types of overlaps may arise involving narrative and argumentation, as well as present a systematic classification of them, according to different criteria.

2. NARRATIVE AND ARGUMENTATION: SPEECH-ACTS AND TEXTS

We understand argumentation as a second order speech-act complex, composed of the constative speech-act of adducing (i.e., the reason) and the constative speech-act of concluding (i.e., the conclusion) (Bermejo-Luque, 2011, pp.60-62).

According to this model, arguments are mere representations of the syntactic and semantic properties of the inferences underlying argumentations or inner reasonings.

Regarding the definition of narrative, Garrido Domínguez (1993, p.2) points out that:

> The difficulties that arise in offering an adequate definition of the narrative text get more complicated as 20th century productions are taken into account. It is known that the romantic ideal of mixing genres appears on them —specially on those in which the narrative mood changes, like Ulysses, The Magic Mountain or In Search of Lost Time—[...]. Inside them, dramatic, lyric and argumentative elements cohabit, as well as the strictly narrative ones, blended in a way that any attempt of isolation could success without attacking against the essence of this kind of stories.

Genette, in his classic *Narrative Discourse: An Essay in Method* (1980) clarifies the concept a bit more: "oral or written discourse that undertakes to tell of an event or series of events". He compares three common uses of the world narrative, to finally differentiate in the following way:

> I propose [...] to use the word story for the signified or narrative content [...], to use the word narrative for the signifier, statement, discourse or narrative text itself, and to use the word narrating for the producing narrative action and, by extension, the whole of the real or fictional situation in which the action takes place.

According to that, we may say that a narrative is a representation of certain events, that might be real or fictional, in a sequence of time. This representation is made by means of certain speech-acts that constitute the narration of the events or, in Genette's, words, the narrating of the story.

In order to illustrate that, let's consider the following text, "The Black Sheep", by Augusto Monterroso:

> In a far-off country many years ago there lived a Black Sheep.
> They shot him.
> A century later, the repentant flock erected an equestrian statue of him, which looked very good in the park.
> From then on, every time Black Sheep appeared they were promptly executed so that future generations of common, ordinary sheep could also indulge in sculpture.

Let's consider the following declarative sentence: "In a far-off country many years ago there lived a Black Sheep". (1)

García-Carpintero (2007) considers fictioning as uttering a sentence with the communicative intention of putting certain addressee in position of imagining this proposition. Following Currie (1990), García-Carpintero states that fictioning can be considered as a speech-act that presents certain constitutive conditions regarding the truth of the expressed propositions.

Regarding the speech-act of non-fiction making, Romero Álvarez (1996) proposes an analysis. In her article, she points out that—in order to make a correct pragmatic interpretation—the considered speech-acts should be considered not only in its contextual frame, but in relation to the previous and successive speech-acts made by the addresser. In this way, a set of speech-acts turns out to be another speech-act, which Romero Álvarez names macro-act, following Van Dijk (1996). In consequence, we can assume that non-fiction narratives are, in a macro level, speech-acts.

According to these definitions of the speech-acts of fiction and non-fiction making, it is clear that the essential feature of narrative is not the single kind of speech-act that generates it, because fiction and non-fiction making speech-acts may cohabit in the same narrative[1].

In relation to the expressions "argumentative text/discourse" and "narrative text/discourse", it is reasonable to understand them as the text or discourse mainly composed, respectively, by argumentative or narrative speech-acts. According to that, the concepts of narrative and of narrative text will be identified[2].

3. CLASSIFICATIONS

In this section, two different classifications about the relationships between narrative and argumentation are presented.

The first one, of a structural nature, is based on the "authorship" of the argumentative speech-act—within the frame of a narrative text. These "authors" will be some character of the story, the narrator or the proper author.

The second classification is based on the analysis of the roles that argumentative speech-acts play in narrative texts, or narrative speech-acts play in argumentative texts.

3.1 Structural classification

Following what has been settled, the first structural type, Type S1, is associated to situations where a character of the story is performing an argumentation, that appears, usually, quoted in the text. This is the case of the argumentation performed by the Owl in the fable by Monterroso that was exposed previously.

There is also an overlap between narrative and argumentation when a narrative text includes an argumentative speech-act, performed by the narrator of the story. This would be the case of the fragment from Montaigne's Essays (I, II: Of the Inconstancy of Our Actions), also previously included. This case will be labeled as Type S2.

Finally, the whole narrative text may be part of an argumentation. This happen when the narrative text is adduced as a

---

1. Olmos (2013, pp.10, 11) also admits that the fictive character of a narrative is not a fundamental criterion in which basis a classification of the relationships between narrative and argumentation would be settled.

2. It might be tempting to define narrative through speech-acts as the product of performing exclusively fiction or non-fiction making speech-acts. However, given the presented theoretical frame, this wouldn't be correct: the representation of events that Genette and Prince mention may be performed through different kinds of speech-acts: expositives (describe, emphasize, affirm), commissives (promise, contract), etc. (Searle, 1979).

reason within the course of an argumentation. As the argumentative speech-act is conceived here as a second order speech-act complex, based on the speech-act of adducing and the one of concluding, it is possible to consider the macro speech-act that the narrative text constitutes as the speech-act of adducing, making it to act as a reason for some conclusion. The paradigmatic example of this situation is a classic fable, when the story is presented as a reason that justifies the conclusion settled explicitly through the moral. This one will be named as Type S3.

The presented way of considering narratives and argumentations as speech-acts, and the narrative and argumentative texts as sets of speech-acts mainly composed by narrative or argumentative speech-acts allows the possibility of analysing each of these structural types as follows:

In Type S1 situations, a character of the story is performing an argumentative speech-act. According to that, what actually appears on the text is the representation of an argumentative speech-act.

In Type S2 situations, the narrator of the story is performing an argumentative speech-act, along with the rest of speech-acts that compound her narration.

In Type S3 situations, the author is adducing the set of speech-acts that compound the narrative text as a reason, with the conventional illocutionary force of trying to justify certain conclusion, that may be explicit or not. As it was mentioned before, the more evident examples of this type are classic fables including an explicit moral. But there are more subtle examples. The compilation of stories The Red Notebook, by the writer Paul Auster, is composed by certain autobiographical stories. As an example, a fragment of chapter 7 is presented:

> Twelve years ago, my wife's sister went off to live in Taiwan. Her intention was to study Chinese (which she now speaks with breathtaking fluency) and to support herself by giving English lessons to native Chinese speakers in Taipei. That was approximately one year before I met my wife, who was then a graduate student at Columbia University. [...]
> It is scarcely possible for two cities to be farther apart than Taipei and New York. They are at opposite ends of the earth, separated by a distance of more than ten thousand miles, and when it is day in one it is night in the other. As the two young women in Taipei marveled over the astounding connection they had just uncovered, they realized that their two sisters were probably asleep at the moment.

In an interview that the journalist Roberto Careaga made, for the Argentinian newspaper "La Tercera", to Paul Auster, he declared the following: "That's why I wrote The Red Notebook: to show, through examples from my own life, how strange life is. We would have to be really stupid and blind to affirm that chance doesn't play a role [...] There are happy consequences, and terrible ones. But we also have the ability of reasoning, taking decisions, have goals and plans. I'm interested in this tension"[3]. (Careaga, 2014) That is, Paul Auster would have narrated a series of facts in his book, adducing it as a reason to justify that the influence of chance is a key point in the development of our own life.

On the other hand, as the author and the narrator's voices often coincide, structural types might get overlapped sometimes. In order to illustrate this phenomenon, a fragment from Michel de Montaigne's Essays (I, I, XXX: Of Cannibals) where types S2 and S3 get overlapped is presented:

> When King Pyrrhus invaded Italy, having viewed and considered the order of the army the Romans sent out to meet him; "I know not," said he, "what kind of barbarians"[...] "these may be; but the disposition of this army that I see has nothing of barbarism in it."

Until this moment, the text has seemed exclusively composed by speech-acts of non-fiction making, which makes it a purely narrative text. However, it continues as follows:

> By which it appears how cautious men ought to be of taking things upon trust from vulgar opinion, and that we are to judge by the eye of reason, and not from common report.

Montaigne, as narrator, is adducing the first narrative piece of text in order to try to justify "how cautious men ought to be of taking things upon trust from vulgar opinion, and that we are to judge by the eye of reason, and not from common report". In a more explicit way: he is concatenating speech-acts: in the first place, speech-acts of narrating, and then, a speech-act of concluding. The compound of those two speech-acts makes the second order speech-act complex of arguing—where the speech-act of adducing is of a narrative nature.

3.2 Functional classification

As it was settled at the beginning of section 3, the proposed functional classification is based on the study of the roles—the functions—that

---

3. In Spanish in the original version. Translated by myself.

speech-acts of arguing play in narrative texts, and that speech-acts of narrating play in argumentative texts.

3.2.1 The role of narrative speech-acts in argumentative texts

On the basis of the previous definition of the argumentative text as the one mainly composed by argumentative speech-acts, the discussion about the roles that narrative speech-acts play in relation to these texts is discussed. This first functional type will be named as type F1.

Given the fact that it is modelled as a speech-act, argumentation presents, along with its illocutionary force (that is, trying to show that some conclusion is true), a perlocutionary force, as a function of the actual perlocutionary effect it provokes on its addressee, that is based on its capacity for inducing the same addressee to make the inferences associated to the argumentation. In order to reach this perlocutionary effect, the addresser has to make the addressee believe both the reason she is adducing and the inference that leads to the desired conclusion from the reason.

Narrative speech-acts in an argumentative text might act as rhetorical devices that provide dynamism, plausibility and vividness to to both the reason and the inference, and, as a consequence of that, increase the justificatory force of the argumentation. A narrative that presents an example about something that is been discussed may assure the existence of something by showing it within the frame of the story.

Montaigne's Essays provide again an example of this usage of narrative resources. As typical structure among the different essays collected on his books, Montaigne use to present, at the beginning of each piece, some topic he will try to justify through the stories he presents after that. This strategy can be seen in the following fragment, from chapter I, I, XXXI (That a man is soberly to judge of the divine ordinances):

> The true field and subject of imposture are things unknown, forasmuch as, in the first place, their very strangeness lends them credit, and moreover, by not being subjected to our ordinary reasons, they deprive us of the means to question and dispute them [...]
>
> In a nation of the Indies, there is this commendable custom, that when anything befalls them amiss in any encounter or battle, they publicly ask pardon of the sun, who is their god, as having committed an unjust action, always imputing their good or evil fortune to the divine justice, and to that submitting their own judgment and reason.

In this fragment, the topic of discussion is presented, at first, through the title—then, it is explained along the first paragraph. It could be reconstructed in an argumentative form in the following way: "their very strangeness lends them credit, and moreover, by not being subjected to our ordinary reasons, they deprive us of the means to question and dispute them, therefore it is good to judge soberly of the divine ordinances". Across the following parts of the chapter, Montaigne provides examples in order to try to justify the conclusion of his argument, by doing so to the reasons and the inference.

3.2.2 The role of argumentative speech-acts in narrative texts

With respect to the function that argumentative speech-acts perform in narrative texts—that will be labelled as type F2—, two options can be considered. As introducing this kind of utterances in narrative texts, the author can try to represent them with rhetorical and non-argumentative intentions (type F2-a) or to try to induce the reader to perform certain inferences (type F2-b). Type F2-a is paradigmatic in relation with representations of argumentations by any character the author want to present as, for instance, evil, weak, ridiculous, eccentric, etc. An example of this usage can be seen on The Magic Mountain, by Thomas Mann, chapter 6:

> On the contrary, Naphta hastened to say. Disease was very human indeed. For to be man was to be ailing. Man was essentially ailing, his state of unhealthiness was what made him man. There were those who wanted to make him "healthy," to make him "go back to nature," when, the truth was, he never had been "natural." [...] They talked of "humanity," of nobility — but it was the spirit alone that distinguished man, as a creature largely divorced from nature, largely opposed to her in feeling, from all other forms of organic life.

The conventional intention with which Thomas Mann introduced Naphta's argumentation is not to convince the reader of that that Naphta is defending, but to help her to create an idea of how Naphta must be: cynical, morbid, complicated, etc.

On the other hand, as it has been shown on Montaigne's Essays, the distance between the author of the text and its narrator can vanish, so the first one is in a position that allows him to argue directly through the text. In these cases, as argumentative speech-acts are not represented, but performed, they keep their typical perlocutionary force: being a tool to try to induce the addressee to make the same inference that the addresser is making on his speech-act. This inference is where the rhetoric import appears, as also happens when the author himself, without any narrator-overlap, argues through the story, as was mentioned before in relation to classic fables or the quoted text by Paul Auster.

3.3 Alternative proposals

On his article "On novels and arguments", Gilbert Plumer (2015) presents some results about the relationships between narrative (in the particular form of novels) and argumentation, with which he tries to justify his proposal about the process of reading a novel.

Despite his main interest concentrates specifically on novels, Plumer distinguish two different types of "narrative arguments", extracted from the work of Ayers (2010), which are a story offering an argument (P1), and a structural type of argument (P2).

Plumer explains that P1 is based on narratives through which creation its author argues, and defends that the argument that constitutes the product of this argumentation can be extracted from the novel. With respect to type P2, it consists on narratives that, despite the fact that arguments and narratives are different kinds of objects, their external structures coincide.

The proposed structural classification of the relationships between narrative and argumentation contains Plumer's distinction. Likewise, Plumer doesn't consider any functional feature in relation with his classification of the relationships between narrative and argumentation—any mention to functions performed by argumentative or narrative speech-acts (or a similar concept) appears on his work. Situations like the presented in Montaigne's Essays, where a narrative is adduced as a reason in an argumentative process (functional type F1) couldn't be described through his model.

Although it shares with Plumer's proposal a Platonic concept of argument, Olmos (2013) distinguish different types of relationships between narrative and argumentation, following a more pragmatic approach. In this way, Olmos explores the possible ways of attributing argumentative character to certain narratives. The first type of relation between narrative and argumentation that she presents is: narratives in which eventually explicit arguments are exposed. The second type of relation between narrative and argumentation refers to narratives "within a context in which, there being facts under discussion, the only visible support or evidence presented for a certain version of them, would be the manifest plausibility of the narrative sequence"(Olmos, 2013: p.13). Distinction based on distinguishing between reproducing an argumentation or performing it, nor presenting an argumentation with rhetorical intentions or just using it to invite to infer certain conclusions couldn't be considered by using Olmos' model.

4. IN DEFENCE OF A FUNCTIONAL ACCOUNT

The thesis defended here doesn't state that a functional analysis — based on the study of the functions performed by narrative or argumentative elements acting in relation with argumentative or narrative structures; in our approach that corresponds to narrative speech-acts in an argumentative text, or by argumentative speech-acts in narrative texts — is, by itself and in solitude, more accurate or preferable than a structural one—that is, one based on form relations among narrative and argumentative elements—. As opposed to that, what is defended here is that structural analysis in exclusive are not able to describe properly all the plausible relations that can hold between narrative and argumentation—and, as it has been illustrated in section 3.3., historically, this one has been the predominant account in which analysis have been based. But a precise functional analysis, combined with form appreciations provided by a structural one, may present the scope and precision needed to successfully describe and analyse these relations.

The previously mentioned "virtues of functional accounts" take relevance in a particular context: literary analysis, in relation with both literary criticism and hermeneutics. The bottom line that is being pursued in this article consist on being able to get the whole meaning of a certain narrative literary text, in the more integral way—through exploring the relationships of the narrative text with argumentation.

4.1 An application: analysis of Vladimir Nabokov's Lolita

It has already been stated that the functional account allows a finer and more precise argumentative analysis of narratives, specially in relation with literary ones.

For instance, and depending on certain pragmatic factors, a type S1 situation, in which an argumentation performed by a character is quoted inside a narrative text, may function as a speech-act indirectly performed by the author (through the character) or as a rhetoric device trying to produce a specific effect on the attendee. An example of that phenomenon is based on Vladimir Nabokov's Lolita. A fragment of chapter 13 is shown then:

> As she strained to chuck the core of her abolished apple into the fender, her young weight, her shameless innocent shanks and round bottom, shifted in my tense, tortured, surreptitiously laboring lap
> [...]

> The implied sun pulsated in the supplied poplars; we were fantastically and divinely alone; I watched her, rosy, gold-dusted, beyond the veil of my controlled delight, unaware of it, alien to it, and the sun was on her lips, and her lips were apparently still forming the words of the Carmen-barmen ditty that no longer reached my consciousness. Everything was now ready. The nerves of pleasure had been laid bare. The corpuscles of Krause were entering the phase of frenzy. The least pressure would suffice to set all paradise loose. I had ceased to be Humbert the Hound, the sad-eyed degenerate cur clasping the boot that would presently kick him away. I was above the tribulations of ridicule [...]

Nabokov is playing with the duality between the author and the narrator. By making Humbert narrate the reader how Lolita moved on his lap as she threw the apple, etc. he uses the narration of these facts in order to try to justify Humbert's own conclusion: "I was above the tribulations of ridicule". The first interpretation for this procedure is that Nabokov is intending to make the reader believe that it is justified that Humbert feels this way (1), while the second one consists on creating a sort of ambivalence between the closeness that may arise between the reader and Humbert as she follows his reasonings and experiences, and the rejection that it may provoke us to know it (2).

This distinction could not be made by using only a structural account—according to the presented model, both interpretations correspond to type S2 overlapped with S1—, but it would be done by using functional criteria: the interpretation (1) can be identified as a case of type F2-b—the author is preparing the reader a solid ground in which basis she can infer conclusions about Humbert—, while interpretation (2) as a case of type F2-a—the rhetorical effect of presenting the argumentation enhances the rhetorical force of the piece—.

5. CONCLUSION

This article has been written with the first objective of presenting a study, as systematic as possible, of the relationships between narrative and argumentation.

In order to do that, two classifications have been developed. The first one is based on structural criteria, i.e., formal features attending to the way in which an argumentative speech-act can appear on a narrative text. The second classification is based on functional features. The rhetorical dimension that representations of certain argumentations in certain context may exhibit is studied.

The second objective of this paper consists on defending the pertinence of using functional classifications in order to properly understand narrative texts in relation to its argumentative character. This thesis shouldn't be understood as a defence of the exclusive pertinence of using functional classifications, but to the one of combining functional and structural —that has been, from an historical point of view, the most predominant in academic literature—criteria.

REFERENCES

Bermejo-Luque, L. (2011). *Giving Reasons. A Linguistic-Pragmatic Approach to Argumentation Theory*. Dordrecht: Springer.
Currie, G. (1990). *The Nature of Fiction*. Cambridge University Press.
Fisher, W. R. (1987). *Human Communication As Narration: Towards a Philosophy of Reason, Value and Action*. Columbia, South Carolina: University of South Carolina Press.
García-Carpintero, M. (2007). Fiction-Making as a Gricean Illocutionary Type. *The Journal of Aesthetics and Art*, 65(2), 203–216.
Garrido Dominguez, A. (1993). *El texto narrativo*. Madrid: Síntesis.
Genette, G. (1980). *Narrative Discourse*. Ithaca, New York: Cornell University Press.
Olmos, P. (2013). Narration as argument. Virtues of Argumentation. *Proceedings of the 10th International Conference of the Ontario Society for the Study of Argumentation (OSSA)*, 22-26 May 2013 (pp. 1–14).
Plumer, G. (2015). On novels as arguments. *Informal Logic*, 35(4), 488–507.
Prince, G. (2003). *A dictionary of narratology*. University of Nebraska Press.
Romero Álvarez, M. L. (1996). El relato periodístico como acto de habla. *Revista Mexicana de Ciencias Políticas Y Sociales*, 41, 9–27.
Searle, J. R. (1975). A taxonomy of illocutionary acts. *Language, Mind and Knowledge* 344-369. Minneapolis, MN: University of Minnesota.
Van Dijk, T. A. (1996). *Estructuras y funciones del discurso*. México: Siglo XXI.

Literary sources

Jacobs, J. (Ed.). (1902). *The fables of Aesop*. Hurst.
Auster, P. (2014). *The Red Notebook*. Faber & Faber.
Careaga, R. (2014) Entrevista a Paul Auster: "Escribir es como una enfermedad, el mundo real no es suficiente". *La Tercera*
Mann, T. (2005). *The magic mountain*. Everyman's Library.
Montaigne, M. (2004). *The complete essays*. Penguin UK
Monterroso, A. (1971). *The black sheep and other fables*. New York: Doubleday & Co.
Nabokov, V. (1955). *Lolita*. Olympia Press

Interactive Discourse Features Supporting Aversive and Existential Acknowledgment In Legislative and Ceremonial Government Apologies

MARTHA S. CHENG
Rollins College
mcheng@rollins.edu

This study identifies and compares two sub-genres of collective government apologies for historical wrongs: legislative and ceremonial. The analysis focuses on discourse features related to interactivity, such as use of pronouns, direct address, and orientation to difference. It finds legislative apologies suppress interactivity while ceremonial apologies are highly interactive, suggesting that the former foreground aversive acknowledgement of the wrongs committed while the latter emphasize existential acknowledgment of the victims.

Keywords: collective apology, acknowledgment, legal language, political discourse, interactional discourse

1. INTRODUCTION

The last three decades have seen the growing practice of governments apologizing for historical wrongs, most often in the form of ceremonial speeches by leaders. Thus, studies of collective apologies have focused on ceremonial speeches. However, at times, a government may offer an apology through a congressional resolution or law—what I call "legislative apologies". This paper seeks to expand our understanding of collective apology by offering an initial investigation into distinguishing the sub-genres of legislative and ceremonial apologies. Specifically, it presents a rhetorical and discourse analysis of two U.S. acts of legislation: Public Law 103-150 (PL 103-150) passed in 1993, apologizing to Native Hawaiians for the annexation of Hawaii and Senate Joint Resolution 14 (S.J. Res. 14) of the 111th Congress passed in 2009, apologizing to Native Americans for past government discriminatory practices. Two ceremonial apologies are employed for comparison: President Clinton's 1997 apology to victims of the Tuskegee Syphilis Study and Australian Prime Minister Kevin Rudd's 2008 "Apology to the Stolen Generations," addressing Australia's Indigenous peoples. The apology to Native Hawaiians and to the victims of the Tuskegee Syphilis study were both issued by or under President Clinton while the apology to Native Americans and to the Stolen generations addressed similar historical injustices.

The analysis finds that the two sub-genres of apologies have dramatically different discourse features impacting their interactional dimension—the degree to which "the presence and nature of the source and the recipients" are made visible (Fahnestock, 2011, p. 278). The legislative apologies suppress this dimension through a highly impersonal tone and collective voice typical of legal language, conveyed through the use of third person pronouns and the absence of naming or addressing individuals. Interactivity can also be demonstrated through an "orientation to difference," Fairclough's term referring to the degree to which a text recognizes and addresses possible differences of attitudes or positions of the audiences (2003, p. 41). Such orientation is absent in the legislative apologies. Instead, the debated histories surrounding the injustices are presented as uncontested facts. Further, victims' expectations appear to be overlooked in the notably brief and restrain statements of mortification and the explicit disclaimers for corrective action. In contrast, the ceremonial apologies are highly interactive texts, with speakers taking on very personal stances toward specific audiences and their concerns. The interactivity and orientation toward victims are established through first-person pronouns, direct address, naming, repeated and emphatic mortification, and explicit corrective action.

These generic and linguistic differences of legislative and ceremonial apologies suggest different functions. That is, while both acknowledge historical wrongs and apologize to victims, they each foreground different aspects of apology: legislative apologies emphasize the historical facts warranting the apology and ceremonial apologies focus more on the victims and repairing relationships. These different functions reflect distinct forms of acknowledgement identified by Trudy Govier (2006): aversive and existential. Aversive acknowledgment recognizes that the "acts in question were wrong" and "those charged with committing them did in fact do so" (Govier, 2006, p. 48) while existential acknowledgment recognizes that "the persons harmed possess human worth and dignity and merit full and equal human rights..." (Govier, 2006, p. 48). Govier points out that the two types of acknowledgments are closely related, but are distinct and can have distinct effects.

Most studies of collective political apologies have investigated cases to help establish typologies of strategies and purposes to define the genre broadly. This paper extends that work by investigating relevant sub-genres, recognizing that collective apologies can take different forms. I argue that legislative collective apologies function more as aversive acknowledgment, due to their lack of interactivity and focus on historical facts, while ceremonial collective apologies foreground existential acknowledgment with their highly interactive and personal stance. Thus, this analysis of legislative and ceremonial apologies encourages a more nuanced understanding of collective apologies for historical wrongs, one that recognizes that the specific sub-genre can significantly affect the apologetic function. This paper first discusses apology in general and collective apology more specifically, second, it reviews the contexts of each studied apology, third, it highlights the varying discourse features between ceremonial and legislative apologies, relating these variations to the kinds of acknowledgement they perform.

2. APOLOGY

At the interpersonal level, Goffman has described apology as a remedial action, a form of corrective action when some social norm has been violated and the offender reassures the offended that they understand the norm and recommits themselves to it (1971, p. 113). Lazare has defined apology as "an encounter between two parties in which one party, the offender, acknowledges responsibility for an offense or grievance and expresses regret or remorse to a second party, the aggrieved" (2004, p. 23). In the public arena, the idea of "image repair" has dominated the scholarship, which draws heavily from the work of Ware and Linkugel (1973) and William Benoit (2014). Image repair focuses on specific strategies, such as bolstering or shifting blame, individuals use to reinstate their reputations.[1] In addition to image repair strategies, scholars have noted that the performative nature of the public apology, usually televised and distributed via mass media, is itself a kind of penance, a *metanoia* (Ellwanger, 2012)—a performance of mortification—required for image repair. Others have discussed the role of cultural traditions in apologetic practices—that leaders draw on cultural resources, such as dominant cultural narratives, in justifying, explaining, or repudiating their actions (Jackson, 2012; Liebersohn, Neuman, and Bekerman, 2004; Suzuki and van Eemeren, 2004).

When governments apologize for historical wrongs, the focus turns from individual image repair to relationship repair among communities: "collective apologies seek to reconstitute, rebuild, and strengthen relationships amongst communities harmed by historical wrongdoing perpetrated by one community against another" (Edwards 2011, p. 75). They do so through acting "as a mediation on past, present, and future relationships with the victimized collective." (Edwards, 2011, p. 75). These apologies often reinterpret history with an eye to affirming shared values, (Villadesen, 2008). They repudiate acts once justified by

---

[1] See Kampf's work (2009) on how strategies for image repair can lead to public non-apologies.

the government, reframing them as morally reprehensible and often provide assurances that such acts will not be repeated.

Whether addressing individual or collect apology, much of the scholarship has revolved around what constitutes an acceptable apology. Many theorists deem acknowledgment of the wrong the most important aspect of any apology. As Lazare notes, "without such a foundation, the apology process cannot even begin" (2004, p. 75). Govier's work, likewise has argued for the fundamental importance of acknowledgment, particularly in efforts at national reconciliation. She defines acknowledgement in the collective context as

> those responsible for committing such wrongs to recognize and admit having done so, and to articulate or represent that admission in a public form so that it becomes an enduring part of the public history of the state and society. The public admission and expression amount to acknowledgment. (2006, p. 48)

Acknowledgement is the opposite of denial and implies that the wrong actions will not be recommitted (Govier, 2006, p. 15). However, acknowledgment must be detailed and specific. Perpetrators or their representatives can be tempted to be vague in describing the transgression (Negash, 2006, pp. 9-10). Thus, "truth telling" and "being transparent about the facts" is essential (Negash, 2006, p. 9). Specific acknowledgment should also recognize the consequences of the wrongs done, particularly to effects on victims. This latter function recognizes the victims' innocence and humanity, reflecting a new way of treating them. A full acknowledgment, then, would detail the wrongs done and their effects on victims, thereby including what Govier termed as aversive and existential acknowledgment.

In addition to acknowledgement, mortification and corrective action are usually necessary for an acceptable apology. Through mortification, the offender accepts responsibility and expresses remorse. It is typically signaled through an explicit Illocutionary Force Indicating Device (IFID) such as the words "sorry" or "apologize" (Augoustinos, Hastie, and Wright, 2011, p. 509) when paired with an accurate portrayal of the transgression. Finally, corrective action entails addressing the source of the injury (Benoit, 2014, p.26) so as to prevent reoccurrences (Edwards, 2010, p. 69). Though distinct from compensation, corrective action might also include some payment or measures to remedy the negative consequences of the transgression and a commitment to prevent future offenses and to repair the damage done by the offense.

The legislative and ceremonial apologies examined here both contain aversive and existential acknowledgment. However, their discourse features suggest that they each foreground one type of acknowledgment more than the other. Also, legislative apologies omit corrective action while ceremonial apologies take pains to detail specific plans for corrective action.

3. CASES

The four apologies examined address historical wrongs perpetrated by governments or their representatives over years and even decades with damaging consequences lingering until today. Prior to 1893, Hawaii was an independent kingdom with a monarch, recognized by other national governments, including the United States, with whom it had diplomatic and trade agreements. In 1893, Hawaii was overthrown by U.S. and other agents, though not at the direction of the U.S. government. In 1898, the United States congress officially annexed Hawaii through the Newland Resolution. It became a state in 1959 through a referendum. PL 103-150, was sponsored by Democratic Senators Daniel Akaka and Daniel Inouye from Hawaii. After passing both Senate and House as a joint resolution, it was signed by President Clinton into law on November 23, 1993 during a White House ceremony. While the Hawaiian congressional delegation was pleased, and Hawaiians celebrated, there is still debate as to the historical accuracy of the law, as well as its impact on the Hawaiian sovereignty movement which is ongoing (Lopez-Reyes, 2000).

On April 30, 2009, the US Senate passed a joint resolution to "acknowledge a long history of official depredations and ill-conceived policies by the Federal government regarding Indian Tribes and to offer an apology to all Native Peoples on behalf of the United States"(S.J. Res. 14). It was sponsored by Senator Brownback and others, first going through the Committee on Indian Affairs before being presented to the Senate for a vote and passed. However, only a small piece of this legislation was included in a defense spending bill and signed into Public Law 11-203 by President Obama. The apology and signing were not publicly announced or accompanied by a ceremony. Due to this lack of publicity, most Native Americans were unaware that this apology had taken place (Longley, 2019; McKinnon 2009).

The two ceremonial apologies considered here are President Bill Clinton's 1997 apology for the Tuskegee Experiment and Prime Minister Kevin Rudd's 2008 apology to Australia's Indigenous Peoples.[2] The Tuskegee Experiment was a program sponsored by the United States Public Health Service and the Tuskegee Institute in Macon County, Alabama in the 1930s into the 1970s to treat African American men with

---

[2] Edwards, J. (2010) analyses Clinton and Rudd's apologies for purpose and rhetorical strategies to establish genre features of collective apologies. Harter, L., Stephens, R.J. and Japp, P.M. (2000) examine the narrative strategies in Clinton's apology.

syphilis. Under the guise of treatment, but without participants knowledge or consent, doctors left the disease untreated to study its progression. Even after effective treatment with penicillin was widely available, they did not treat the participants, essentially using them as unknowing guinea pigs. The study lasted 40 years, until 1972. In 1973 Congressional hearings found the experiment deeply unethical. Victims and their families won a $10 million law suit and the U.S. government set up the Tuskegee Health Benefits Program to provide medical and support services to victims and their families. On May 16, 1997, President Clinton apologized through a speech at a White House ceremony during which other government officials and survivors of the study also spoke. Survivors and families of victims were present. Survivors who spoke at the ceremony thanked the president for his apology[3].

In one of his first acts as Labor Prime Minister, Rudd, on behalf of Parliament, apologized to Australia's Indigenous Peoples for decades of discriminatory and oppressive policies. His lengthy apology ,given before Parliament and televised, contained a condensed version—a motion that was then passed by Parliament. His speech received a standing ovation in the parliament and generally positive coverage in the press (Augoustinos, Hastie, and Wright, 2011). However, the issue of apologizing had been the topic of public debate with the previous ruling Liberal party opposed such an act (Davidson, 2014).

4. LEGISLATIVE APOLOGIES AS AVERSIVE ACKNOWLEDGMENT

The legislative apologies are bound by legal genre conventions, including an impersonal style and formality. The formality comes from the legal written tradition of British law (Williams, 2005, p. 30), and helps to signal that the law has been through a process of negotiation and careful, considered construction and its contents are no longer preliminary (Stinchcombe 2001, p. 9). The impersonal style also supports the authoritativeness and archival nature of law—reflecting an impartial and collective stance that endures beyond any particular person, political administration, or social-historical context. Also, the author and audience are both collective entities leading to the impersonal style. Thus, the language of legislative texts represses features signaling their interactive dimension. Instead, the texts foreground the historical events warranting the apology. The majority of the legislation is spent chronicling the events, laying out the facts of the case in the highly formal, authoritative

---

[3] The current U.S. Center for Disease Control website offers a detailed overview of the Tuskegee Experiment and compensation efforts, as well as the transcript of the presidential apology (Tuskegee Study and Health Benefits Program, 2019). Video of the White House ceremony, including victim speeches, are available (*Apology to Survivors*, 1997).

style. These features point to a more aversive acknowledgment of the wrongs done.

4.1 *Legal Language: Impersonal Style*

One of the ways of making a text personal or interactive is through the use of pronouns, specifically the first person "I," "we," "us," "our" and the second person "you." Thus, the impersonal style of the legislative apologies is most obviously reflected through the almost complete use of the third person throughout, never referring directly to a specific speaker or audience. First person and second person pronouns are absent.[4] Despite PL 103-150 being sponsored by Senator Daniel Akaka and signed into law by President Clinton, and S.J. Res. 14, being sponsored by Senator Brownback, any reference to or voice of these individuals is absent from the texts. Instead, the actors in the text are "the Senate and House of Representatives of the United States of America in Congress…" (THE APOLOGY, 1993) and "The United States, acting through Congress…" (S.J. Res. 14, 2009).[5] Such language suggests that the apologies are not contingent of individual politicians or historical circumstances and are backed by the U.S. government.

Also, the victims, to whom apologies are directed, are referred to by collective nouns, rather than named more specifically, reinforcing the impersonal style. In his discussion of representing social actors, Fairclough states, "The opposite extreme to impersonalization is naming—representing individuals by name" (2003, p. 150). Fahnestock notes that "the first method of bringing participants into a text is by referring or naming" (2011, p. 303). The legislative apologies avoid such individual naming, relying on simply "Native Hawaiians" and "Native Peoples" or "Indian tribes." More specific groups, such as the many distinct Native American tribes, or individuals or populations, such as children, who were victims are not identified. One exception exists in PL 103-150 in which the Preamble chronicles the mistreatment of Queen Liliuokalani who ruled Hawaii at the time of annexation. However, this naming is provided in the context of providing historical background; thus, she is not spoken *to*, but rather *about*.

4.2 *Legal Language: Orientation to Difference*

In addition to the use of third person, the legislative apologies' low interactivity is demonstrated by their ignoring possible counter perspectives and audience stances. When discussing orientation to

---

[4] With one exception in S.J. Res. 14 where the text refers to the "…land we share;"
[5] Throughout the paper, excerpts of PL 103-150 are taken from THE APOLOGY, 1993; excerpts of S.J. Res. 14 are taken from S.J. Res. 14, 2009.

difference, Fairclough differentiates various levels of orientation, from highly oriented and embracing difference, to ignoring difference. (2003, pp.41-42). In the mid-range level, a speaker may try to "resolve or overcome difference" or "focus on commonality" (2003, p. 42). Making definitive pronouncements in law suppresses counter perspectives. This lack of orientation to difference is first demonstrated by the portrayal of the transgressions in the Preambles. The history surrounding historical injustices are often ignored and if not, then highly contested, but the legislative apologies elide these issues in their representations. The lack of dialogism is further demonstrated by the brief and restrained mortification and the explicit disclaimer, denying corrective action, which overlooks victims' needs.

These apologies follow the structure of typical statutes, beginning with a Preamble before moving on to the Resolutions containing the apology and other performatives. The Preamble "sets out the context in which the text was drawn up and what the purpose and scope of the law...is meant to be"(Williams, 2005, p. 39). Preambles can be long, using a series of "citations" beginning with a "whereas clause," often separated by semicolons to form a single sentence. In the legislative apologies, the Preambles represent the transgressions warranting the apologies. These statements elide the contested history by using unmodalized language and never acknowledging counter histories; the statements are presented categorically as statements of fact, with no hedging or qualifiers. PL 103-150 begins with the following "whereas" statement and continues with 34 more:

> *Whereas*, prior to the arrival of the first Europeans in 1778, the Native Hawaiian people lived in a highly organized, self-sufficient, subsistent social system based on communal land tenure with a sophisticated language, culture, and religion;
>
> *Whereas*, a unified monarchical government of the Hawaiian Islands was established in 1810 under Kamehameha I, the first King of Hawaii;
>
> *Whereas*, from 1826 until 1893, the United States recognized the independence of the Kingdom of Hawaii, extended full and complete diplomatic recognition to the Hawaiian Government, and entered into treaties and conventions with the Hawaiian monarchs to govern commerce and navigation in 1826, 1842, 1849, and 1887;

S.J. Res. 14's Preamble is structurally and linguistically similar, with 20 "whereas" statements.

In the legislative apologies the Preambles dominate the texts. In PL 103-150, the Preamble makes up 83% of the words of the legislation, while in S.J. Res. 14, it makes up 69%. Thus, the laws provide more "presence" (Perelman and Olbrechts-Tyteca, 1969, pp. 115-120) to detailing the events, and do so with specific dates, locations, actors, etc. As noted earlier, apologies for historical wrongs become part of the story of these injustices and the efforts at reconciliation—they become part of a nation's history. By apologizing through law, with the authoritative, collective voice these acts offer a singular sanctioned view of history. Thus, while the legislative apologies take pains to acknowledge the historical wrongs, they do so by silencing contesting voices.

A lack of orientation to difference also exists in the legislations' mortification and disclaimers, which appear in the Resolutions following the Preambles. While acknowledgment is a necessary condition for apology, mortification, as the acceptance of responsibility and expression of remorse, is the heart of apology. Given the extent of the historical wrongs—the vast numbers of peoples affected, the decades of harm done to them, and the long-term consequences—the mortifications in the legislative apologies are quite restrained. PL 103-150 states,

> The Congress...(1) Apologizes to Native Hawaiians on behalf of the people of the United States for the overthrow of the Kingdom of Hawaii on January 17, 1983 with the participation of agents and citizens of the United States, and the deprivation of the rights of Native Hawaiians to self-determination;

S.J. Res. 14's apology is equally brief:

> The United States, acting through Congress...apologizes on behalf of the people of the United
> States to all Native Peoples for the many instances of violence, maltreatment, and neglect inflicted on Native Peoples by citizens of the United States.

While the acknowledgment of wrongs in the Preambles is long and detailed, the mortification is notably singular and brief. Such brief expressions of mortification almost seem almost anticlimactic after pages detailing decades of mistreatment. Thus, while the mortification is present and explicit, (with the IFD "apologizes"), its brevity signals a lack of orientation to difference. After generations of suffering and being ignored, victims of such crimes may need a more extensive and emphatic mortification to be persuaded that the apology is sincere and the wrongs will not be repeated.

Finally, successful apologies often contain promises of corrective action—that the perpetrator will take concrete action to prevent offending again and will make efforts to mitigate the harm done, perhaps through some compensation. Such promises reassure victims that the perpetrators have changed and the apology is sincere. The legislative apologies are decidedly ungenerous in this regard. They make gestures toward corrective action, but avoid committing to any specific action. PL 103-150 "commend[s] efforts at reconciliation" and "expresses its commitment to acknowledge the ramifications of the overthrow of the Kingdom of Hawaii...". S.J. Res. 14 "expresses its commitment to build on the positive relationships of the past and present to move toward a brighter future...". However, these statements are quite general, with no specific corrective action. Further, each legislative apology ends with a Disclaimer preventing victims from seeking compensation based on the apologies. For example, S.J. Res. 14 states, "Nothing in this Joint Resolution (1) authorizes or supports any claim against the United States; or (2) serves as a settlement of any claim against the United States". The lack of discernable corrective action, and explicit ban on seeking such action ignore victims' desire for reparations and changed behavior.

Overall, the legislative apologies lack orientation to difference—they do not recognize any possible counter positions in their portrayals of history and present that history a categorical truth; the restrained, brief statements of mortification and lack of corrective action overlooks victims' desires for reassurance or their possible skepticism that the apology is sincere and that similar wrongs will not reoccur.

This lack of orientation to difference, together with the formal, impersonal language indicates a text with little interaction between speaker and audience. Instead, the texts, as legislation, speak for the record, establishing an authoritative and timeless written catalogue of the historical events warranting the apologies and they archivally record the act of apology itself. Thus, the analysis suggests that the legislative apologies serve more as aversive acknowledgement of the wrong rather than existential acknowledgment of the victims, though the latter is present.

5. CEREMONIAL APOLOGIES AS EXISTENTIAL ACKNOWLEDGEMENT

In contrast to the legislative apologies, the ceremonial apologies are marked by highly interactive discourse. This textual addressivity is accomplished through the frequent use of 1st and 2nd person pronouns, directly naming and addressing audience members, and a high degree of orientation to difference.

5.1 Ceremonial Language: Personal Interaction

While the legislative apologies avoided 1st and 2nd person pronouns, Clinton and Rudd use these throughout their speeches, particularly when expressing mortification. When Clinton describes the transgressions and apologizes, the 1st and 2nd person pronouns create a personal stance and address the audience:

> It is a time when *our* nation failed to live up to its ideals, when *our* nation broke the trust with *our* people that is the very foundation of *our* democracy...*we* can make amends and repair our nation.
>
> ...what the United States government did was shameful, and *I* am sorry.
>
> The American people are sorry—for the loss, for the years of hurt. *You* did nothing wrong, but *you* were grievously wronged. *I* apologize and *I* am sorry that this apology has been so long in coming.
>
> ...to the doctors who have been wrongly associated with the events there, *you* have *our* apology as well. To our African American citizens, *I* am sorry that *your* federal government orchestrated a study so clearly racist. (Tuskegee Study-Presidential Apology, 1997; emphasis added)

Rudd's apology is similarly personal and even more emphatic:

> ...the laws that *our* parliaments enacted made the stolen generations possible. *We*, the parliaments of the nation, are ultimately responsible,...
>
> *We* apologise for the laws and policies of successive Parliaments and governments...
> *We* apologise especially for the removal of Aboriginal and Torres Strait Islander children...
> For the pain, suffering and hurt...*we* say sorry.
> To the mothers and fathers, the brothers and sisters...*we* say sorry.
> For the indignity and degradation...*we* say sorry.
>
> To the stolen generations, *I* say the following: as Prime Minister of Australia, *I* am sorry. On behalf of the government of Australia, *I* am sorry. On behalf of the parliament of Australia, *I* am sorry. *I* offer this apology without qualification. *We* apologise for the hurt, the pain and suffering...*We* apologize for the indignity...*We* offer this apology to the mothers, the fathers, the brothers, the sisters...(Rudd, 2008; emphasis added)

Opponents of collective apologies argue that current individuals and generations bear no responsibility for past wrongs. With their use of personal pronouns, Clinton and Rudd reject this distancing of past and present responsibilities. Instead, the responsibility remains with the collective entity, which they and their colleagues and fellow citizens constitute. Also, although they are apologizing as government representatives, they take an individual stance that is absent in legislative apologies.

The excerpts above also illustrate ways in which Clinton and Rudd directly address various audience members. Clinton notes the doctors caught up in the scandal, as well as the African American victims, and Rudd differentiates family members torn apart—mothers, fathers, brothers, sisters and children. They name and directly address other members of the audience as well. In his opening remarks, Clinton speaks directly to the members of the audience:

> I would like to recognize the other survivors who are here today and their families: Mr. Charlie Pollard is here. Mr. Carter Howard. (Applause.) Mr. Fred Simmons. Mr. Simmons just took his first airplane ride, and he reckons he's about 110 years old, so I think it's time for him to take a chance or two. I'm glad he did. And Mr. Frederick Moss, thank you, sir. (Tuskegee Study-Presidential Apology, 1997)

He also names congressional leaders present and thanks them for their roles in addressing the Tuskegee experiment and the needs of the survivors and their families. Rudd also names various members of the audience: He directly addresses the various groups that were victimized:

> ... I would also like to speak personally to the members of the stolen generations and their families: to those here today, so many of you; to those listening across the nation—from Yuendumu, in the central west of the Northern Territory, to Yabara, in North Queensland, and to Pitjantjatjara in South Australia. I know that, in offering this apology on behalf of the government and the parliament, there is nothing I can say today that can take away the pain you have suffered personally. (Rudd, 2008)

Through these techniques, Clinton and Rudd make their speeches highly personal and interactive, recognizing their various audiences and speaking directly to them.

5.2 Ceremonial apology: Orientation to difference

Clinton and Rudd signal a high orientation to difference in several ways. First, they are attuned the victims' need for believable and sincere mortification. As noted earlier, the wrongs being addressed were perpetrated over years and generations of victims. Thus, victims might easily expect or need a forceful "sorry" statement. As seen in the above excerpts, both Clinton and Rudd use IFIDs repeatedly, using "sorry" and "apologize" repeatedly. Of course, repetition is a classic stylistic feature of speeches to provide rhythm and coherence, but it also can serve as a means of conceptual amplification. Fahnestock points out that "To amplify an element means to endow it with stylistic prominence so that it acquires conceptual importance in the discourse and salience in the minds of audience" (2011, p. 390). Given the severity and extent of the wrongs done, the repeated and emphatic expressions of remorse are more than justified and demonstrate recognition of victims' needs.

Yet, no matter how emphatic a verbal apology, in the case of historical injustices, it is not enough to right the wrongs. From the victims' perspective, compared to the injustices they suffered, mere words are inadequate. Because of this inadequacy, victims may not necessarily forgive the perpetrators. Clinton and Rudd recognize their apologies' inadequacy in the eyes of their audiences. Clinton states,

> To the survivors, to the wives and family members, the children and the grandchildren, I say what you know: No power on Earth can give you back the lives lost, the pain suffered, the years of internal torment and anguish. What was done cannot be undone...
>
> But you have the power, for only you — Mr. Shaw, the others who are here, the family members who are with us in Tuskegee — only you have the power to forgive. Your presence here shows us that you have chosen a better path than your government did so long ago. (Tuskegee Study-Presidential Apology, 1997)

Rudd makes a similar appeal for forgiveness:

> I know that, in offering this apology on behalf of the government and the parliament, there is nothing I can say today that can take away the pain you have suffered personally. Whatever words I speak today, I cannot undo that. Words alone are not that powerful; grief is a very personal thing.
>
> My proposal is this: if the apology we extend today is accepted in the spirit of reconciliation in which it is offered, we can today resolve together that there be a new beginning for Australia. (Rudd, 2008)

By requesting that the victims accept the apology, rather than expecting it, Clinton and Rudd recognize that their audiences may view the apology as woefully inadequate compared to the harmful effects they suffered. By empowering the victims to accept or deny the apology, these discourse moves signal "a new way of treating the victim, which itself reverses a prior way of mistreating the victim that began with the initial wrongdoing" (Helmreich, 2015, p. 76).

Finally, successful apologies usually require corrective action—a tangible demonstration of the remorse and a sign to the victims that the offender will not reoffend. While the legislative apologies encouraged or supported "efforts at reconciliation," they did not specify any corrective action and actively prevented claims in their disclaimers. The ceremonial apologies, in contrast, contained specific commitments to corrective action. In addition to the compensation already given by Congress, Clinton lists five initiatives including a grant to establish a bioethics research and training center and directing the Secretary of Health and Human Services, together with higher education to develop training materials for researchers on bioethics. Rudd, likewise proposes initiatives to help victims, such as having every Indigenous child attend early education schools, all Indigenous children having access to health care, and forming a joint commission to ensure adequate housing for remote communities. With these commitments to corrective action, Clinton and Rudd, recognize the expectations of the victims for some concrete acts that demonstrate a commitment to not reoffending. Overall, the ceremonial apologies are highly interactive in significant contrast to legislative apologies. This stance signals more attention to the victims and their needs, foregrounding existential acknowledgement.

6. CONCLUSION

Studies of collective apology for historical wrongs have sought to understand the genre broadly defined. They have productively identified shared purposes, functions, and characteristics. However, collective apologies can take different forms such as legislation or ceremonial speeches. These sub-genres, not surprisingly, reflect drastically different discourse features. The analysis suggests that the contrasting interactive dimension of these sub-genres can have different effects on the type of acknowledgment foregrounded.

PL 103-150 and S.J. Res. 14 follow the genre constraints of U.S. legislation with an impersonal, objective style, and authoritative stance while Clinton and Rudd's apologies reflect personal and performative qualities, oriented to their audiences. Legislation is supposed to transcend individual politicians or political moments and establish guiding principles by which to govern. Thus, the legislative apologies suppress interaction with the audience, instead focusing on chronicling the wrongs committed to establish an uncontested record of historical fact. In doing so, they perform aversive acknowledgement. In contrast, in the ceremonial apologies the speakers take highly personal, interactive stances and are attuned to audience expectations and needs. They foreground the victim, treating them with the respect and deference previously lacking, thereby enacting existential acknowledgment.

These findings encourage a more nuanced understanding of collective apology, with attention to ways in which governments choose to apologize to victims. At the same time, these cases raise questions about apology dissemination and reception: legislative apologies may or may not be publicized, while ceremonial apologies are typically mass media events. Surely, the degree of dissemination would likely influence efforts at reconciliation. Finally, research into how victims receive either type of apology would be productive to evaluate the effectiveness of the two sub-genres of collective apology.

REFERENCES

Apology to Survivors of the Tuskegee Syphilis Experiment. (1997, May 16). Retrieved from
https://www.youtube.com/watch?v=F8Kr-0ZE1XY

THE APOLOGY - United States Public Law 103-150. (1993). Retrieved July 13, 2016, from
http://www.hawaii-nation.org/publawall.html

Augoustinos, M., Hastie, B., & Wright, M. (2011). Apologizing for historical injustice: Emotion, truth
and identity in political discourse. *Discourse & Society*, *22*(5), 507–531.

Benoit, W. L. (2014). *Accounts, Excuses, and Apologies, Second Edition: Image Repair Theory and*
Research. SUNY Press.

Davidson, H. (2014, September 22). John Howard: there was no genocide against Indigenous
Australians. *The Guardian*. Retrieved from
https://www.theguardian.com/world/2014/sep/22/john-howard-there-was-no-genocide-against-indigenous-australians

Edwards, J. A. (2010). Apologizing for the Past for a Better Future: Collective Apologies in the
United States, Australia, and Canada. *Southern Communication Journal*, *75*(1), 57–75.

Ellwanger, A. (2012). Apology as Metanoic Performance: Punitive Rhetoric and Public Speech.
Rhetoric Society Quarterly, *42*(4), 307–329.

Fahnestock, J. (2011). *Rhetorical Style: The Uses of Language in Persuasion*. Oxford. Oxford
University Press.

Fairclough, N. (2003) *Analyzing Discourse: Textual Analysis for Social Research*. New York:
Routledge.

Goffman, E. (1971). *Relations in Public: Microstudies of the Public Order*. New York. Basic Books.

Govier, T. (2006). *Taking Wrongs Seriously: Acknowledgment, Reconciliation, and the Politics of*
Sustainable Peace. Amherst: Humanity Books.

Harter, L.M., Stephens, R.J., and Japp, P.M. (2000) President Clinton's Apology for the Tuskegee
    Syphilis Experiment: A Narrative of Remembrance, Redefinition, and Reconciliation. *The Howard Journal of Communication, 11*, 19-34.
Helmreich, J. S. (2015). The Apologetic Stance. *Philosophy & Public Affairs, 43*(2), 75-108.
Jackson, J. L. (2012). "God's law indeed is there to protect you from yourself": The Christian
    personal testimonial as narrative and moral schemata to the US political apology. *Language & Communication, 32*, 48-61.
Lazare, A. (2004). *On apology.* Oxford ; New York: Oxford University Press.
Liebersohn, Y. Z., Neuman, Y., & Bekerman, Z. (2004). Oh baby, it's hard for me to say I'm sorry:
    public apologetic speech and cultural rhetorical resources. *Journal of Pragmatics, 36*, 921–944.
Longley, Robert. (2019, August 22). Did You Know the US Apologized to Native Americans?
    Retrieved March 14, 2019, from ThoughtCo website: https://www.thoughtco.com/the-us-apologized-to-native-americans-3974561
Lopez-Reyes, R. (2000). Hawaiian Sovereignty. *Peace Review, 12*(2), 311–318.
McKinnon, J. D. (2009, December 22). U.S. Offers An Official Apology to Native Americans. Retrieved
    March 14, 2019, from WSJ website: https://blogs.wsj.com/washwire/2009/12/22/us-offers-an-official-apology-to-native-americans/
Negash, G. (2006). *Apologia Politica: States and Their Apologies by Proxy.* Lanham: Lexington Books.
Perelman, C., & Olbrechts-Tyteca, L. (1969). *The New Rhetoric: A Treatise on Argumentation.*
    Notre Dame: University of Notre Dame Press.
Rudd, K. (2008). APOLOGY TO AUSTRALIA'S INDIGENOUS PEOPLES. Retrieved July 17, 2017, from
    http://parlinfo.aph.gov.au/parlInfo/search/display/display.w3p;query=Id%3A%22chamber%2Fhansardr%2F2008-02-13%2F0003%22
S.J.Res.14 - 111th Congress (2009-2010): A joint resolution to acknowledge a long history of official
    depredations and ill-conceived policies by the Federal Government regarding Indian tribes and offer an apology to all Native Peoples on behalf of the United States. [Legislation]. (2009, August 6). Retrieved February 18, 2017, from https://www.congress.gov/bill/111th-congress/senate-joint-resolution/14/text
Suzuki, T., & van Eemeren, F. H. (2004). "This Painful Chapter": An Analysis of Emperor Akihito's
    Apologia in the Context of Dutch Old Sores. *Argumentation and Advocacy, 41*, 102–111.
Tuskegee Study - Presidential Apology - CDC - NCHHSTP. (May 16, 1997). Retrieved September 14, 2018,

from https://www.cdc.gov/tuskegee/clintonp.htm
Villadsen, L. S. (2008). Speaking on Behalf of Others: Rhetorical Agency and Epideictic Functions in
Official Apologies. *Rhetoric Society Quarterly*, *38*(1), 25–45.
Ware, B. l., & Linkugel, W. A. (1973). They Spoke in Defense of Themselves: On the Generic Criticism
of Apologia. *Quarterly Journal of Speech*, *59*(3), 273–283.
Williams, C. (2005). *Tradition and change in legal English: verbal constructions in prescriptive texts*.
Bern: Peter Lang.

Devil's Advocates are the Angels of Argumentation

KATHARINA STEVENS
University of Lethbridge
katharina.stevens@uleth.ca

DANIEL H. COHEN
Colby College
dhcohen@colby.edu

Is argumentation essentially adversarial? The concept of a devil's advocate – a cooperative arguer who assumes the role of an opponent for the sake of the argument – serves as a lens to bring into clearer focus the ways that adversarial arguers can be virtuous and adversariality itself can contribute to argumentation's goals. It also shows the different ways arguments can be adversarial and the different ways that argumentation can be said to be "essentially" adversarial.

KEYWORDS: argument-as-war, adversarial argument, cooperative argument, devil's advocate

1. INTRODUCTION

At risk of being overly dramatic, we would like to point out that there seems to be a bit of a war going on in argumentation theory. However, if it *is* a war, it is neither very destructive nor especially adversarial because it is an argument about the value of thinking about arguments in terms of wars, with a specific focus on the role of *adversariality* in argumentation. One of the questions that this little war is fought over is: *Is adversariality essential to argument?*

The broader discussion about adversariality in argument can be frustrating because every component of the claim that *adversariality is essential to argument* is ambiguous, beginning with what is meant by "argument", continuing with how we are supposed to understand "essential" and ending with some very serious confusion over what is to count as "adversarial". We can agree with the Pragma-dialecticians that at the *start* of every critical discussion there is a *difference*[1] and admit with Govier, that at the *heart* of every argument there is a at least some *opposition*,[2] but still have no argument with Bailin and Battersby that *conflict* does not have to be an important *part* of argumentation because deliberation does not need it.[3]

Like many of the best academic arguments, the argument about adversariality in argument is a tangle of several arguments. It is an overgrown thicket of different perspectives, insights, and concepts, which connect to adversariality in various ways. The result is more of a cacophonous controversy than a fruitful critical engagement that can integrate all those perspectives, insights, and concepts.

We will not even try to adjudicate all those debates, but we will argue that adversariality really is essential to argumentation – in a specified sense of *adversariality* and in an appropriate sense of *essential*. We begin by distinguishing several things that might be meant by saying, "Arguments are essentially adversarial." We then introduce the kind of arguer who embodies the kind of adversariality that is essential: The Angelic Devil's Advocate. We use this figure to highlight the relevant kind of adversariality, and to explain both how and why it is essential.[4] We are, however, mindful of ideal-theory problems, so our faith in these angels does not extend to belief in their existence.

Let us sidestep the last of the three ambiguities in the claim "*adversariality is essential to argument*" by stipulating what we mean by "argument". We are interested in argumentation as a process, and a joint activity between two or more parties.[5] We will not enter debates about whether arguments understood as abstract sequences of inferentially

---

[1] Some difference in standpoint is necessary for the opening stage of a pragma-dialectical critical discussion, but the existence of a difference is not sufficient for argumentation.
[2] Govier (1999).
[3] Bailin and Battersby (2017). For example, among Walton and Krabbe's six suggested models for different ways to engage in argumentation, we find models that pit arguers against each other such as persuasion dialogues and negotiations, but also models that unite arguers in the common pursuit of the answer to a problem or question, like deliberations and inquiries. So not all dialogue types that count as arguments pit arguers against each other in an adversarial relation.
[4] We have not made this figure up, at least not entirely. She has had many prior incarnations in argumentation theory. She appears as the ideal interlocutor in Johnson's *Manifest Rationality*; she is part of the universal audience in Perelman and Olbrecht-Tyteca's *New Rhetoric*; and we find her embodied as the opponent in an idealized critical discussion in *pragma-dialectics*. Since she may be most fully realized in Wohlrapp's pragmatic theory, we will begin there. We suspect that in another guise and gender, she may also be the legendary *Argumensch* of the oral tradition.
[5] Arguments2, in O'Keefe's vocabulary (O'Keefe, 1977).

structured propositions must somehow contain an implicit adversarial component to count as arguments. Nor will we address the argumentative status of solitary reasoning.[6] With these limits in place, we can begin to disentangle various concepts of adversariality in argument.

2. VARIETIES OF ADVERSARIALITY EXPERIENCE

In response to some feminist critiques of adversarial argumentation, Trudy Govier distinguished "ancillary" from "minimal adversariality" (Govier, 1999). *Ancillary* adversariality in argumentation is characterized by "lack of respect, rudeness, lack of empathy, name-calling, animosity, hostility, failure to listen and attend carefully, misinterpretation, inefficiency, dogmatism, intolerance, irritability, quarrelsomeness, and so forth" (Govier, 1999, p. 245). While acknowledging that ancillary adversariality is common in argument, Govier maintained it is neither necessary nor even central to argumentation. By contrast, *minimal* adversariality which casts arguers as opponents in arguments is both necessary and central. However, the opposition of argument opponents need not extend beyond the argument any more than the opponents in a friendly game of chess need be enemies outside the game. Minimal adversariality, she says, originates from the "bipolarity of "for and against" [that] seems to be inherent in thought itself" so is a necessary ingredient in argumentation. Thus, while making an argument tacitly acknowledges that others might disagree about the conclusion and explicitly implies that they would be mistaken,[7] it does not require anything more confrontational than just that.

Govier's distinction sparked a discussion on adversariality. Some criticisms targeted her concept of minimal adversariality as more than what was absolutely necessary for argumentation; others thought it

---

[6] What we have to say will be relevant for those dramatic instances in which a reasoner really does take up (or experience) different voices in her mind on behalf of different standpoints. What makes these phenomena relevant is that many of the most important cognitive benefits to be gained from argumentation come directly from the engagement, i.e., from *arguing*. You are fortunate, then, if, like Socrates, you have an inner *daimon*. For the rest of us, it helps to be on speaking terms with our own inner Socrates – who, for the record, sounds suspiciously like another manifestation of an Angelic Devil's Advocate. But with worse people-skills.

[7] For current purposes, both explicitly rejecting and simply not accepting a claim when sufficient reasons are available would count as "mistaken" from the proponent's standpoint.

needed to be fleshed out.⁸ Our take-away from the literature is that thinking of arguments as either adversarial or not, or even as more or less adversarial, is inadequate because when it comes to argumentation, "adversarial" can mean many things. Fortunately, the existing literature provides the conceptual means to develop a vocabulary for all the requisite distinctions.

As a start, we differentiate (1) the *adversarial attitude*, (2) an *adversarial stance*, (3) *adversarial functions*, and (4) the *persuasive-adversarial effects* of argument. This is surely not the only taxonomy for adversariality, but it builds on existing distinctions and it turns out to be useful.⁹

2.1. The adversarial attitude

Arguers with an adversarial attitude argue primarily *to win*, because there are obvious benefits from winning an argument, including making other people think or act to our advantage. In contrast, an arguer bringing a cooperative attitude to an argument is more concerned with bringing it to an *optimally successful conclusion* – a conclusion based on a fair representation of the balance of reasons applicable to the issue of the argument and available to the arguers. She argues not to win, but to *get it right*. Unlike Govier, we do not think that aggression necessarily accompanies an adversarial attitude – rather, the adversarial attitude will predispose the arguer to do what is necessary to win, even if that means suppressing a correct evaluation of the available balance of reasons (open aggression is one way to do so, but there are others).

It is the adversarial attitude that has deservedly attracted the most criticism: it encourages partisanship, elevates tactics over strategy and means over ends, and generates the negative behaviors that give argumentation a bad name.¹⁰

---

[8] For contributions trying to find an adversarial core, see, e.g. (Aikin, 2011, 2017; Casey, 2018). For contributions critiquing such attempts, especially Govier's, see e.g. (Hundleby, 2013; Rooney, 2004).
[9] This is the only way the concept of adversariality in argumentation could have been disambiguated. Another (for his purposes very useful) way to distinguish different kinds of adversariality in argumentation has been suggested by Casey (2018). However, we think that a careful reading of the work already done supports the distinctions we suggest and that making them will help structure the broader discussion about the place of adversariality in argumentation and remove some of the confusion that we at least felt when we first started reading about it. We will here describe each of these different kinds of adversariality and link them to the contributions that provided the grounds for them.
[10] Govier (1999, 245) cites, "lack of respect, rudeness, lack of empathy, name-calling, animosity, hostility, failure to listen and attend carefully, misinterpretation, inefficiency, dogmatism, intolerance, irritability,

2.2. The adversarial stance

"What does it mean for a practice to be adversarial?" Govier asked, and answered, "It means that in this practice people occupy roles which set them against each other, as adversaries or opponents," citing law, politics, and debates as institutions in which conventionally defined, oppositional roles are assigned. We shall call this taking an *adversarial stance*.

In assuming an adversarial stance, an arguer is committed at a minimum to finding reasons and formulating arguments for the associated standpoint, defending that position from objections, and, often, to raising objections to contrary alternatives. A *conscientious* court-appointed attorney would be an example of someone arguing this way without necessarily adopting an adversarial attitude or engaging in any of the aggressive behavior associated with it.[11] In a paper defending war- and sports-metaphors for argumentation, Aikin (2011) points out that the adversariality in competitive sports actually presupposes rather than precludes an underlying cooperative basis, and that same "cooperative adversariality" can inform argument. We have argued, similarly, that a globally cooperative attitude strategy can lead an arguer to adopt a tactical adversarial stance to further the chances for successfully resolving an argument by more fairly representing the balance of reasons[12] Indeed, a common justification for the Dominant Adversarial Model for arguments – the so-called "DAM account" – is that structuring arguments with arguers in adversarial stances enhances the prospects for optimal resolutions of difference: make sure each side has a champion presenting its case in the strongest light so we can better judge between them.[13] Many countries use this reasoning to justify their adversarial models for adjudication.[14]

---

quarrelsomeness, and so forth", but see Hundleby (2013) for reasons for hesitating before signing on to Govier's praise of politeness: the weight of norms is never gender-equitably distributed.

[11] The distinction between aggressive behavior in argument and argumentative adversariality as arguing against each other has been accepted an integrated by several authors, especially those aware and critical of Govier's contribution e.g.Aikin, (2011), Hundleby (2013), Rooney (2004). However, apart from Hundleby, we have seen little awareness that open aggression is not the core of the problem with Govier's ancillary adversariality, so that simply eliminating aggression will not solve the associated problems. We hope that our distinction between an adversarial attitude and an adversarial stance, instead of the distinction between minimal adversariality (as arguing against each other) an ancillary adversariality (as arguing aggressively), preserves Hundleby's important insight.

[12] Stevens and Cohen (2018).

[13] Zarefsky 2012 is a recent endorsement of this view.

[14] See, e.g. Fuller and Winston (1978), Luban (1988), and Sommaggio (2014).

As often noted, argumentation does not actually require arguers in opposing roles because arguers need not argue *against* each other.[15] Hundleby (2013) highlights this by pointing out that people may argue without even being committed to a claim, e.g., when pooling reasons to solve a problem or as a deliberative tool. This motivated some early criticisms of the DAM account, but in retrospect, we can see those objections conflated (not without some justification) the proponent-opponent structure of the DAM account with aggression and the argument-as-war metaphor, and thus were directed against the adversarial attitude not the adversarial stance.[16] Later criticisms on the basis that not all arguers can fulfill the tasks associated with an adversarial stance where more successfully directed against this form of adversariality (Burrow, 2010; Hundleby, 2013).

2.3. The adversarial function

Our refinement of Govier's (1999) distinction between adversariality as an attitude and taking an adversarial stance is meant to help clarify where exactly the distinction lies that Govier drew attention to. But these two are not the only kinds of adversariality we can identify in argument. A further distinction can be made between them and fulfilling *an adversarial function* in an argument, perhaps by merely serving as a sounding board for another or even just temporarily raising hypothetical objections against oneself. Of course, good sparring partners land real punches!

Aikin (2017), in response to Rooney's and Hundleby's criticisms that Govier's minimal adversariality is unnecessary, proposed an even more minimal version: *dialectically minimal adversariality*. He starts from the insight that arguing is necessary only when a view is or might be controversial, to mitigate its controversiality. Arguments target audiences who may have doubts about the view (possibly including the proponents themselves). Arguing is directed at critical challenges and objections. Aikin writes: "The thought is that without the role-related duties of critical dialogue, there are moves of critical probing that must be performed that are, in their dialectical function, *oppositional*" (Aikin, 2017, p. 16). Aikin's important insight is that even when arguers are not adversarial in the sense of taking thoroughgoing adversarial *stances*, there are still adversarial *functions* to be performed such as formulating

---

[15] See, e.g. Bailin & Battersby (2017), Gilbert, (1994, 1997), Hundleby (2013), and Rooney (2004).
[16] E.g. Cohen, (1995) and Moulton (2003 first publ. 1983). Rooney (2010) argues, helpfully, that the line between Govier's minimal adversariality and Govier's ancillary adversariality is porous.

objections, raising questions[17] – functions that show where arguments are broken, in contrast to functions belonging to what Hundleby calls "argument repair".[18]

Since the adversarial function can be filled by such helpful and cooperative arguers as sounding boards and sparring partners, the term *adversariality* should be understood according to its conventional usage for opposition of any kind, including the opposition of ideas, without any connotations of personal animosity or the toxic combative aspects of arguing that reinforce the DAM account's stranglehold on the imagination of argumentation theorists.

2.4. Persuasion as an adversarial effect

John Casey (2018) suggests that argumentation's "essential" adversariality can be found in the effects of reason-giving as such, without reference to the opposition of either ideas or arguers. Our beliefs are not entirely subject to our willful control: the causal nexus of belief-production includes everything we experience and hear – including reasons given in argumentation. From this perspective, arguing appears as an attempt to *cause* changes in our cognitive systems. In Casey's view, then, argument is adversarial because, it tries to *impose* change, fighting the epistemic inertia of the status quo, thereby amounting to an "attack" of sorts on personal autonomy.[19]

However, Casey (2018) points out that arguments can be between consenting adults who may welcome or even seek the change that comes with understanding the reasons presented in arguments.[20] What makes even those arguments *adversarial* is that they work *against* the arguer's cognitive *status quo*, although neither coercively nor aggressively. Because they happen whether they are welcome or not, the change-invoking effects of persuasive argumentation are not above moral considerations. For example, when our attempts at rationally persuading others include arguing with an adversarial attitude, we are committed to changing the beliefs despite their own preferences. And

---

[17] In the context of thinking of argumentation as calling for different tasks rather than different roles, Bailin and Battersby (2017) draw a similar distinction between Govier's minimal adversariality and "the confrontation of ideas". In Stevens and Cohen (2018), we argue for the value of thinking in terms of roles rather than tasks, but agree that the distribution of tasks can happen in different ways and need not result in oppositional roles.
[18] Hundleby (2010) is the source for this contribution to the discourse.
[19] Nozick (1981) reaches a similar conclusion. It also resonates with the claim, albeit for very different reasons, from some early feminist epistemologists claim that every attempt to persuade is an act of violence. See, e.g., Foss and Griffin (1995), Gearhart (1979), or Nye (1990).
[20] Fulkerson (1996) and Govier (1999), *inter alia,* also raise this point.

even if we have their epistemic betterment in mind, arguing could be unjustifiably paternalistic if it happens without consent.[21]

3. BUT IS IT ESSENTIAL?

The literature on whether and what kind of *adversariality* is essential for argumentation largely ignores the question of what *kind* of essential presence adversariality might have. But the ambiguity of *adversariality* is nearly matched by the ambiguity of *essential*. Is the claim that adversariality, of whatever sort, is essential to argumentation meant *conceptually*, so that adversariality is a necessary part of arguing; is it meant *descriptively*, so that it is universally present, or nearly so, and cannot be ignored by theory; or is it meant as a *normative* evaluation, saying that it is an *important* part of argumentation, regardless of whether it necessary or pervasive? Casey (2018) argues the conceptual claim while Zarefsky (2012) addresses the normative point. Others touch on all three in heterogeneous discussions. Critics of adversariality often stress its normative dangers before arguing that it is not conceptually necessary.[22] Defenders of adversariality often combine arguments that it is necessary with reasons why it is positive.[23]

Our earlier distinctions go a long way to answering the descriptive and conceptual questions of whether adversariality is essential to argument. Adversarial effect may be inevitable but only its possibility could be a pre-condition for arguing. Neither an adversarial attitude nor an adversarial stance is conceptually *necessary* for arguments, although both are sufficiently pervasive to be necessary for describing arguments. In practice, arguers tend to be torn between adversarial and cooperative attitudes: they want to win *and* to be right. We might say that anyone with an exclusively adversarial attitude is not genuinely arguing: since she is uninterested in the balance of reasons, she is not really engaged in the reasons-giving and reasons-responsive enterprise, and she no need for anyone to fill the adversarial function. Nonetheless, arguers engaged in genuine argument overwhelmingly often have an at least partially adversarial attitude. That leaves only the adversarial *function* as a candidate in need of investigation for being an essential ingredient of arguments. Is it? As noted, Hundleby (2013) showed how arguments can occur even without the adversarial function.

---

[21] See Tsai (2010). Davis (2017) adds the important caveat that this is the case mainly or only when the argument is *unwanted*. Neither author seems to be aware of the literature on adversariality in argumentation, so they do not explain how they see their arguments interacting with the broader discussion on adversariality.

[22] Bailin & Battersby (2017), Foss & Griffin (1995), Hundleby (2013), Moulton (2003, first publ. 1983), and Rooney (2004).

[23] Both Govier (1999) and Aikin (2011) fit here.

Admittedly, this depends on where the line is drawn between arguing and other dialogue types. Others may draw the line differently. We will simply recognize the adversarial function as *central* to argumentation – especially to *good* argumentation – and sidestep the larger question of whether it is *necessary* to all argumentation to avoid having to define argumentation precisely.

The important question for us is the one whether it is essential in the *normative* sense: Is adversariality *indispensably important* for argumentation? Would eliminating it from arguments imperil what it is that makes argumentation valuable? It is, after all, the intuition that adversariality *is* valuable, despite good reasons for thinking it is *detrimental* to good argumentation, that motivates these debates about adversariality and argumentation. Here is where the taxonomy of adversarial kinds helps because it makes it possible to say *this* kind of argumentative adversariality – the adversarial function – is essential in the normative sense, but *those* kinds – the adversarial stance and its accompanying attitude, which are nearly universally present, descriptively – are potentially toxic sources for the negative effects. Thus, if the adversarial attitude is descriptively universal, but not conceptually necessary, while the normative function is *normatively* essential, but sadly not universally present, then what argumentation theorists need is normative models that will motivate arguers to perform adversarial functions while inhibiting our natural tendencies to let the adversarial attitude crowd out cooperation in arguments. This model, we think, can be represented in the ideal figure of the *Angelic Devil's Advocate*.

4. THE ANGELS OF ARGUMENTATION

Competitors, rivals, and opponents are all adversaries but they are not all enemies. A punch from a sparring partner feels the same as one from a match opponent but there is a difference that makes all the difference.[24] A sparring partner serves as an opponent but is really an ally. She is there to make the boxer *better*. She may try to win but winning is not her motivation.

The same thing holds for arguments: opposing arguers need not be enemies. They can be allies. When making difficult decisions or considering controversial claims, it pays to find someone to provide the input that comes only from opposition. We need others to fill the

---

[24] Similarly, a football coach chooses the starting quarterback from rival *teammates*, not from *enemies*. The successful candidate has a very different relation to the now-back-up quarterback than he does to quarterbacks from other teams. Barack Obama used exactly this analogy in the 2008 Democratic primary to prevent the competition between him and Hillary Clinton from becoming hostile (and he did indeed keep her on his team).

adversarial function, not to *be* adversaries. We want an arguer who *opposes* us to *help* us: an *advocatus diaboli,* a Devil's Advocate.[25]

A devil's advocate is not merely a useful interlocutor: in the idealized version described here, she is the ideal *other* who embodies what is best and most important about argumentation.[26] She is the opponent we need because her overall goal is to enhance the prospects of *successful* argumentation, i.e., getting it right.

Harald Wohlrapp highlights this role in his epistemic approach to argumentation.[27] Our ability to figure out the world is limited by our subjectivity – our prior beliefs, opinions, interests, and prejudices. Even when we can reason our way to a conclusion, perhaps even identifying reasons for and against it, we should not fully trust ourselves. The reasons we consider are *our* reasons; the inferential paths we follow are one *we* build; and the conclusions we reach have to be acceptable *to us.* This is where we can use a cooperative opponent: someone to help us transcend our limits by criticizing our argument in order to strengthen it, not to defeat it.

Along with our beliefs, experiences, and perspectives, we also bring a motley of biases to our reasoning. We are good at incorporating these biases into our arguments – indeed, the more skilled we are at arguing, the better we are at rationalizing those distortions, the harder it is for us to detect them, and the easier it is for us to be taken in by them.[28] Argumentation is – and *needs to be* – about more than just giving reasons. Arguing as a proponent *for* a conclusion is only part of it. Argument have other roles and there are other skill sets for those roles, including *hearing* reasons, *evaluating* inferences, *asking* the right questions, and raising good *objections,* and then *answering* those questions and *responding* to those objections.

---

[25] *"In 1587, Pope Sixtus V established a process involving a canon attorney in the role of Promoter of the Faith or Devil's Advocate. This person argued against the canonization (sainthood) of a candidate in order to uncover any character flaws or misrepresentation of the evidence favoring canonization."* Wikipedia

[26] Of course, if we see argumentation only as a tool to help us reach our practical goals by changing people's beliefs and actions in our favor, our ideal is an audience that uncritically accepts our every word.

[27] Wohlrapp does not think, though, that arguing alone can produce knowledge. Knowledge can only arise when theory proves itself reliable in practice. Instead, arguing produces the *trust* in new theory to rely on it in practice (and give it a chance to become knowledge). He describes argumentation as an intersubjective activity aimed at the testing of theses and proposed solutions. It is especially useful for epistemic gaps where our knowledge and well-grounded, established opinions run out, and where our well-rehearsed theories do not provide us ready solutions.

[28] This point is emphatically made in Kornblith (1999).

The deck is stacked against solitary reasoners, but arguing *with others* helps the odds. It is not the only way to succeed, but it is better than discovering our epistemic flaws through practical failures.[29] Wohlrapp envisions an opponent who is adversarial insofar as her job is to find *flaws* in our reasoning. She tries to undermine our reasoning as part of her adversarial function, not from an adversarial attitude. It may hurt when opponents raise objections we cannot answer, but that is because our subjective view of the world is on the line: we are invested in our arguments. When they are revealed as flawed, so are we. But we should be able to recognize the short-term, apparent loss as a genuine long-term gain. Wohlrapp, like the outspoken critics of adversariality, realizes that losing an argument can be an epistemic gain.

Wohlrapp's opponent functions as a Devil's Advocate: an arguer brought in to save an arguer from his own subjectivity. She argues *against* an argument *for* the benefit of the argument. She is a Guardian Angel of Argumentation.

5. ANGELS AND VIRTUES

Mercier and Sperber provide a context for understanding why the social dimension of argumentation is integral to reasoning, and empirical data showing that some opposition helps us reason.[30] Wohlrapp complements this with a description of good opponents that identifies their specific skills and virtues, and an explanation of how they benefit arguments. So is the *Devil's Advocate* the ideal interlocutor? We think that she represents *one* ideal: adversariality scrubbed clean to preserve only its normatively essential aspects. Nonetheless, she is not all we need in an interlocutor. The DA's focus is too much on the response to proponents either by raising objections or asking questions. What is overlooked in the ideal of the DA are such argumentative moves as initiating new lines of reasoning or proposing improvements to the standpoint. The DA is essentially *reactive;* to be truly *angelic*, Devil's Advocates need to be *proactive*. The original Devil's Advocates were called in by papal courts to fulfill their roles; Angelic Devil's Advocates – true Guardian Angels of argumentation – wouldn't wait for the call.

The thing to consider is *argumentative engagement*. An arguer can conduct himself impeccably whenever he finds himself in arguments, but if he is confrontation-averse and consistently avoids arguments, he would be hiding that light under a bushel: an able advocate, but not angelic. Alternatively, an arguer might argue cogently once in an

---

[29] Quine (1970) p. 48„ defending the use of inductive reasoning, put the point rather more dramatically: "Creatures inveterately wrong in their inductions have a pathetic but praiseworthy tendency to die before reproducing their kind."
[30] Mercier and Sperber (2011).

argument, astutely pushing back where she should, but have no idea how or when to *disengage*: a devil of a Devil's Advocate. There is an art to engaging in arguments and an art to disengaging from arguments. A full account of ideal arguers should address the conduct, skills, and virtues associated with entering and exiting arguments in addition to the conduct, skills, and virtues associated with all of the different roles in ongoing arguments.

We identify four aspects of argument engagement and the virtues associated with them. In each case, the virtue can be located, following Aristotle's lead, as a mean between extremes.

The first form of failure for would-be guardian angels of argumentation is failure to report for duty. Colleague who have not served as sounding-boards for the drafts of your latest manuscript have not contributed and do not deserve acknowledgment; a friend who wasn't there to argue you out of a foolish course of action, was not a friend in deed. There are two different failures here: those who *don't* engage and those who *won't* engage. Some would-be arguers might miss an argument because they are unaware of them; others might hear them but deliberately choose not to enter. The former failure might be a lack of empathy or an inability to pick up on conversational markers that signal that a dialogue has become an argument. The latter could be due to being confrontation-averse, a condition that might not be blameworthy but it is evidence that an important argumentative virtue is lacking. However, sometimes the refusal to engage is culpable: for example, filibustering in order to prevent critical engagement is a transgression, even though it is technically not a "fallacy" in the sense of being a mis-step *in* an argument. Not-arguing is not a kind of bad arguing, but non-arguers can be vicious on account of the missing virtue.[31]

Second, there are corresponding failures at the other ends of those spectra: sometimes engagement is ill-advised and even the angels of argumentation should dare not tread. We are all too familiar with "argument provocateurs" who are too eager to argue and manage to turn every communicative exchange into an argument.[32] We also have too much experience with diehard arguers from whom we cannot disengage. Beating a dead horse need not involve any logical fallacies, but it is still an objectionable form of arguing. (The categories are not exclusive. Provocateurs who are also diehards are best avoided!)

Argumentative engagement is generally a yes-or-no situation, but sometimes it makes sense to consider degrees of engagement. Half-hearted engagement can sabotage argumentation. At the other end, half-

---

[31] See Cohen (2003) for a taxonomy of fallacies, transgressions and sins in argumentation beyond just inferential failures.
[32] The character of the "argument provocateur" was introduced in Cohen (2005) along with a bestiary of other arguers notably lacking in argumentative virtues.

hearted *dis*-engagement in the form of lingering hostilities can prevent us from processing and learning from an argument. However, for the purposes of triangulating in on the qualities that make Angelic Devil's Advocates the embodiment of what is essential to good argumentation, it is more helpful to ask about *how* arguers engage rather than *how much*. Again, the extremes are instructive. On one spectrum, we find an arguer who is so invested in his position that he takes any criticism as a personal attack contrasted with a parody of academic objectivity, someone so disinterested as to be unaffected by stomach-churning atrocities, uninspired by breath-taking beauty, or unmoved by mind-numbing injustice. On another spectrum, we find career contrarians at one end whose inability to bring closure to the role of a Devil's Advocate disqualifies them from being angelic, while the other end finds overly amenable enablers whose agreement encourages exaggeration and radicalization.[33] She, too, fails as an Angelic Devil's Advocate, but not for a lack of angelicism.

In each case, there is a golden mean that represents an argumentative virtue: the willingness, ability, and skill to enter into argument; the willingness, ability, and skill to dis-engage from argument; the willingness, ability, and skill to genuinely engage without becoming inappropriately invested in the topic; and the willingness, ability, and skill to agree or disagree as needed. Taken together, they define an angelic devil's advocate.

And, we submit, she embodies the essence of good argumentation. Sadly, she is not real.

REFERENCES

Aikin, S. F. (2011). A Defense of War and Sport Metaphors in Argument. *Philosophy and Rhetoric, 44*(3), 250-272. doi:10.5325/philrhet.44.3.0250

Aikin, S. F. (2017). Fallacy Theory, the Negativity Problem, and Minimal Dialectical Adversariality. *Cogency, 9*(1), 7-19.

Bailin, S., & Battersby, M. (2017). DAMned if you do, DAMned if you don't: Cohen's "Missed opportunities". *OSSA Conference Archive, 90*. doi: https://scholar.uwindsor.ca/ossaarchive/OSSA11/papersandcommentaries/90

Burrow, S. (2010). Verbal Sparring and Apologetic Points: Politeness in Gendered Argumentation Contexts. *Informal Logic, 30*(3), 235-262.

Casey, J. (2018). Revisiting the Adversary Paradigm. *Presented at the 2018 International Society for the Study of Argumentation Conference.*

Davis, R. W. (2017). Rational Persuasion, Paternalism, and Respect. *Res Publica, 23*(4), 513-522. doi:10.1007/s11158-016-9338-x

---

[33] Sunstein (2000) provides the empirical data related to these phenomena.

Foss, S. K., & Griffin, C. L. (1995). Beyond persuasion - a proposal for an invitational rhetoric. *COMMUNICATION MONOGRAPHS, 62*(1), 2-18. doi:10.1080/03637759509376345

Fuller, L. L., & Winston, K. I. (1978). The Forms and Limits of Adjudication. *Harvard Law Review, 92*(2), 353-409. doi:10.2307/1340368

Govier, T. (1999). *The philosophy of argument*. Virginia: Vale Press.

Hundleby, C. (2013). Aggression, Politeness, and Abstract Adversaries. *Informal Logic, 33*(2), 238-262. doi:10.22329/il.v33i2.3895

Kornblith, H. (1999). "Distrusting reason." *Midwest Studies in Philosophy*, XXII, 181-196.

Luban, D. (1988). *Lawyers and Justice: An Ethical Study*. Princeton: Princeton University Press.

Moulton, J. (2003 first publ. 1983). A paradigm of philosophy: the adversarial method. In S. Harding & M. Hintikka (Eds.), *Discovering reality: feminist perspectives on epistemology, metaphysics, methodology, and philosophy of science*. Dordrecht: Springer.

O'Keefe, D. J. (1977). Two Concepts of Argument. *Argumentation and Advocacy, 13*(3), 121-128. doi:10.1080/00028533.1977.11951098

Quine, W.V.O. (1970). "Natural kinds". *Essays in Honor of Carl G. Hermpel* , N. Rescher, ed. Dordrecht: D. Reidel. P 41-56.

Rooney, P. (2004). Feminism and Argumentation: A Response to Govier. *Presented at the 2003 Ontario Society for the Study of Argumentation Conference*. doi:http://web2.uwindsor.ca/faculty/arts/philosophy/ILat25/editedrooney.doc

Rooney, P. (2010). Philosophy, Adversarial Argumentation, and Embattled Reason. *Informal Logic, 30*(3), 203-234. doi:https://doi.org/10.22329/il.v30i3.3032

Sommaggio, P. (2014). The Socratic Heart of the Adversarial System. *University of Leicester School of Law Research Paper No. 14-21*.

Sunstein, C. (2000). Deliberative trouble? Why groups go to extremes. *Yale Law Journal, 110*(1), 71-119.

The semantic mechanisms underlying disagreement. An argumentative semantics approach

ANA-MARIA COZMA
University of Turku
anacoz@utu.fi

This paper approaches disagreement from a lexical and argumentative perspective. I first give a brief presentation of Galatanu's model, the Semantics of Argumentative Possibilities. Then I examine disagreement by identifying the argumentative orientations that are manifest in discourse. Finally, based on the illustrations, I propose a list of semantic mechanisms inherent to disagreement. The illustrations are drawn from an online participatory consultation on issues related to bioethics.

KEYWORDS: disagreement, Semantics of Argumentative Possibilities, argumentative semantics, discourse analysis, meaning potential, semantic mechanism, bioethics, participatory democracy

1. INTRODUCTION

This paper will shed light on disagreement from a complementary perspective to the one that is generally adopted in argumentation studies. Within the framework of argumentative semantics[1], I will approach disagreement with the analytical tools provided by the model of the Semantics of Argumentative Possibilities (SAP). Thus, in this paper, the cases of disagreement[2] will be analysed via the argumentative orientations triggered by lexical meaning.

---

[1] The term has lately been used for a range of theories that derive from the Argumentation within Language (Anscombre & Ducrot, 1983), including the AWL theory itself. Ducrot's article on 'rhetorical argumentation and linguistic argumentation' (2004) provides a synthesis of this view compared to the one that is specific to argumentation theories.

[2] Disagreement is understood here in a broad sense as an 'utterance that comments upon a pre-text by questioning part of its semantic or pragmatic information (sometimes its format structure as well), correcting or negating it (semantically and formally)' (Sornig, 1977, p. 363).

The corpus on which this study is based comes from an online consultation that took place in France in 2018, namely the Estates General of Bioethics 2018, a nine-topic consultation from which I retain the discussion on the topic of 'Artificial intelligence and robotisation'. Due to lack of space, no general conclusions for the whole corpus will be drawn here; instead, I will simply use the corpus in order to illustrate my approach to disagreement in terms of lexical meaning, and leave the general considerations on the whole corpus for a future study.

Disagreement is an inherent feature of democratic debate and, in the consultative setting of the Estates General of Bioethics, there are no negative connotations attached to it. It is simply part of the democratic participation, without any other implications than everyone being able to express their views. However, in other settings, for instance a deliberative one, disagreement has a different significance and broader implications. These are beyond the scope of this paper.

I will start by providing a short description of the SAP model and by introducing the key concepts of this study. The model will then be illustrated with the selected corpus put into practice in order to present a new way of looking at discursive cases of disagreement. The examples that I will analyse, namely the comments left by the participants in case of disagreement, will also serve as a preliminary basis for the classification of the mechanisms of disagreement.

2. ARGUMENTATIVE MEANING IN DISCOURSE AS A RESULT OF THE ARGUMENTATIVE POTENTIAL OF LEXEMES

The SAP is a theoretical model developed by Olga Galatanu, and is in line with Hilary Putnam's perspective on lexical meaning and Oswald Ducrot's view of argumentation[3]. The model is positioned at the interface between lexical semantics and discourse analysis, as one of Galatanu's

---

[3] Olga Galatanu has been developing her theoretical model for over twenty-five years and the name 'Semantics of Argumentative Possibilities' has been used for the model since 2002 (see Galatanu, 2004). A comprehensive view on the motivations and design of the model can be found in the book she published in 2018, in French. A presentation in English of this model can be found in the paper published in 2009. As for the English name of the theory, several other translations can be found – for instance, Semantics of Argumentative Potentials –, but the one that was finally favoured is *Semantics of Argumentative Possibilities*, which is also the closest to the French name of the theory. Although the expression 'argumentative possibilities' might sound strange to someone who is not used to it, 'Semantics of Argumentative Possibilities' is in fact the most accurate translation of the French name *Sémantique des possibles argumentatifs*, since the French expression '*les possibles argumentatifs*' can also be initially puzzling.

main concerns is to show what is the role that lexical meaning plays in the construction of discursive meaning and, vice versa, to show how lexical meaning – and therefore the vision and values underlying meaning – are reconstructed through discursive mechanisms.

I will give a very brief description of this complex theoretical model, in accordance with the objectives of my study. I will limit this summary to the three following aspects:

    i.   the description of lexical meaning according to the SAP;
    ii.   the description of discursive meaning based on the pre-established lexical meaning;
    iii. the methodology for discourse analysis provided by the SAP model.

i. Galatanu's model postulates a three-layer semantic representation: **lexical meaning** is described in terms of core, stereotypes and argumentative possibilities (APs). The core stands for the stable part of the meaning, while the stereotypes are understood as an open and evolving list of culturally motivated representations. Both the core and the stereotypes have an argumentative dimension, which is represented by the structures 'A hence B' or 'A nevertheless B'. Thus, the core consists of an argumentative chain, whose elements are limited in number, for example 'X hence Y hence Z'. As for the stereotypes, they associate only two elements (argument-conclusion), the first of which corresponds to one of the elements in the core (in this case, X, Y or Z). In other words, the stereotypes are extensions of the core elements. Together, the core and the stereotypes account for the meaning potential of the lexeme. This potential is also represented in the form 'A hence/nevertheless B', but in this layer of the semantic representation, A stands for the word itself. Here is an illustration of the three layers of the word *robot* (the semantic representation must be understood as the linguist's abstract construct, in this case my construct based on the dictionaries):

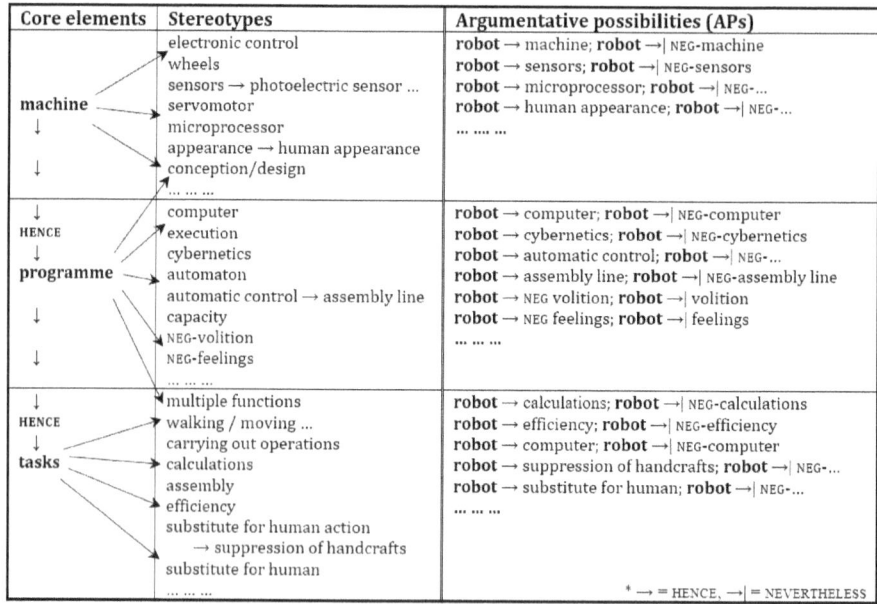

Figure 1. The meaning of 'robot' according to the SAP model

The argumentative association between the meaning elements is represented by means of the abstract connectors HENCE and NEVERTHELESS[4]. Galatanu describes this argumentative association as relying on a natural link such as cause-consequence, symptom-phenomenon, intention-means, whole-part etc., in other words, a link that has some intrinsic necessity to it.

This part of the model will be used to a lesser extent in this paper, but is nevertheless essential for the understanding of the approach used in this study.

ii. To describe the **discursive meaning** according to the SAP model means to identify the argumentative associations of the form 'A hence/nevertheless B' occurring in that discourse, and to show in what way they are related to the meaning potential of the occurring lexemes, i.e. to show if they correspond or not to the activation[5] of a part of the meaning potential understood in terms of core-stereotypes-argumentative possibilities. These argumentative associations occurring in discourse are considered to be 'discursive deployments of the meaning potential'. When the discursive deployments are not in line with the

---

[4] The French connectors DONC and POURTANT play this role in the Argumentation within Language theory and its developments (see, for instance, Ducrot, 2002).
[5] 'In SAP, the core and the stereotypes constitute a device for the generation of argumentative discursive sequences, 'the argumentative potentials', which can be activated within discursive occurrences, or deconstructed, even inverted, by co-contamination or contextual phenomena.' (Galatanu, 2009, p. 283).

meaning potential, they serve as a means to de/reconstruct the lexical meaning of the given words. Without going into detail (see Galatanu 2018b, p. 226), I will refer to all argumentative associations occurring in discourse as being 'discursive deployments' (DDs).

This facet of the model will be central in this paper. However, it should be noted that my approach in terms of DDs makes sense insofar as we describe the idea of the meaning potential in terms of core-stereotypes-argumentative possibilities.

iii. When **using the SAP model for discourse analysis**, the analyst focusses generally on one or several lexemes. First, they provide a description of the meaning, i.e. of the meaning potential, of those lexemes, by establishing the core, stereotypes and APs in accordance with the semantic representations shared by the speakers (this can be done either by using the language dictionaries or by submitting questionnaires to the speakers). Second, they examine the occurrences of the lexemes in a corpus, they identify the corresponding DDs and explain the mechanisms of (de/re)construction of the meaning in the discursive context. This methodology is made explicit in Galatanu's work (2009, p. 284).

However, the SAP model will be used in a new manner in this study. Instead of focussing on the DDs of some given lexemes, I will consider all the DDs that are present in the pieces of discourse I analyse (the CCNE's proposal and the participants' comments). The result will be a network of DDs, or more precisely, an oriented graph. This approach to the implementation of the SAP model can be productive for two reasons: the proposal and comments that constitute the corpus are brief, and the interaction in the corpus is limited to the participants' comments on proposals (each proposal is accompanied by a set of comments that are linked directly to it).

In addition to these three aspects, I would like to specify that the SAP model analyses words by means of other words. Meaning is not necessarily described through a set of minimal or primary elements, but rather through other words that are understood as meaning elements. It is the meaning elements *and their organisation*, i.e. the oriented links between them, that constitute the meaning of the word. Behind a word, there are other words, more precisely a specific configuration of other words. That is why Galatanu points out that her semantic theory is an associative, holistic, encyclopaedic and dynamic approach to lexical meaning. Modality also plays a key role in her theory.

3. DISCOURSE AS A NETWORK OF ARGUMENTATIVE ASSOCIATIONS

The methodology of this study has already been outlined in the previous section. I will now illustrate it and provide practical details. But first, I will introduce the corpus from which the illustrations originate.

The topic 'Artificial intelligence and robotisation' was one of the nine topics displayed on the website of the Estates General of Bioethics 2018, an online consultation that took place from January 18th to April 30th, 2018, and was organised by the CCNE, the French governmental advisory council on issues related to bioethics[6]. The general question was 'How should robots be integrated into medicine in order to improve treatment and health care?'. According to the numbers still on display on the consultation's website, the topic drew the attention of 4,514 participants and totalled 238 proposals (out of which 11 emanated from the CCNE). The participants had several options: to vote for or against existing proposals, to comment upon them, and to add their own proposals. The comments to a proposal are listed under it, in two columns, according to whether they are in favour of or against the proposal. The comments do not always focus on the topic; they also discuss and criticise the way the consultation is organised or the way the proposals are formulated. These forms of disagreement occurring in the corpus could be easily analysed from a pragmatic perspective, for example using Grice's conversational maxims, and for this reason I will disregard them. For the purpose of this paper, I will consider the comments on only one CCNE's proposal, titled 'Developing social robots', which received not one negative comment on a pragmatic level (see below). This proposal received a total of 91 comments, but for some reason only 87 are visible on the website, out of which 72 are against the proposal. For practical reasons, I use the term 'comment' for the participants' reactions, but on the website they are referred to as 'arguments, reasons' (in French, *argument*).

> Développer les 'robots sociaux'
> Les robots sociaux sont aujourd'hui utilisés dans certaines maisons de retraite au Japon, et permettent des interactions relationnelles avec les patients, en complémentarité avec les soignants. Le développement de ce type de robot, à dimension affective et pratique, pourrait être interrogé afin de pallier les problèmes, par exemple, relatifs aux zones dans lesquelles le besoin de personnels médico-sociaux est important ; et à la solitude des patients.
> Developing 'social robots'
> Social robots are now used in some retirement homes in Japan, and allow for relational interactions with patients, in complementarity with caregivers. The development of this type

---

[6] The website of the consultation is https://etatsgenerauxdelabioethique.fr, and CCNE stands for 'Comité consultatif national de bioéthique'.

of robot, with an affective and practical dimension, could be questioned in order to overcome the problems concerning, for example, the areas in which the need for medical-social staff is considerable; and the loneliness of patients.[7]

The corpus I analyse here is in French and the semantic analysis is based on the French lexemes. I will therefore translate everything into English in order to make my approach also clear to non-French speakers. Given that lexical meaning is culturally motivated, it goes without saying that the French 'maisons de retraite', 'Japon', 'personnels médico-sociaux' and the English 'retirement homes', 'Japan', 'caregivers' are not equivalent; they most probably do not have the same stereotypes and meaning potential. I hope my readers will be able to cope with this inconvenience and, if they are familiar with French, they will look up the original words in my analysis.

I turn now to the first illustration of my methodology. The diagram below (Figure 2) consists of the DDs contained in the text of the abovementioned proposal. As a reminder, a DD of a word, or discursive deployment, is an argumentative association between the word itself and another semantic representation; the arrow indicates the argumentative orientation. In this illustration, [Japan HENCE social robots] and [Japan HENCE retirement homes] are DDs of 'Japan'; there are four DDs of 'retirement homes'; and 'social robots' has the highest number of DDs. I will also use the term 'argumentative orientation' when talking about the DDs of a word, for instance the argumentative orientations of 'Japan' are 'robots' and 'retirement homes'.

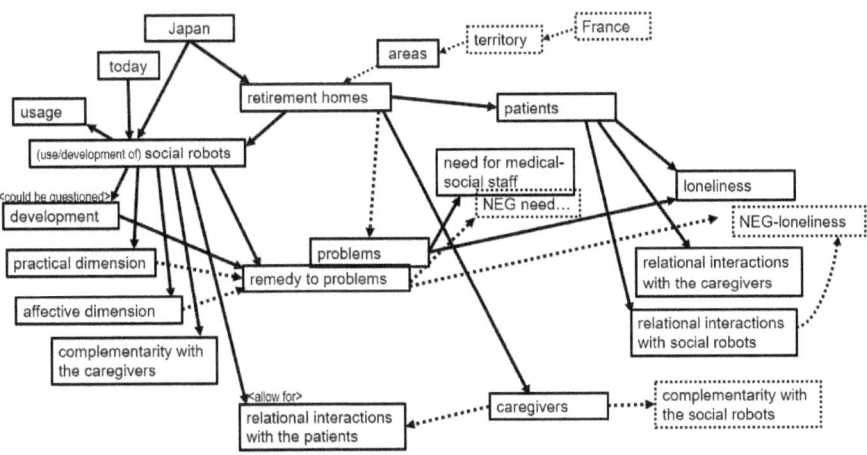

Figure 2. The argumentative orientations contained in the proposal

---

[7] All translations in the text are mine.

The arrows in the diagram represent the abstract connector HENCE – there are no associations based on the connector NEVERTHELESS, but if there were any, they would be represented in a distinctive manner. The dotted arrows indicate a presupposed, implicit, or even inferred association. The angle brackets are used for modalisation: <could be questioned> (alethic and epistemic modality), <allowed for> (alethic modality). In fact, modality is also present in the semantic representations put in the boxes of this diagram, for instance 'problems' (pragmatic modality), loneliness (affective modality) or 'development' (volition and pragmatic modality). As will become clear below, the axiological modalities play a particularly important role in disagreement.

Regarding the way in which the DDs and their argumentative orientations are identified, the argumentative association relies, as I said, on a natural link (cause-consequence, symptom-phenomenon etc.). In discourse, this link can either be expressed through lexical items (*allow, concerning, for example*), or connectors (*in order to*), prepositions (*in*: social robots are now used in some retirement homes; retirement homes in Japan; *of*: this type of robot: robot HENCE type), appositions (*this type of robot, with an affective and practical dimension*), attributes (*social robots*: 'robot HENCE social'; *the development of this type of robot*: robot HENCE development) etc. The argumentative association is thus retrieved from the syntax-semantics interface. Some words have not been taken into account in this diagram: *some, considerable* (these belong to the class that Ducrot, 2002, calls 'operators') and *type* (as an anaphoric expression for 'social robots').

The diagram is intended for visualising the discursive representations constructed in the proposal. The title provides the central element, 'developing social robots', and the text specifies the argumentative orientations associated to it: 'remedy to problems', 'practical dimension', 'affective dimension' etc. The DDs [social robots HENCE affective dimension] and [social robots HENCE relational interactions] are definitional in nature, since their conclusion 'affective dimension'/'practical dimension'/'relational interactions' explicates core features of the phrase 'social robots', while the DDs [social robots HENCE development] and [social robots HENCE remedy to problems] correspond to the activation of stereotypical features[8]. The core meaning

---

[8] See Galatanu (2018b, pp. 226 sq.) for the different types of DDs. A similar distinction is made by Anscombre and Ducrot regarding their concept of 'topos': intrinsic topoi, for instance *Pierre is rich: he can buy himself whatever he wants*, are such that the second segment (*he can buy whatever he wants*) is simply the explication of the meaning of first segment (*Pierre is rich*); extrinsic topoi such as *Pierre is rich: therefore he is miserly* work differently, since the second segment brings a real conclusion, not a mere explication of the first segment (Anscombre, 1995). In the SAP model, however, discursive deployments are viewed as

of the phrase 'social robot' can be described by the argumentative chain [relational machine HENCE programme HENCE relational tasks], based on the one for the lexeme 'robot': [machine HENCE programme HENCE tasks]. At the same time, 'social robot' can also be described as being an AP of 'robot': [robot HENCE sociability/sociality/social skills]. This argumentative association could also have been displayed in the diagram, since it is presupposed; it has not been done for reasons of efficiency, but it should be noted that all the argumentative orientations of the phrase 'social robots' are also shared by the lexeme 'robot' itself. The comments in Section 4 will show that the participants most of the time simply use the lexeme 'robot', and that some explicitly reject the use of the phrase 'social robot' (in fact, what they reject is the AP [robot HENCE sociability] and *a fortiori* the internalisation[9] of this association that leads to the core meaning [relational machine HENCE programme HENCE relational tasks]). The diagram also constructs a certain representation of 'retirement homes', 'patients', 'Japan' etc. Moreover, it can be read as a set of argumentative chains, for instance 'Japan hence social robots HENCE remedy to problems HENCE NEG-loneliness', 'retirement homes HENCE patients HENCE loneliness' etc.

The advantage of such a diagram, that accounts for all the DDs, is that it does not privilege one element over another; instead, it details all the argumentative orientations, even the less salient. That said, the salience can be determined from the number of associations in which one element is involved, either as the argument or the conclusion: 'social robots', '(remedy to) problems', 'retirement homes' and, to a lesser extent, 'patients' are the most salient, while 'today', 'usage' are the less salient, even insignificant. The participants in the consultation can theoretically show disagreement with any of the argumentative associations in the proposal, even the ones that are marginal (like [today HENCE social robots]).

The figure 2 is thus meant to account for all the argumentative orientations that are contained in the above-quoted proposal. I will now use this diagram in order to analyse a few comments showing disagreement with the proposal.

---

explications of the meaning in both cases: the core meaning in the first case (*rich* HENCE *capable of buying*), and the stereotypical meaning in the second case (*rich* HENCE *miserly*).

[9] I borrow the term from Ducrot (2002), but my use of this term does not correspond to his.

4. Analysing disagreement from a lexical perspective

The participants' comments below express disagreement with the proposal. Since they are displayed in the 'arguments against' column, the comments do not need to name the difference of opinion by using words such as 'don't agree', 'disagree' etc. However, there are linguistic markers that indicate the difference of opinion, for instance various forms of negation, contrastive structures, general statements, expressions showing indignation or negative appreciation etc.

For each comment, I will draw a new diagram that corresponds to the argumentative orientations occurring therein. In order to see how the comments are related to the proposal, each diagram will be built into the one shown in Figure 2. The elements from Figure 2 that are not mentioned in the comment will be greyed out, while the new ones stirred by the comment will be displayed in red.

> Comment n°1
> Le robot n'aura jamais de sentiments et l'inter-relation homme/robot est une illusion. Attention à conserver notre Humanité.
> The robot will never have feelings and the human/robot inter-relation is an illusion. Be careful to preserve our Humanity.

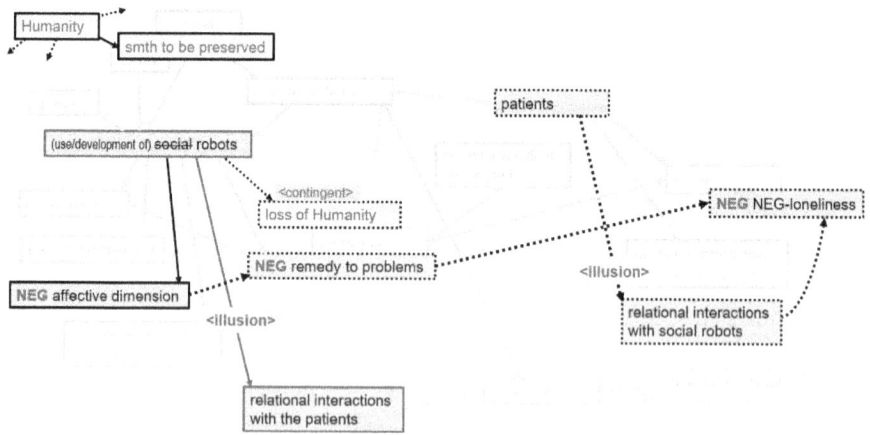

Figure 3. Comment n°1

In this comment, the difference of opinion is manifested in three ways. First, by saying that 'the robot will never have feelings', the author of the comment rejects and opposes one of the stereotypes of 'robot' introduced by the proposal, that is 'affective dimension'. S/he does so by making a prediction (*will never*) based on the commonly shared linguistic stereotype 'machine HENCE absence of feelings'. Second, by saying that 'the human/robot inter-relation is an illusion', the author rejects the DD [robots HENCE relational interactions], given that this argumentative association corresponds to the meaning of 'human/robot inter-relation' itself. The modal element <illusion> adds a doxological evaluation (belief) to the negation that it implies. Third, in the added semantic representation 'Humanity' argumentative orientations are left implicit, except for 'something to be preserved'. The warning 'Be careful to preserve our Humanity' implies that there is a risk of losing humanity because of the use of robots, HENCE the association [robots HENCE loss of humanity]. This association leaves a negative imprint on the representation of 'robot', i.e. a negative axiological evaluation (namely pragmatical and ethical: unfavourable and unethical).

> Comment n°2
> Un robot n'a pas de présence et de chaleur humaine : dimension fondamentale dans le soin à la personne. Le risque est la suppression des postes de soins, l'augmentation du chômage et l'accentuation du délitement des liens sociaux.
> A robot has no presence and no human warmth: a fundamental dimension in human care. There is a risk of job cuts in healthcare, rise of unemployment and intensification of the disintegration of social bonds.

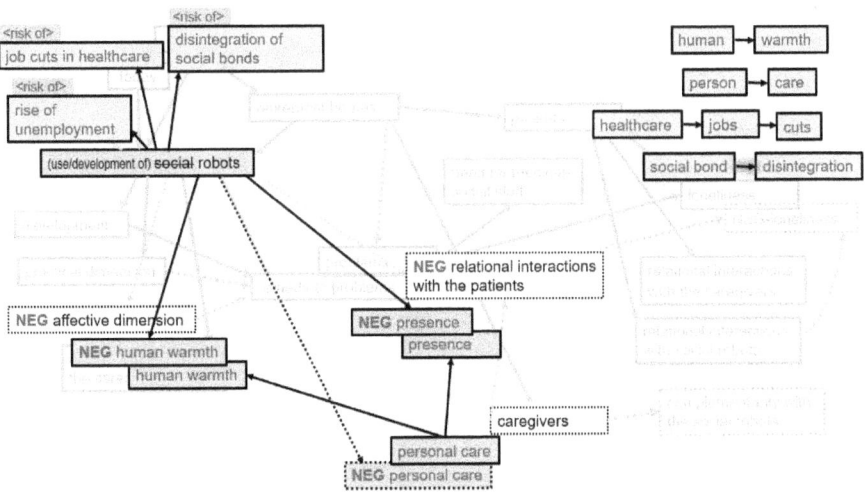

Figure 4. Comment n°2

This comment, just as the previous one, constructs a semantic representation of 'robot' that diverges from the one in the proposal. In contrast to the positive associations mentioned in the proposal, this comment triggers a series of negative ones: the comment contradicts the associations in the proposal ('NEG-human warmth' and 'NEG-presence' are contrary to 'affective dimension' and 'relational interactions') and introduces new associations that correspond to pragmatically negative evaluations ('job cuts', 'unemployment' and 'disintegration of social bonds'). Another negatively oriented DD is inferred: [robots HENCE NEG-personal care]. Overall, the comment rejects the positive representation of 'robot' without taking into consideration other semantic representations mentioned in the proposal; it also marginally presents a certain vision of representations such as 'healthcare', 'jobs' etc.

> Comment n°3
> L'utilisation de robots sociaux est une déresponsabilisation des proches et du personnel soignant dans les souffrances existentielles des patients. La société ne doit pas déléguer ce qui fait la base de son fondement qui est le lien interpersonnel. Les robots pourraient ainsi être utilisé de manière complémentaire comme une distraction, mais ne doivent pas être vu comme un remplacement du lien humain qui demeure aujourd'hui un critère prépondérant du bonheur.
> The use of social robots deresponsibilises relatives and caregivers with respect to the existential suffering of the patients. Society must not delegate what makes the basis of its foundation – the interpersonal bond. Robots could thus be used in a complementary way as entertainment, but must not be seen as a substitute for the human bond which remains today a dominating criterion for happiness.

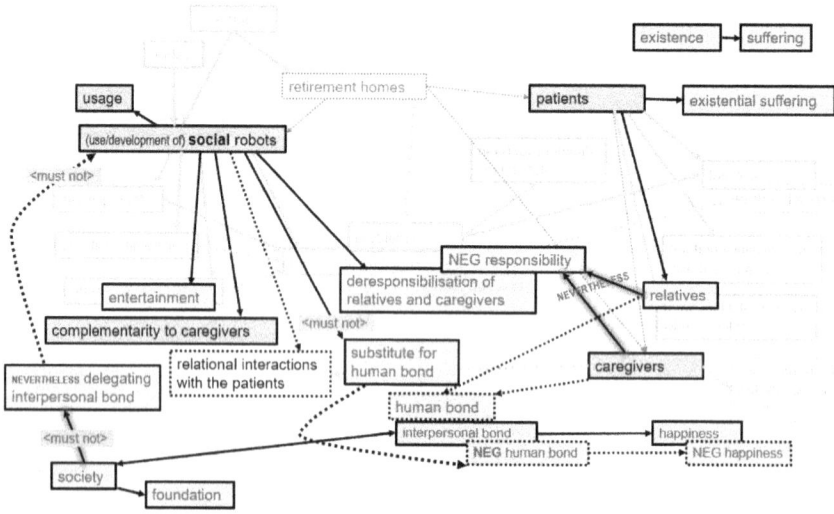

Figure 5. Comment n°3

Unlike the second comment, this one does not rely only on the representation of 'robot', but also takes into consideration other representations from the proposal ('patients', 'caregivers'), while also introducing new ones ('society', 'interpersonal bond', 'relatives'). The comment accepts some of the DDs of 'robot' in the proposal ([robot HENCE usage/complementary to the caregivers/relational interactions]), as well as the phrase 'social robot', which was not the case with the previous comments. In terms of evaluation, it adds a hedonistic orientation to robots ('entertainment') and an unethical one ('deresponsibilisation'). The argumentative association [robots HENCE substitute for human bond] is excluded (*must not be seen as*), since it would lead to unhappiness caused by the absence of human bond. Three DDs can be distinguished for 'society', out of which one is excluded (*must not*) and thus blocks the conclusion 'use of social robots'. The idea of responsibility is central, and it appears in two argumentative associations: [relatives NEVERTHELESS NEG-responsibility] and [caregivers NEVERTHELESS NEG-responsibility]. These transgressive associations (*nevertheless neg-*) construct an ethically negative representation of 'patients' and 'caregivers'.

I would like to make a remark related to rephrasing and anaphoric expressions such as *human bond*, that rephrases *interpersonal bond*, or *human warmth* that rephrases *affective dimension* in the comment number 2. They are considered equivalent in my analysis, but actually they are different semantic representations and have a slightly different meaning potential (i.e. a different set of core elements, stereotypes and APs). Rephrasing can also involve an argumentative association taken as a whole, as in the comment number 1: *the human/robot inter-relation* rephrases the sentence *social robots allow for relational interactions with patients* and, therefore, stands for [robot HENCE relational interactions...]. For these reasons, rephrasing and anaphora would need to be discussed in greater detail than I am allowed to do within the limits of this paper.

The following and last comment is the opportunity to make another remark concerning the argumentative orientations that are presupposed, implicit or inferred.

> Comment n°4
> je rappelle que grace a tous leurs robots, le Japon est le pays ayant le plus bas taux de natalité... Un robot ne remplacera jamais le contact, ni l'intelligence humaine.
> Be reminded that thanks to all their robots, Japan is the country that has the lowest birth rate... A robot will never replace contact, nor human intelligence.

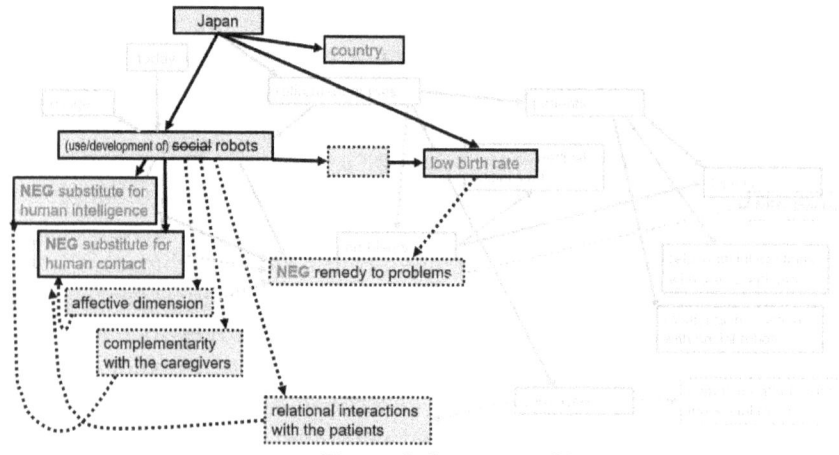
Figure 6. Comment n°4

The diagram identifies three explicit DDs of 'Japan' and another three for 'robots'. However, the DD [robots HENCE low birth rate] is a shortcut, and one or even several intermediary nodes can be added to explain the whole path from the argument 'robot' to the conclusion 'low birth rate': for example, 'NEG-human contact' could fill the gap in between. In the same way, the other two DDs of 'robots', [robots HENCE NEG-substitute for human intelligence/contact], constitute standalone associations, and at the same time, they imply a longer path from the argument 'robot' to the conclusions 'NEG-substitute...', as shown by means of dotted arrows. Finally, there is an implicit orientation 'NEG-remedy to problems' that can be retrieved from the comment as a reaction to the proposal. Moreover, part of the dotted elements in the diagrams simply arise from the context, i.e. the proposal. Presupposition, implicit meaning and inference would deserve an in-depth description that cannot be provided here. I therefore bring this analysis to an end by referring to Galatanu (2018b) for an argumentative approach to implicit meaning.

5. CONCLUSION

In the analysis above, I provided the description of disagreement by looking into the argumentative orientations of the words used in the proposal and comments. This gave insight into the various semantic mechanisms that are used in situations of disagreement at a micro-level. My approach to disagreement is thus only one among many, and needs to be combined with other approaches that focus, for instance, on speech acts, polyphony, linguistic markers etc. It also needs to be integrated into macro-level approaches that consider the context in which the disagreement takes places, since the virtues and vices of disagreement cannot be apprehended outside the implications it has within a particular field of activity.

The illustrations discussed in this paper indicate that disagreement relies on semantic mechanisms of rejection. These are:

a) the stereotype on which a DD relies is rejected
    - either the DD is replaced by its opposite ('robots HENCE NEG-affective dimension') (Comment 1),
    - or a new one is added that is very close to the negation of the stereotype ('robot HENCE NEG-human warmth' opposed to 'robots HENCE affective dimension') (Comment 2);
    - the negation will spread to the whole argumentative chain based on that stereotype, if there is such a chain ('NEG-affective dimension HENCE NEG-remedy to problems HENCE NEG-neg-loneliness') (Comment 1);
b) an absent but plausible argumentative orientation is rejected
    - the rejection affects an argumentative association that is both absent from the proposal and authorised by it ('caregivers NEVERTHELESS NEG-responsibility') (Comment 3);
    - this is based on a particular type of negation: the argumentative association is the same, but its normative form 'caregivers HENCE responsibility' is undermined and replaced by the transgressive form 'caregivers NEVERTHELESS NEG-responsibility';
c) an implicit argumentation is rejected
    - for example, by adding a new association that both explicates and negates the implicit association in the proposal ('robots HENCE NEG-substitute for human contact' as opposed to 'robots HENCE affective dimension HENCE substitute for human contact') (Comment 4);
d) the whole argumentative association is rejected
    - for example, by means of *'illusion'* and *'must not'* (Comments 1 and 3);
e) the axiological orientation is rejected
    - a new association is added that has an opposite axiological orientation (*'job cuts'* or *'low birth rate'* as opposed to 'remedy to problems') (Comments 2 and 4).

These semantic mechanisms co-occur and their list is likely to be extended.

The texts analysed in this paper are short because this was the only feasible way to fully illustrate my approach. It goes without saying that this approach could not be applied to the same extent to longer discourses. However, I still claim that the view offered by the SAP model, i.e. apprehending discourse in terms of lexical argumentative associations, can be profitably integrated into macro-level approaches of disagreement.

REFERENCES

Anscombre, J.-Cl. (1995). Topique *or not* topique: formes topiques intrinsèques et formes topiques extrinsèques. *Journal of Pragmatics*, 24 (1-2), 115-141. https://www.sciencedirect.com/journal/journal-of-pragmatics/vol/24/issue/1

Anscombre, J.-Cl. & Ducrot, O. (1983). *L'argumentation dans la langue*. Brussels: P. Mardaga.

Ducrot, O. (2002). Les internalisateurs. In H. L. Andersen & H. Nølke (Eds.), *Macro-syntaxe et macro-sémantique* (pp. 301-323). Berne: Peter Lang. http://semanticar.hypotheses.org/files/2018/09/Ducrot-2002-Les-Internalisateurs.pdf

Ducrot, O. (2004). Argumentation rhétorique et argumentation linguistique, In M. Doury & S. Moirand (Eds.), *L'argumentation aujourd'hui* (pp. 17-34). Paris: Presses Sorbonne Nouvelle. https://books.openedition.org/psn/756?lang=en

Galatanu, O. (2004). La sémantique des possibles argumentatifs et ses enjeux pour l'analyse de discours. In M.J. Salinero Cascante & I. Iñarrea Las Heras, *El texto como encrucijada: estudios franceses y francófonos. Actes du Congrès International d'Études Françaises, La Rioja, Croisée des Chemins, 7-10 mai 2002, vol. 2* (pp. 213-225). Logroño: Université de La Rioja. https://dialnet.unirioja.es/descarga/articulo/1011551.pdf

Galatanu, O. (2009). Semantic and discursive construction of the 'Europe of knowledge'. In E. Suomela-Salmi & F. Dervin (Eds.), *Cross-Linguistic and Cross-Cultural Perspectives on Academic Discourse* (pp. 275-293). John Benjamins.

Galatanu, O. (2018a). *La sémantique des possibles argumentatifs. Génération et (re)construction discursive du sens linguistique*. Bruxelles: P.I.E. Peter Lang.

Galatanu, O. (2018b). Les fondements sémantiques de l'implicite argumentatif. *Corela*, HS-25. https://journals.openedition.org/corela/6577

Sornig, K. S. (1977). Disagreement and contradiction as communicative acts. *Journal of Pragmatics*, 1, 347-374.

Strategic Maneuvering in Implicit an Pseudo-explicit Advertising Discussion

HÉDI VIRÁG CSORDÁS
Budapest University of Technology and Economics
hedi.csordas@filozofia.bme.hu

Advertisements can be analyzed in pragma-dialectical terms as contributions to discussions but they are not discussed in detail. I will distinguish implicit and pseudo-explicit advertising discussions and define them. How can they be analyzed as a critical discussion? Which PD rules moderate these special debates? Strategic maneuvering and the majority of the PD rules are applicable with some modifications.

KEYWORDS: pragma-dialectics, strategic maneuvering, commercial communication, stages, implicit discussion, pseudo-explicit discussion, validity of the ten PD rules, violence of the ten PD rules

1. INTRODUCTION

In my paper, I will discuss the applicability of the pragma-dialectical (PD) and strategic maneuvering (SM) approach to advertising communication. Similarly to everyday communication situations, advertising communication also presents us with cases of argumentation. I will therefore argue that different approaches to argumentation based on communication domain and genres of communication, and then make some points regarding the genre of advertising communication and promotion. Afterwards, I will identify the discussion stages in advertising communication, and analyze the validity and presence of the ten standard rules of critical discussion.

2. DEFINITIONS OF THE COMMERCIAL COMMUNICATION

The PD and SM approach does not distinguish between types of arguments: all arguments can be analyzed and evaluated on the basis of the ideal model of critical discussion. On the other hand, Van Eemeren considers arguments to be conventionalized communicative practices that can be distinguished by area, genre, and communicative act. He identifies eight major domains of communicative acts, attributing them to various genres and communication practices. Van Eemeren discusses only four of these domains: adjudication, decision-making, mediation, and negotiation. However, he presents this system as open, meaning that the categories, domains, and genres of communication can be further expanded and combined. To analyze advertisements from a PD and SM perspective, a thorough understanding of commercial communication is necessary; without this, I would only go so far as to define the genre of promotion.

We can examine advertising communication from the perspective of implicit argumentation, with only one of the parties actually participating in the discussion. In the case of advertising, the active (arguing) party is the service provider or manufacturer, while the passive party is the target audience. Whilst most advertisements fall into this category, we can encounter special cases where two competing companies conduct pseudo-explicit discussions[1]. Here, they make seemingly interconnected arguments, responding to the other party's propositions, but their actual discussion partner is still the customer.

Van Eemeren discusses the typology of argumentation on the basis of communication domain, genre, and activity. I will use a similar system to identify the categories of advertising communication and promotion, as well as to present the types of informative, persuasive, reminder, and reinforcement advertising. Since the PD and SM approach does not discuss advertising communication as a *communication domain*, I will use Kotler & Keller's definition from the glossary of classical marketing communication:

> "Advertising is any paid form of nonpersonal presentation and promotion of ideas, goods, or services by an identified sponsor. Ads can be a cost-effective way to disseminate messages, whether to build a brand preference or to educate people."
> (Kotler & Keller, 2006. p. 740.)

There are two things I point out in this definition: firstly, marketing theory does not consider advertising communication to be a personal discussion; and secondly, it is used for influencing the consumer's decision. Both of these factors underline our hypothesis that advertising can be viewed as implicit argumentation.

---

[1] The most emblematic case for the pseudo-explicit discussion is the Mercedes vs. Jaguar commercial battle from 2013.

Van Eemeren considers the pragma-dialectical analysis of advertisements a contribution to critical discussion. His commentary is not so much a definition of advertising communication, rather an addendum (footnote 41) to support the analysis of advertisements as a form of argumentation.

> *"In a pragma-dialectical analysis advertisements are, just like other specimens of argumentative discourse, viewed as contributions to a critical discussion."*
> *"41. A dialectical analysis of the advertisement is certainly relevant because listeners and readers will demand faithful information and good reasons for buying the advertised product, even if the advertisers cannot be expected to make an attempt at critical dispute resolution."* (Van Eemeren, 2010. p. 235.)

This definition shows us that besides influencing them through persuasion techniques, advertisements also provide customers with reliable information and solid arguments. In Van Eemeren's view, commercial actors cannot be expected to resolve the critical discussion; it must be done by the customer. He believes that the only difference of opinion is between the product distributor and the customer. However, as the above-cited footnote cannot stand in as a definition for advertising communication as a communication activity, I will make an attempt at a definition in consideration of the marketing and legal terminology.

Advertising communication is a paid form of nonpersonal communication, information, or presentation, which is directed at presenting or promoting a concept, product, or service as well as increasing revenue.

After defining advertising communication, let us take a closer look at its *communicative genre*, promotion. Instead of a single genre, marketing uses the 4P (product, price, promotion, and place) of the marketing mix to present and promote products, services, and concepts, while increasing the turnover of any given company. On closer examination, the category of promotion includes sales promotion, advertising, sales personnel, public relations, and direct marketing (Kotler & Keller, 2006. p. 54.) Promotion is a process of information transfer between seller and customer that is intended first and foremost to facilitate the decision to purchase. As Van Eemeren assigns specific characteristics to the genre of decision-making, I will examine which of these genre-specific traits can be applied to the genre of promotion.

In Van Eemeren's categorization, decision-making falls mainly into the domain of political communication, but its characteristics give us reason to assume that some of its specifics can also be assigned to the genre of promotion. Therefore, I propose that when discussing advertising communication, we also consider the genre-specific characteristics of decision-making, because it is eventually the consumer's decision which advertiser they will believe. My point of departure is the five criteria regarding decision-making: (1) there should be a confrontation between the parties; (2) an equal and appropriate amount of time should be available; (3) both parties should be present during the discussion; (4) the discussion should concern one specific issue; (5) the parties should argue in order to facilitate the audience's decision. (Van Eemeren, 2010. p. 142.)

Let me begin by analyzing the criterion of *confrontation between the parties*. Promotion does not necessarily require such confrontation. The difference of opinion rarely surfaces; the parties do not establish the topic of discussion, and they do not consult on their viewpoints. While confrontation is often identified as rivalry between companies, this is a misconception, given that the implicit argumentation is actually between the seller and the customer. The real difference of opinion is whether the customer should buy the product or not. As I have mentioned in my introduction, we can also encounter seemingly explicit, but in fact implicit discussions between competing companies: however, in this case, instead of having an actual discussion, their rivalry is still aimed at communicating their arguments towards potential customers[2].

The second characteristic of decision-making is that the parties have an *equal and appropriate amount of time* to make their arguments. The genre of promotion does not guarantee an equal amount of time for argumentation. The parties often have to consider factors which limit the length of their utterances. For television commercials, for example, the length of commercial spots will determine the length and quality of the arguments these companies can make. Neither nor is the response to promotion always immediate: a customer may choose to buy a product or a service long after the promotion was seen or heard. In implicit argumentation, the criterion of equal and appropriate amount of time cannot be fulfilled, as the reaction and arguments of the passive participant in the discourse are not available. They may consider the

---

[2] This situation is similar to Walton's forensic debate, where two participants debate each other, but their main purpose is to convince a third neutral party. (Walton 1989. p. 4)

advertising message intended for them, and may come up with counterarguments, but they may just as well refuse to argue, and therefore the utterances of the parties cannot be compared the basis of this criterion.

The third criterion is that *both parties should be present* for the decision-making. In implicit argumentation, this condition cannot be fulfilled, since there is only one party – the advertiser – who is present for the time of the discourse. Not even the advertiser is directly participating in the discussion; they make their argument in motion picture format. If we encounter a pseudo-explicit discussions between two rivals, we will see that the advertisers make their arguments independently of each other; they do not appear in the same place for the discussion. While rival television commercials may be aired within the same commercial spot that does not equalize the actual appearance of the competing parties.

There are only two of the criteria defining the genre of promotion which are met in full: one that the *discussion should concern one specific issue*; and two, that *the parties should argue in order to facilitate the audience's decision*. The issue in question is whose product or service is better – more desirable – for the customers. Manufacturers are understandably convinced that their product is the best. And companies choose arguments which are suitable for gaining the sympathy of the audience.
Based on the above, I am proposing the following definition for promotion as a communication genre: promotion is a (mostly) implicit discussion between the seller and the customer with the primary function to raise arguments concerning one specific issue to facilitate the customer's decision to purchase.

Finally, let us consider the *types of communicative activities*, staying strictly within the categories identified in advertising communication. Kotler & Keller describe four such categories: informative, persuasive, reminder, and reinforcement advertising. (Kotler & Keller, 2006. p. 740.)

3. STAGES OF THE COMMERCIAL COMMUNICATION IN THE TYPE OF IMPLICIT AND PSEUDO-EXPLICIT VERSIONS

As a type of argumentation, advertising communication is also different from critical discussion in its confrontation, opening, argumentation, and closing stages. Here, I will give a short recap of the function of these stages, and then I will present the characteristics of an implicit or pseudo-explicit discussions within the framework of advertising communication.

In the *confrontation stage*, the difference of opinion between the parties comes to the surface. In an implicit discussion within advertising communication, only the advertiser participates in the argumentation, so the advertiser and their imaginary opponent will never consciously acknowledge their differing opinions. However, the customer's personal presence is not required for the advertiser to be able to imagine the potential arguments against their viewpoint, and respond to them in the argumentation stage. Similarly, the competing companies are not present in a pseudo-explicit discussions, yet the confrontation stage is inapplicable to the genre of promotion, as their opposing viewpoints are natural results of their market positions.

In the *opening stage*, whether it is an implicit or explicit discussion, the parties do not agree on the subject of the discussion, and they do not set a point of departure, because that would also be alien to the genre of promotion. Nor do they agree on the acceptable premises and arguments to be used in the discussion, lay down rules, or set preconditions to moderate the process[3]. In an implicit discussion, the role of the protagonist is assigned automatically to the advertiser who argues in favor of their product or service. However, the potential customer – who is not participating directly in the discussion – will not automatically become an antagonist – they rather have the role of a passive skeptical listener. In pseudo-explicit discussions, the roles are different. While the advertiser is still defending their standpoint (as the protagonist) and the customer will remain the skeptical listener, the competitor joining the discussion will have the role of antagonist. An example of the missing opening stage and lack of agreement on the topics to discuss would be the pseudo-explicit discussion *Mercedes versus Jaguar* argument, where the parties constantly change their viewpoints. For example, when the Mercedes presented the shock absorber system of their car, the rival Jaguar responded by adding a new aspect – speed – to the discourse.

In the *argumentation stage* of advertising communication – whether it is an implicit or pseudo-explicit discussions – the advertiser makes arguments in favor of their standpoint. In an implicit argument, the customer does not participate in the discussion, and does not necessarily perform any argumentative activity. But even if they did

---

[3] Except the institutional conventions and the rules of Competition Authority etc.

make counterarguments, the advertiser would not find out, since they are not present in the discussion situation simultaneously. Analyzing a pseudo-explicit discussion is easier, given that the parties do respond to each other's arguments. Notably, due to the genre-specific limitations of promotion, the arguments in a pseudo-explicit discussion may not be presented at the same time: the response may sometimes come months later. In some cases, the antagonist party may change the form of communication (responding, for example, to a commercial spot with a still image, like in *Mercedes versus Jaguar*). The platform of the debate may move from Youtube advertisements to billboards or social media. The arguments of advertisers show a tendency of focusing less on adhering to dialectical norms than maximizing rhetorical effect, because advertising as a genre is less conducive of discussing complex arguments and more biased towards rhetorically efficient persuasion techniques. Therefore, these arguments are effective rather than strong. In a pseudo-explicit discussion, the customer does not necessarily perform any argumentation activity.

In advertisements, the *closing stage* of discussions is generally missing. In the closing stage of critical discussion, the parties summarize which standpoints have been defended or disproved. In implicit discourse and explicitly behaving implicit debates, such a simultaneous assessment and a potential admission of the other party's dominance is made unfeasible by the fact that neither of the parties is present in the discourse. Considering Van Eemeren's view that in the closing stage of implicit discussions, even if there is only one debater, they must still assess their own argumentation performance and withdraw the statements which are less well-founded, we can maintain that this is an unrealistic expectation in the case of advertising communication. First, the limitations of advertising as a genre do not allow for expressing long, multi-level arguments. Second, advertisers cannot be expected to evaluate and withdraw a strong argument. Notably, while identifying the stages is a normative requirement just as much as conducting a critical discussion, is rarely fulfilled, because the parties are too insistent on their own standpoints. Van Eemeren states that in advertising communication, *"in an advertisement, the difference of opinion that is to be resolved is whether or not the appraised product should be purchased."* (Van Eemeren, 2010. p. 235.) While we might presume that Van Eemeren refers to the closing stage here, implying that the inactive party in the implicit discussion also provides an evaluation alongside the purchase, we would be mistaken. First, these events take place at different times; second, since we have no access to the thoughts of the customer, we cannot be sure that they actually purchase a product or a service because of the arguments seen in the advertisement rather than any other reason. We should also mention a case when the discussion is "extended": that is, when the authorities start regulatory proceedings, and a legal case unfolds. From a PD and SM perspective, we consider this legal discourse to be a separate discussion. In such cases, the advertiser must present scientific evidence to prove their statements made in the media sphere, and argue in favor of their standpoint. This discussion is characterized by the specifics of legal communication and adjudication: an authorized figure makes a deliberation based on the factual argumentation of the parties.

4. THE VALIDITY CRITERIA OF THE TEN RULES IN THE IMPLICIT AND PSEUDO-EXPLICIT VERSIONS OF THE COMMERCIAL COMMUNICATION

Besides the basic characteristics of argumentation in advertisements, we must also determine the applicability of the ten rules of critical discussion. In this case, I consider advertising communication as multimodal argumentation with visual elements, since advertisements with only visual arguments fall under a different interpretation of the ten PD rules. Let us consider the ten rules in the context of advertising communication. First, I will examine the validity of each rule, and then the fulfilment criteria: whether the parties adhere to these restrictions.

Van Eemeren makes no distinction between explicit and implicit discussions when setting the normative criteria of validity. Based on his categories, I argue that the rules are always valid in explicit discussions, whereas the criteria of validity are not fully met in implicit ones. In his paper, J. Anthony Blair asserts that there are five cases in implicit discussions where PD rules are not valid. (Blair, 1998. p. 335.) While I admittedly agree with him with regard to implicit discussion in advertising communication, there is one more rule that I consider to be problematic.

Below, I will only discuss the rules which fail the validity criterion in the case of implicit discussions.

Freedom rule: In an implicit discussion, the limited presence of the parties allows only certain standpoints to be expressed. Statements in advertisements are one-sided; the advertiser cannot receive counterarguments and cannot respond to them. Blair thinks that arguers in implicit discussions narrow down the range of possible arguments, ignoring them if they are not closely related to their standpoint, they would take too long to elaborate, or they would derail the discourse. (Blair, 1998. 335.)

Burden of proof rule: In advertising communication, the advertiser cannot respond directly to counterarguments, and so they cannot defend their standpoint even if asked to do so. In the heterogenous audience, there may be people who would dispute their arguments (although potential customers rarely engage in argumentation), but the characteristics of the implicit argumentation prevent them from directly discussing their misgivings. Or, if their counterarguments are unrealistic, the arguer can decide to ignore them. (Blair, 1998. 335.)

Unexpressed premise rule: In an implicit discussion, the arguer is isolated, arguing without direct contact with the audience, so they cannot be held responsible for denying the unexpressed premises presented in their argumentation. (Blair, 1998. 335.) Furthermore, the customer does not necessarily respond to the advertisement with argumentation – so even if we disregard the lack of direct presence, we can presume that the advertiser cannot falsify the implicit premises presented by the customer.

Starting point rule: The audience of an implicit argumentation is not available to agree on the starting point and original premises, and they cannot challenge the arguer for attempting to falsely present them. (Blair, 1998. 335.) Let me note that advertising communication makes it highly unlikely that changes are made to the starting points and premises, given that the advertisement is created long before the audience could even voice their concerns.

Closure rule: Blair thinks that in an implicit argumentation, the lack of direct feedback from the audience may make the arguer reluctant to withdraw their statements or to accept failure at defending their standpoint. As I have already mentioned with regard to identifying stages, in advertising communication, the parties cannot be expected to summarize the argumentation process together: the customer is typically not present in the discourse, and does not necessarily perform argumentation activities.

Although Blair does not question the validity of the standpoint rule, in my view, this criterion is also problematic in implicit discussions.

Standpoint rule: In implicit argumentation, the arguer imagines the potential counterarguments brought by the other party in order to make their statements in opposition to them: in this case, attacking the standpoint means attacking the opposition's standpoint. In implicit discussions, the antagonist cannot respond directly to this argumentation, so it cannot be decided whether they are actually responding to the original standpoint.

All the other rules (the rules of relevance, validity, argument scheme, and usage) are considered to be valid in implicit argumentation.
I will also examine the validity criteria of the rules in pseudo-explicit discussions. Since such discussions are based on implicit argumentation,

I will start with the five plus one rules previously identified as invalid.

Freedom rule: While implicit discussions did not meet this criterion, our pseudo-explicit ones do. If we break the discourse into two parts, we can see that the rule is inapplicable to the implicit argumentation between the advertiser and the consumer: only one party's standpoint is represented in the argument, and the other party cannot express their concerns. But if another rival company responds directly to the argumentation, the standpoint can be debated, giving validity to the criterion.

Burden of proof rule: Similarly to the previous rule, implicit argumentations do not meet the validity criterion, while the explicit sections of the discourse do. As I have pointed out, the implicitness of the discussion means that the advertiser cannot respond directly to counterarguments, and thus cannot be expected to defend their standpoint if asked to do so. In an explicit setting, however, the rival company often asks the protagonist to defend their position.

Unexpressed premise rule: This rule also presents a duality: the criterion is not met by the implicit argumentation, but when an explicit discussion unfolds, it sometimes becomes applicable. In implicit argumentation, the advertiser is the only participant, and therefore cannot be challenged for denying unexpressed premises. However, in a pseudo-explicit discussion, the rival company entering the discussion may point out that the advertiser has presented an unexpressed premise inappropriately.

Starting point rule: Here, we can see the first difference regarding the validity criteria: neither the implicitly nor the pseudo-explicit sections of the discourse meet this condition. The advertiser cannot directly contact the customer, which means that they cannot agree on the starting premises and they cannot point out if there are any changes throughout the discourse. This rule is similarly invalid in explicit advertising communication, because the advertisers participating in the argument conduct their part of the discourse at different places and times, meaning that they are unable to settle the starting conditions of the discussion

Closure rule: Similarly to the previous one, this rule is also inapplicable in both the implicit and explicit part of the discussion. The advertiser cannot evaluate the argumentation process with the audience, since the customer is not present in the discourse and does not necessarily engage in argumentation. But while in the pseudo-explicit discussions both rivals put forward their standpoint, their market position makes it impossible for them to openly admit that their competitor won the debate and to retract their standpoint in favor of the opposition. I reiterate here that advertising communication is aimed at selling a product or a service for a profit, which makes the closure rule inapplicable here by definition.

Finally, I look at the validity of the standpoint rule – the one that J. Anthony Blair has found unproblematic – with regard to the pseudo-explicit discussion in advertising communication.

Standpoint rule: The rule is not applicable to implicit argumentation, but it has validity in explicitly behaving implicit ones. In implicit discussions, the advertiser imagines the counterarguments of potential customers and responds to them, and therefore attacking the standpoint means attacking the other party's standpoint. However, in implicit argumentation, the customer cannot respond to the arguments, and thus cannot clarify whether the advertiser is attacking their actual, intended standpoint. In a pseudo-explicit discussions, we have no such problem, since the advertiser can make it transparent to the rival company if they are not attacking the advertiser's actual standpoint.

What we can conclude regarding pseudo-explicit discussions in advertising communication is that there are six rules which are not applicable to the discussion between the consumer and the advertiser, and there are two which are not valid to the argumentation between the advertiser and their competitor.

5. THE APPLICATION OF TEN RULES IN THE IMPLICIT AND PSEUDO-EXPLICIT COMMERCIAL COMMUNICATION

While the ten rules of pragma-dialectics provide a normative framework for argumentation, in practice, the conditions tend to be violated, which consequently infringes on the dialectical norms. I will now discuss the application of these rules to implicit and pseudo-explicit discussions. I will observe their application only where their validity has been established – if we have excluded their validity, I will not consider how they are applied. First, I am going to discuss the application of the valid rules in implicit argumentation.

Relevance rule: valid, but often violated. In an implicit discussion, the advertiser can defend the point made by the advertisement in other ways, not only by putting forward arguments in favor of their standpoint. The original relevance of the rule is to exclude non-argumentative persuasion techniques from the discussion. If we observe advertising communication, we can conclude that it mostly applies such non-argumentative persuasion techniques that means derailment.

Argument scheme rule: unlike the previous one, this rule is not only valid but also applied to the practice of implicit advertising communication. In implicit discussions, the arguer has no possibility to agree on the general argument schemes with their partner, and therefore arguers are solely responsible for putting forward sound arguments.

Validity rule: similarly, the validity rule is also valid and applied to implicit argumentation. In an implicit discussion, the arguer is only responsible for arguing in favor of their own standpoint, and therefore they must exercise caution in putting forward correct and formally valid arguments.

Usage rule: valid for implicit argumentation but violated in practice. Advertisements are not known for using clear and unambiguous statements; advertisers tend to use multi-modal arguments, which are reconstructed in various individual ways.

Now, I am going to discuss the application of the valid rules to pseudo-explicit discussions.

Freedom rule: valid and applied for pseudo-explicit discussions. If an explicit discourse unfolds in the media sphere, the parties cannot prevent new standpoints or related concerns from emerging.

Burden of proof rule: valid, but this type of argumentation is often violated. The protagonist can be asked to defend their standpoint in the media sphere (as they were, for instance, in the Mercedes versus Jaguar advertisement series) but it is atypical for them to actually respond to the request and prove their standpoint.

Standpoint rule: valid but rarely violated in pseudo-explicit discussions. Advertisements reflecting on each other may attack a distorted opinion instead of the other party's actual standpoint. Violating the standpoint rule usually serves the interest of the advertisers, who tend to focus on gaining the sympathy of customers in a variety of ways instead of identifying the appropriate standpoint.

Relevance rule: valid and partly violated. Similarly to implicit argumentation, advertisers tend to use unclear phrasing here; their toolset includes persuasion techniques derived from social psychology.

Unexpressed premise rule: although an applicable criterion for pseudo-explicit discussions, it is regularly violated. The unexpressed premises that can be deducted from the standpoints put forward by advertisements are usually reconstructed by the opposing party in an inaccurate, exaggerated, or superficial way.

Argument scheme rule: valid for pseudo-explicit discussions, but sometimes violated. A norm for critical discussion is that an accepted argument cannot be (logically) invalid. In advertising communication, the opening stage of the discussion is missing, and therefore the parties cannot be expected to agree on the valid forms of argumentation. On the other hand, advertisers have a general knowledge of what we consider to be valid arguments, and therefore – without an opening stage – we can expect them to use that knowledge.

Validity rule: valid and partially violated in pseudo-explicit discussions. When rival companies are conducting a discussion in the media sphere, their main goal is to respond to the counterarguments of the potential customer, but they do it in a way as if they were attacking the opposition's standpoint. Therefore, the parties typically fail to adhere to the valid forms of argumentation.

Usage rule: valid for pseudo-explicit discussions, and rarely violated. Instead of using clearly phrased statements, advertisers communicate through visuals (open to individual interpretation) combined with short text (multi-modal arguments). In some cases, they attack their rivals through enigmatic ambiguous messages.

6. SUMMARY

We can conclude that while the standards of critical discussion are only partially met by implicit and pseudo-explicit discussion in advertising communication, the analytical criteria are valid. I argue furthermore that the validity and application criteria regarding the rules of implicit argumentation should be considered valid not only for advertising communication, but for all other forms of implicit argumentation as well.

REFERENCES

Blair, J. Anthony (1998). The Limits of the Dialogue Model of Argument. *Argumentation 12.* (pp. 325-339). Kluwer Academic Publishers. Printed in the Netherlands.
Eemeren, F. H. van (2010). Strategic Maneuvering in Argumentative Discourse. *John Benjamins Publishing Company.* Amsterdam
Eemeren, F. H. van & Grootendorst & Henkemans (2002): Argumentation: Analysis, Evaluation, Presentation. *New Jersey, Lawrence Erlbaum Associates, Publishers.*
Kotler, P. & Keller, K. L. (2008). Marketing Management. *Pearson Education, Inc, publishing as Prentice Hall*
Walton, D. N. (1989). Informal Logic A Handbook for Critical Argumentation. *Cambridge University Press, Printed in USA*

Reasons for Rational Disagreement from Dialectics
The Van Inwagen Cases

ISTVAN DANKA

Budapest University of Technology and Economics, Hungary
John von Neumann University, Hungary
<u>danka.istvan@filozofia.bme.hu</u>

> This paper investigates faultless disagreements by seeking for success criteria in argumentative situations derived from 'van Inwagen cases'. It argues that van Inwagen's definition of a successful argument implies that success is impossible in faultless disagreements. Analysing gradually more complex argumentative situation types shows that with increasing complexity, disagreement is increasingly more rational because of the evasion techniques available for the party in charge of defence.
>
> KEYWORDS: knockdown arguments, faultless disagreement, peer disagreement, pragma-dialectics, successful arguments, van Inwagen

1. INTRODUCTION

Some debates are rationally irresolvable because in the lack of successful counterarguments, parties are dialectically obliged to keep committed to their respective views (Danka 2018). Following van Inwagen (2006), this paper investigates success criteria in such debates. While in simple cases, the reasons for rational irresolvability are simple, they are even simpler in gradually more complex cases because complexity opens further and further possibilities for rational evasion. This implies that the possibilities of successful rational arguments in fields describable by the 'van Inwagen cases' are very limited.

Van Inwagen defines successful arguments as follows: an "argument is a success if it starts with premises almost no sane, rational person would doubt, and proceeds by logical steps whose validity almost no sane, rational person would dispute, to the conclusion" (van Inwagen 2006, p. 3). Later he considers this first approach too strong because the success of an argument is context-sensitive: it depends on purposes (e.g. persuasion or raising doubts), the audience (standards of acceptance in their field of expertise) and other circumstances. For these reasons, there is no formal definition for successful argumentation but the best criterion available is pragmatic: an argument is successful if it convinces an expert audience to change their minds on the issue. E.g. if experts believe that p, an argument against p is successful if after considering the argument carefully, experts either come to believe that *non-p* or they suspend judgments about p.

Though there are good reasons that this modified version of van Inwagen's definition should be further weakened (Hanna 2015), for the present purposes, it is sufficient. So let us accept for the rest of this paper that

> **Successful argument**$_{def}$: An argument for p is a success if under ideal circumstances, it alters the doxastic attitudes of an expert audience so that if audience disbelieved p, they come to believe p (or at least suspend their judgment about p), if they suspended judgment about p, they come to believe p, and if they believed p, their credence is strengthened.

A related notion often applied in this paper is '*prima facie* successful argument'. This is an argument that the target audience is expected to find convincing but will nonetheless reject because it would refute their standpoint for which they have even more convincing arguments. How this is exactly done will be explained below by analysing what I shall call as 'van Inwagen cases'.

2. THE VAN INWAGEN CASES

Following van Inwagen (2006), nine types of argumentative situations will be discussed in order to see how parties can react to prima facie successful arguments against their views. The first three are directly derivable from van Inwagen's two examples (by dividing his second case to two), the rest is an extension of his examples. Van Inwagen discusses philosophical debates but as far as I see, there is nothing specifically philosophical in his cases (other than a lack of empirical evidence that is nonetheless applicable to several non-philosophical fields too). Hence, the cases – and the consequences following from them – can be generalised (presumably) to all empirically unsupported debates.

The van Inwagen cases build on the idea of epistemic peers debating while having all relevant evidence shared, approx. the same epistemic capacities, virtues and background knowledge, a similar level of commitment to truth-seeking, etc. If two epistemic peers disagree under these ideal circumstances, their disagreement is called as 'faultless' because neither commits any mistake but still, they disagree. In faultless disagreements, both views are equally well-supported and no new (decisive) evidence can ever emerge. Hence, faultless disagreements are supposed to be irresolvable, and the case types below demonstrate why it is so (in similar cases at least).

Disagreements can be irresolvable on other grounds too. For the present purposes, the most interesting is when peers are not idealised so that they are allowed to commit mistakes but their positions are nonetheless faultless: both are equally well defendable from the epistemic perspective of the peers because they have no epistemic access to decisive evidence on the matter. On the supposition that the principle of bivalence applies to the matter of their disagreement, one of the parties must be right and the other must be wrong. But they cannot ever decide which of them is right, no matter the time and effort they invest into the debate. This is perhaps a more realistic sense of faultless disagreements because peers are not expected to get fully prepared for the debate in advance, but nonetheless, they cannot face with successful counterarguments because of their epistemic access to the matter of their disagreement (i.e., they can evade all *prima facie* successful counterarguments in one way or another).

The (types of) cases below demonstrate that in dialectically less complex cases, avoiding an acknowledgement of a defeat due to facing with a *prima facie* successful argument is relatively easy. But in dialectically more complex cases, an avoidance is even easier because the defendant has more advanced dialectical tools for her avoidance. This has a disjunctive consequence: either the pragmatic definition of success is implausible or the possibilities of successful argumentation in non-empirical fields is very limited. As there is some anecdotal evidence and common experience supporting the latter, I take the consequences of van Inwagen cases to be an argument for that horn.

2.1 The 'equal opponents' case

First let us see if two epistemic peers (i.e., fully equal opponents) disagree over a matter of their shared expertise. By starting the debate, they demonstrate that they do not find any successful argument against their views because if they found any, they would not join the debate. Given that all their evidence and arguments are shared due to their being epistemic peers under ideal circumstances, the opposing party has no successful argument *for* their view too (because it would be a successful argument against the view of their opponent). If their disagreement is faultless, no new evidence can ever emerge. Hence, no successful argument can ever be developed on either side, and the dispute cannot be rationally resolved.

This does of course not imply that in a debate between epistemic peers, no successful argument can emerge. They may not be in a faultless disagreement. Many debates end with an acknowledgement that one or the other party has not considered an argument in advance which is proved to be successful later in the debate. But these debates do not run between ideal parties highly committed to their standpoint. Debating about minor issues, it can even be the case that one or the other party gives up their position following (relatively) weak arguments against their point. It does not imply that a weak argument can be successful in a strict sense. It only implies that the circumstances were not ideal in that debate.

The Equal Opponents Case also applies to argumentative situations in which parties are equally un(der)prepared. Insofar as they do not have an epistemic access to decisive evidence on the matter of their disagreement (as the weaker definition of a faultless disagreement implies), no successful argument can ever emerge that must be rationally accepted by the them. This makes their dispute rationally irresolvable.

2.2 The 'protagonist-antagonist' case

In the second scenario, parties are epistemic peers but they are unequal in dialectical terms: one of them called as the 'proponent' or 'protagonist' defends a standpoint, and the other party called as 'opponent' or 'antagonist' aims at attacking that position. Their dialectical roles are therefore asymmetric or complementary. The burden of proof is on the protagonist, and hence her task is *prima facie* harder than the antagonist's: the former has a great attack surface she must defend, whereas the latter does not need to care about possible implications of her attack, as all she needs to do is hitting the protagonist's position.

Hence, while the protagonist's arguments are successful (in the sense defined above) only if the antagonist comes to believe her views, the antagonist's arguments are successful if the protagonist comes to disbelieve her views *or* suspend her judgment about her original position. But even though raising doubts (the task of the antagonist) is seemingly easier, it still requires a doxastic attitude change in the protagonist. If they are faultlessly prepared, it is impossible for the same reason as above: the protagonist would have not joined the debate if there is a successful argument against her position on their shared basis of evidence. If they are not faultlessly prepared but their positions are faultless from their epistemic perspectives, novel arguments can occur but they cannot be successful. The debate is irresolvable, even though preliminary odds are biased towards the antagonist.

2.3 The judicial case

In the Judicial Case, the goal of the protagonist is to convince a neutral party about the truth of her standpoint, whereas the goal of the antagonist is keeping the jury uncommitted to the protagonist's standpoint (either by keeping them neutral or convincing them to disbelieve the protagonist's standpoint). If the jury consists of epistemic peers/superiors of the protagonist and the antagonist, under ideal circumstances, the protagonist should have to convince the jury to give up their *equally well-grounded* neutrality towards the protagonist's position (or, in the weaker scenario, the parties and the jury have no epistemic access to facts that are decisive regarding the debate). This alone would be an impossible task for the protagonist if the jury considered all relevant evidence and arguments in advance (that is, as before, implied by the ideal circumstances). The antagonist needs to do nothing but recite sceptical arguments about the protagonist's view supporting the neutrality of the jury. Insofar as the jury has still been neutral, presenting arguments that support neutrality is much easier than changing their doxastic attitudes. Given that the protagonist and the antagonist are epistemic peers, the protagonist has no chance to win the debate. All she can do is hopelessly defending her position as long as the jury declares that the debate is over.

Hence, a combination of the Judicial Case with the Protagonist-Antagonist Case always results in a win of the antagonist. But first, this of course makes no difference in the protagonist's doxastic attitudes. Second, judgment suspension does not help in deciding *whether* the protagonist is right and hence as a yes/no question, whether p or not remains undecided. Third, the antagonist does not generate a *doxastic attitude change* and hence her argument is not successful in the sense above. The value of her victory is dubious at least.

Combining the Judicial Case with the Equal Opponents Case, both parties are in the same situation: they should have to change the doxastic attitude of a neutral jury which is at least as prepared as the parties. Now the chances get back to equal: under ideal circumstances, the jury will remain unconvinced about both positions, and hence the debate goes endlessly (or stops in an irrational way). In either case, if an alteration in the jury's doxastic attitudes is required for success, the dispute cannot be resolved.

3. IMPROVED VAN INWAGEN CASES

The scenarios above were built on standard van Inwagen cases (with some modifications). Their complexity gradually increases in order to seek for scenarios that are less abstract and ideal on the one hand and may have some details that help a rational resolution on the other. But as we shall see, increased complexity does not alter van Inwagen's conclusion: a rational resolution of dispute is impossible in these cases because complexity provides further possibilities for the defensive party to evade attacks on their standpoint.

3.1 The 'inapparent conflict' case

In the Inapparent Conflict Case, epistemic peers P_1 and P_2 disagree over p. While P_1 holds that p, P_2 also represents the (pro)position that p but by p, she means q (where p and q are slightly different – e.g. two partly conflicting interpretations of a paragraph from a classical philosophy text). Their disagreement is inapparent because their positions equivocate. As far as no direct contradiction arises from their interpretations, their disagreement will remain inapparent. If the difference between their understandings of the same paragraph is slight (and the paragraph is not central in their interpretation of the whole of the text), they could possibly never discover their disagreement. An inapparent disagreement cannot be resolved because the disagreement is not even on the table.

3.2 The 'apparent conflict' case

One may say that the inapparent conflict is not interesting in terms of an epistemology of disagreement precisely because it is inapparent. But its inverse case can be interesting even for them. In the inverse, Apparent Conflict Case, P_1 holds that p and P_2 openly holds that q (p and q are, again, slightly different interpretations of the same (complex of) proposition(s)). By q, P_2 means p (e.g. in a different wording) but this fact is inaccessible to one or both parties. So their disagreement could be said to be superficial, linguistic or terminological only, whereas one or both of them think(s) that their disagreement is substantial. If they lack a linguistic basis for resolution, they cannot even agree in what they disagree. If they could, they would shortly arrive at the acknowledgement that they do not disagree over p at all. Therefore, their disagreement is only apparent (but for some reason, its apparency is unrecognisable and hence the problem is irresolvable for them).

What can possibly count as successful arguments in these cases? In the Inapparent Conflict Case, they seem to represent the same position. Hence, they do not even aim to convince one another as they think there is no need for that. In the Apparent Conflict Case, their positions are the same but their arguments are different as they think they defend different positions. If they provided an argument against the other's position that is successful against the other party's point, the very same argument would be applicable against their own point too. Hence, successful arguments could be provided by self-refutation only (even if self-refutation would be not necessarily recognised by parties).

3.3 The equivocation case

Talking about linguistic misunderstandings, the Equivocation Case comes to the fore. In this scenario, P_2 provides a *prima facie* successful argument that *non-p*. Now P_1, holding that *p*, claims that P_2 in fact argues that *non-q*, where *q* is a substantially different equivocation of *p*. Hence, P_1 reasonably thinks he has nothing to do with P_2's argument as it does not even touch upon her standpoint that *p*. She may argue e.g. that *p* is to be understood at a metalevel, in a different domain of discourse, etc. than the target of P_2's argument. This can be a standard strategy for P_1 to avoid any *prima facie* successful arguments on the basis of some (alleged) linguistic misunderstandings.

The strategy may seem as illegitimate if applied to cases where the other party cannot identify a linguistic misunderstanding (whether there is one or not). But once again, as long as the parties cannot agree on the linguistic grounds of their disagreement, the disagreement between them is even deeper than it seemed when it was taken as a disagreement over a standpoint (rather than what the standpoint really consists in). It does also not matter whether there is a linguistic misunderstanding or not. As far as they cannot come to an agreement about linguistic issues, they cannot even start meaningfully discussing substantive issues and hence their disagreement cannot be overcome. Whether it is an intentional avoidance strategy of P_1 or an involuntary misunderstanding, it is irrelevant if the pragmatic definition of successful arguments is applied. There can be no agreement over a success of an argument if there is no agreement what the argument is about.

3.4 The 'evasion by clarification' case

Let us assume that P_1 and P_2 somehow come to know the linguistic grounds of their (in)apparent disagreement. The most probable strategy they follow is clarifying their respective positions so that misunderstandings can be eliminated. So when P_2 provides a *prima facie* successful argument against *q*, and P_1 identifies this argument as a strawman fallacy, she 'clarifies' that *p* is not equal to *q* or 'refines' what is to be understood by *p* so that it will be no longer equal to *q*. If she succeeds, their apparent disagreement disappears, and the dispute is resolved.

Is this a realistic scenario? They as experts under ideal circumstances would have known most probably that their disagreement was linguistic prior to the arguments presented. A realistic reason why they did not is that p and q are not once-and-for-all fixed propositions but in-progress positions that, because of counterarguments occurring in the debate, can be slightly modified (clarified or refined).

In Strategic Manoeuvring, such 'clarifications' may be taken as derailments (van Eemeren & Houtlosser 2002): an in-progress modification of a standpoint of the protagonist can be considered as giving up the original debate and starting a new one about a conflict between the new standpoints. But there *can* be legitimate clarifications, even if the protagonist had not even thought her position through, if the content of the modified position is contained by the content of the original.

For example, P_1 replaces a predicate F_1 of her original position p_1 with F_2 in her clarified position p_2 so that the semantic extension of F_2 is a subset of the semantic extension of F_1. Let us assume that the protagonist holds a general claim that $\forall x. F(x)$. The antagonist provides an argument by example, $\sim F(a)$. Now the protagonist distinguishes two senses of predicate F so that the domain of x in $F_1(x)$ makes F_1 inapplicable to a. As the antagonist argues that $\sim F(a)$, she must understand $F(x)$ as $F_2(x)$. But $\sim F_2(x)$ is no counterexample for $\forall x. F_1(x)$. This evades P_2's attack.

This may seem to be *ad hoc* but if the protagonist has independent reasons for the distinction, she may legitimately say that he did not exclude *a* from the domain of *x* above simply because it was not important for the line of argumentation before, or she did not even think that *a* might have been seen as relevant for the antagonist because for the protagonist it had been so evident that F is inapplicable to *a* that it did not seem to be worth mentioning that.

One may remain unconvinced by the argumentation above. Under ideal circumstances, standpoints must be perfectly clarified prior to (the argumentation stage of) the debate. If it had not happened, and the protagonist modified her standpoint in the argumentation stage, she would lose the debate anyway.

Two responses should be considered to this objection. First, a dynamic model (with standpoint modifications by clarification and refinement) necessarily loses the aboriginal omniscience of arguers but in exchange, it implies a more realistic scenario because the debate can finally have serious consequences on the positions of (however idealised) parties. But similar to earlier cases, even if parties can face with novel arguments during the debate but since their positions are supposed to be faultless, these novel arguments cannot be successful.

Whether such faultless cases exist is an issue not to be discussed here; the present question is what to do with them if they exist and someone happens to be in one of them.

Second, dissolving the parties' original disagreement would make nothing with their *newly emerged* disagreement over the modified version of the protagonist's view. From an argumentation theoretical point of view, a closure of the original debate may count as important. But from a dialectical viewpoint, it is more important that the disagreement would remain irresolved. In the original debate or in another, the protagonist *can* legitimately present her modified position, and the debate goes further on, even if $\sim F(a)$ formerly seemed to be a potentially successful argument against $\forall x. F(x)$.

Still, if their disagreement was newly emerged, and hence they had finished their original debate, it could be said that their original dispute was resolved: it would have been shown that the original position of the protagonist (as formulated in that debate) could not be defended further. But this is not necessarily the case. First, the original position possibly could be further defended *even if* its original holder is no more interested in defending her original position because the modified position seems to her preferable to the original, *and* there still remains a conflict between the modified position and some position opposing to the original.

3.5 The 'false dilemma' case

The False Dilemma Case builds on the Apparent Conflict Case but in a manner recognised by one or both parties. Along with the Question-Begging Case to be discussed in the next section, this is one of the most powerful evasion techniques. The motivation behind these techniques is as follows. If from any set of premises, an unacceptable conclusion validly follows, the only thing one needs to do in order to avoid accepting the unacceptable is challenging one (or more) of the premises. This can make debates endless, as far as premises that are unfounded can always be challenged and in empirically unsupported fields, a foundation of premises is always relative to further premises. (Anti-foundationalists may wish to extend this line to empirically supported fields as well but this would go too far from the present issue.)

When P_1 (now as antagonist) presents her *prima facie* successful argument against p (that P_2 as the protagonist holds), P_2 replies that P_1's arguments rely on *non-p*. Since *non-p* is inacceptable for P_2 because of the dialectics of the debate, *non-non-p* is to be taken. Now P_2 can follow two different lines: either claiming that from *non-non-p*, *p* follows, or claiming that *non-p* can be evaded by seeing 'either *p* or *non-p*' as a false dilemma. The former is the Question-Begging Case to be discussed in the following section. The latter is the False Dilemma Case, when P_2 turns towards a common ground between *p* and *non-p* (let us call it as *q*), and in the rest of the debate, she argues for *non-q*.

Note that this is an antagonist-only strategy: the protagonist is expected to defend her position in all possible dialectical situations (as far as it can be done rationally). The antagonist, in contrast, only must attack the position of the protagonist. So in this case, P_1 as a protagonist must defend her standpoint *non-p* that she is doing by arguing against *p*. Since *p* is not a standpoint the antagonist must defend, she can manoeuvre away, arguing against both *p* and *non-p*. The protagonist cannot do the same (in a reversed situation) because she must keep committed to *p*. She would preferably choose to follow the Question-Begging Case or she would give up the debate about *p*. In the latter case, the protagonist can start a new debate about *q* (taken as a common ground for *p* and *non-p*), becoming an antagonist by arguing against *q*. Hence, the False Dilemma Case can provide the opportunity to the original protagonist to set the burden of proof to her opponent. By (temporally) giving up her position that *p,* she can take an advantage over the original antagonist, and come back to the original problem whether *p* or *non-p* only if the antagonist wins the modified debate about *q*.

3.6 The question-begging case

The protagonist-only alternative of the False Dilemma Case is the Question-Begging Case. P_1 holds that *p*. Let us assume that P_2 provides, on the ground of *q,* a deductively valid, *prima facie* decisive argument that *non-p*. There is hardly ever a better candidate for a successful argument than a deductive proof. Now P_1 starts arguing that *q* begs the question *precisely because* from *q, non-p* deductively follows. Insofar as in a deductive argument, all information contained by the conclusion must also be contained by the premises, by accepting the premises, the antagonist *presupposes* that *p* is false. That is, she begs the question.

In order to avoid circularity, the antagonist can formulate her premises as independent clauses that separately do not presuppose the conclusion. But in this case, she just makes the job of the protagonist easier, as the latter will be provided a list of propositions among which only one should be rejected so that the whole of the argument would collapse. Circularity can be avoided by an infinite regress only: as argued above, premises can always be questioned, demanding further syllogisms which support them, and taking premises as conclusions of a further syllogism results in even more premises from which only one should be questioned by the protagonist.

As a consequence, the burden of proof will be reversed, as it is the original antagonist who has to develop further and further defensive arguments in order to show that she avoids circularity in her original attack. The more complex an argumentation, the more probable it is that one or more of the premises can be effectively attacked, and due to the infinite regress, practically all argumentation can be extended to infinite complexity. It depends on the antagonist's capacity to provide more and more argumentation for further and further syllogisms.

Now there are two strategies available for the (once) antagonist: either she endlessly defends that her original counterargument against *p* was not a *petitio principii,* or she counterattacks by demanding further argumentation for one or more premises of the original protagonist's counter-counterattack. In the first case, she is forced to be constantly defensive. In the second case, parties constantly pass the burden of proof back and forth. In either case, successful arguments will never emerge.

4. CONCLUSION

The list of cases could be extended further but some consequences follow from this brief list too. Even if there is perhaps no linear relationship between complexity and dialectical possibilities, but as a general consequence, it can be said that the more complex an argumentative situation, the more dialectical possibilities arguers have in order to avoid the acknowledgement of a defeat. It has also been argued that routes for escape are always available for the losing side by giving up their position and start a new debate with a 'refined' or 'clarified' position or resetting the burden of proof.

One may object that these manoeuvres end up the original debate and hence a disagreement has been dissolved, even if only at the price of meeting another. Hence, a successful argument can be identified after finishing the original debate – namely, the one that has made the losing party giving up the debate. But it is not necessarily the case. The loser of the original debate may only temporarily give up her position because another debate can give her better chances to show that her position is secure in the original debate. She may also give up her original position because due to the confrontation of *pro* and *contra* arguments, she has found a better position that is still in a conflict with the position of the opposing party and she is no longer interested in defending her original position (that could be nonetheless defended from the counterarguments arisen if there were parties interested in doing so). This leaves the disagreement *between the parties* irresolved, and that is central for a pragmatic approach central to van Inwagen's definition of a successful argument. Furthermore, it leaves open the question whether the original position can be defended further. Hence, no conclusive refutation is provided, even in pragmatic terms.

Finally, it can always be the case that by passing the burden of proof back and forth, results of a later debate lead the parties back to one of the previous debates. A debate, if closed by opening another debate, is not fully closed. Depending on the outcome of the debate(s) following it, it can be possibly reopened anytime. Hence, a final argument in a debate (if followed by further debate(s)) is not a successful argument but a tool for a (possibly) temporal win only. Successful arguments should result in closing all debates relevant for the original because it should make the opposing party be convinced that the successful arguer is right. It seems to be too demanding for them to expect such a decisiveness. But it seems that demanding less would trivialise successful arguments to turning points in the chains of endless debates.

ACKNOWLEDGEMENTS: Support from the Measures of Rationality Research Group (Dept. of HPS, Budapest University of Technology and Economics, Hungary) is acknowledged.

REFERENCES

Danka, I. (2018). Irresolvable Rational Disputes. In S. Oswald & D. Maillat (Eds.), *Argumentation and Inference: Proceedings of the 2nd European Conference on Argumentation, Fribourg 2017. Vol. II.* (pp. 191-204). London: College Publications.
Hanna, N. (2015). Philosophical success. *Philosophical Studies,* 172(8), 2109–2121.
van Eemeren F. H., Houtlosser P. (2002). Strategic Maneuvering in Argumentative Discourse: A Delicate Balance, in F. H. van Eemeren, P. Houtlosser (Eds.), *Dialectic and Rhetoric: The Warp and Woof of Argumentation Analysis* (pp. 131–159). Dordrecht: Kluwer Academic.
van Inwagen, P. (2006). *The Problem of Evil.* Oxford: Oxford University Press.

Arguing Brexit on Twitter: a corpus linguistic study

NATALIE DYKES
Friedrich-Alexander-Universität Erlangen-Nürnberg
natalie.mary.dykes@fau.de

PHILIPP HEINRICH
Friedrich-Alexander-Universität Erlangen-Nürnberg
philipp.heinrich@fau.de

STEFAN EVERT
Friedrich-Alexander-Universität Erlangen-Nürnberg
stefan.evert@fau.de

We study argumentation on social media by applying discourse and corpus linguistic methods to a corpus of 6 million tweets containing "Brexit". We identify common argumentation strategies through manual annotation and develop corpus queries to find further instances of those schemes. Methods from discourse linguistics help to identify salient aspects shared by the various argument scheme realisations. Through modularity and iterative abstraction, the queries lend themselves to logical formalization and further automatic processing.

KEYWORDS: argumentation mining, corpus linguistics, defeasible, quantitative, social media, Twitter

1. INTRODUCTION

Argument Mining is a relatively new field in Natural Language Processing and Computational Social Science. It is concerned with the automatic extraction and representation of arguments found in large collections of electronic texts. The most foundational tasks in Argument Mining include the identification of argumentative utterances in corpora and the classification of particular structural elements like premises or conclusions. The main outcome of this process is a mapping of argumentative text sequences to their automatic analysis; which then lends itself to further processing in a computational setting.

In many cases, contributions focusing primarily on natural language processing perspectives tackle corpora of relatively well-structured and explicitly argumentative texts, such as debates (Cabrio and Villata 2013) or student essays (Peldszus and Stede 2015).

However, argumentation is often considerably less straightforward when it comes to texts from other genres. While argumentation plays a role in many settings of everyday discourse, its automatic processing faces additional challenges in some text types. Firstly, arguments tend to follow the type of defeasible logic as proposed by researchers including Kienpointner (1992) and Walton et al. (2008). In defeasible settings, the relationship between premises and conclusions is less clear than in the case of traditional argumentation schemes like Modus Ponens. For instance, consider the argument from correlation to cause (Walton et al. 2008, pp. 328–329):

> "Premise: There is a positive correlation between A and B.
> Conclusion: Therefore, A causes B."

While this is a widespread interpretation of correlation – we hope that readers will excuse our own defeasible reasoning – the phrase *Correlation does not imply causation* has gained enough traction to be the title of a Wikipedia page.[1] In this sense, even when the premise holds true, this does not necessarily mean that the conclusion is true as well.

Moreover, it can be challenging to identify premises or conclusions in the first place, because they are often left implicit in everyday argumentation. This holds especially true for social media, where texts are often extremely short. Therefore, Bosc et al. (2016) consider every tweet as argumentative if it can be interpreted to contain either a premise or a conclusion; regardless of whether the line of argumentation is present. The obscured logical structure is connected to a third aspect that makes argumentation mining difficult on less structured texts: persuasion is usually not achieved by strict logical relations, but rather by persuasion through rhetoric strategies such as selection, arrangement, and phrasing of argumentative units (Wachsmuth et al. 2018).

In this contribution, we focus on argumentative discourse on Twitter, where all of these aspects play a particularly prominent role: As Nigmatullina and Bodrunova (2018) suggest for the case of discussion threads, tweets are strongly fuelled by emotions, particularly anger and indignation, and often contain ironic elements. Moreover, a high degree of implicitness is to be expected. As our tweets were collected in 2016, all posts have a maximum length of 140 characters, which was the limit at

---

[1] https://en.wikipedia.org/wiki/Correlation_does_not_imply_causation

the time. Apart from this restriction, there are rather few guidelines: users can post pictures, links or GIFs and write about whatever they please, which makes it a more or less unmediated environment. Finally, tweets contain a large amount of non-standard language, as is usually the case in computer-mediated communication.

Considering these aspects, it is hardly surprising that automatically finding and classifying arguments in social media has led to relatively modest success (cf. Goudas et al. 2014, Dusmanu et al. 2017).

2. RELATED WORK

In corpus linguistics, several approaches have been taken to study argumentation. Degano (2007) studies presuppositions and dissociations in a corpus of British press texts. The analysis starts from a predefined list of explicit argumentation markers; including clause types, comparisons, descriptions or the use of particular vocabulary (cf. Levinson 1983, pp. 181–184). For instance, stating that somebody *did not manage* something indicates the presupposition that they have tried and been unsuccessful (Degano 2007, p. 366).

Another possible strategy is to combine qualitative and quantitative methods. O'Halloran (2011) analyses argumentation in transcripts of oral conversations. In a first step, he uses manual coding to identify utterances as claims and challenges. The following part consists of calculating statistical keywords, key POS tags and key semantic domains (Rayson 2008) to explore the linguistic realisation of arguments.

Al-Hejin (2015) works closer to argumentation schemes in a study of British press reports on Muslim women. He focuses on macro-propositions: "global" motives of the discourse topic (van Dijk 2008, p. 16). For example, when mentioning a woman's choice to wear a hijab, particular arguments tend to follow, e.g. that they are unwilling to integrate into Western culture (Al-Hejin 2015, p. 40). Macro-propositions were identified by keywords, which were grouped manually and verified by calculating key semantic domains.

Baker (2004) compares pro- and anti-reform speeches in a political debate on the age of consent for homosexual men in Britain. The speeches differ not only a lexical level, but also logically: opposers tend to form chains of individual arguments building upon each other, while proponents' arguments were less intertwined and "more straightforward" (Baker 2004, p. 104).

3. THE BREXIT 2016 CORPUS

The basis for our analysis is a corpus of approximately 6 million tweets

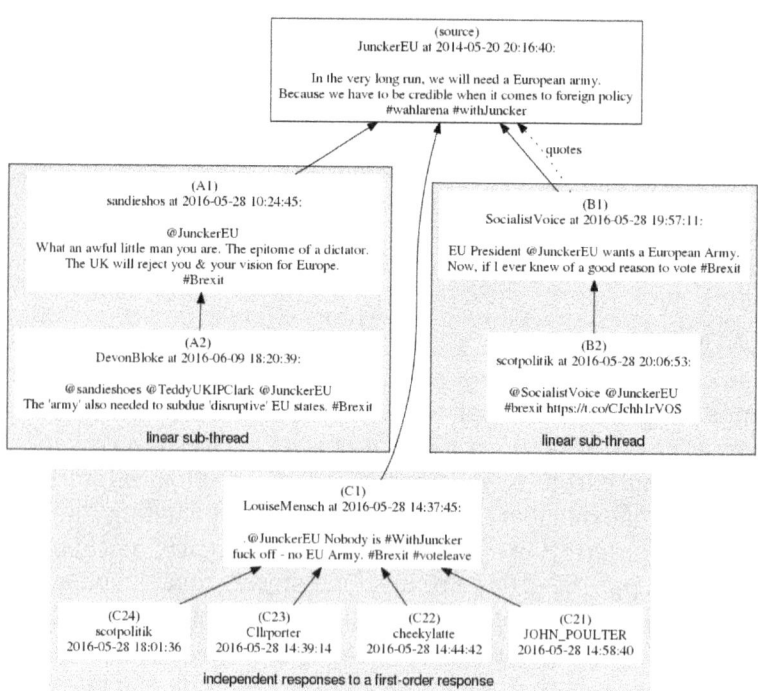

Figure 1: reply thread example

containing the character string "Brexit" (16/05/05–16/08/24)[2] and their associated reply threads. Reply threads were constructed by retrieving all available tweets for which there is a reply in our initial database. One example for such a reply thread can be seen in figure 1. Note that the initial source tweet by @JunckerEU is from 2014, hence before the actual start of the database. For mining arguments, we limit ourselves to tweets where the post containing the relevant string was written before the Brexit referendum, since we assume that this sample will have higher consistency in argumentation. This leads us to a collection of approximately 2.5 million tweets.

As shown in examples (1) to (3), tweets in the corpus may contain all of the features mentioned as challenging for processing; such as sarcasm, non-standard language and misspellings, emojis and implicitness. Moreover, many tweets are likely not directly argumentative; such as example (3).

> (1) Brexit will result in Zombie Hitler returning leading an army of demonic nazi crabs, marching sideways across Europe to devour our brains!

---

[2]

https://web.archive.org/web/20171121195029/http://www.eecs.qmul.ac.uk/~dm303/brexit/

(2) German : Acknowlege significant UK contribution 2 the EU

(3) if we Brexit is this our last Eurovision? Or do we get a guest spot? Like Australia 😄

3.1 Preprocessing

Before the corpus was ready for analysis, several pre-processing steps were taken in order to allow for elaborated searches and more valid results. The first necessary aspect concerned deduplication: Following Schäfer et al. (2017), we removed posts that were likely generated by social bots; thus reducing our dataset to approximately 1.8 million tweets and 32 million tokens. The data was then automatically processed using several pipelines for tokenization, part-of-speech tagging (Owoputi et al. 2013), lemmatization (a simple rule-based lemmatizer based on tokens and POS tags combined with a word lookup), phrase chunking and named entity recognition (Ritter et al. 2011).

3.2 Indexing

In order to make the corpus usable for extracting argumentative sequences, the data was indexed with the IMS Open Corpus Workbench (CWB). This allowed us to combine all levels of annotation and thus formulate complex queries (cf. Evert and Hardie 2011). Moreover, the CWB includes several features that have proven useful to our approach; including macro definitions and wordlists, which will be described below.

4. MODELING ARGUMENTS: PATTERNS AND SCHEMES

4.1 Corpus Queries

Unlike most traditionally corpus linguistic approaches, the central structuring elements of our study are not wordlist of keywords or collocations, but CWB queries, which are intended to provide the input for an inventory of logical formulas. These formulas, which are being developed by another group of our ongoing project, cover contents that is typically to be expected in everyday argumentation, for instance:

(A): $\forall \{?0 : entity\} \in \{?1 : entity\} : \{?2 : property\}(?0)$

In this formula, there are three empty slots, corresponding to an entity (0), a group (1) and a property (2) which, applies to all group members. It summarises aspects found in several argumentation schemes, including Common Folks ad Populum and Position to Know (Walton et al. 2008). The schemes, in turn, manifest in various linguistic forms, including the following examples for *Common Folks ad Populum*, with the relevant parts denoted by brackets:

(4) Study shows <ordinary folk are losers under the EU while the wealthy prosper>

(5) @DrAlanGreene <I'm as against #Brexit as the next man> but this is nonsense

In both cases, an appeal to "regular" people is the core of the argument; framing the speaker's standpoint as the commonsense thing to think. The passages in bold are the part of the respective corpus query that was matched in a given tweet. The queries are designed to capture linguistic patterns, which we expect to reflect particular argumentative strategies. Consider the query that found hits like 5):

```
<np>@0:[pos_simple="L|N|P|Z|#"]       []*</np>       @1:[::]
/region[vp] (/region[np])? "as" (/ap[])* "as" "all|the|any"
"next|old|other" @2:[::](/ap[])* [lemma=$nouns_person_common |
pos_simple = "Z"]+@3:[::];
```

The central semantic element defining the argumentative potential of this query is the notion of normality. This is represented by the specified adjectives variable $nouns_person_common, which in this case is specified to be present within the second noun phrase of the hit. The variable reads from a wordlist consisting of nouns that might be considered "generic" references to people, i.e. fellow, human, person, ppl etc. The rest of the query mostly consists of phrase chunk and part of speech restrictions.

Finally, parts marked with an @ are anchor points, marking particular points where the query matches can be extracted to use as input to the logical formulas. For instance, the sequence between @0 and @1 – *I'm* in the case of example 5) – can be specified to fill the entity definition slot {?0: entity} in the given formula. Thus, whatever takes up the NP slot in the query is assumed to be equivalent to the entity being assigned particular characteristics based on their group membership (in this case, "normal" people, according to the tweeter).

The main goal in developing these queries is to maximise precision, while recall is of secondary concern. In this way, we aim to generate input to the formulas, which are suited to incorporate arbitrary linguistic structure and reflect its logical relation, provided that the query is precise enough. For instance, if the restriction to the wordlist

$nouns_person_common is omitted, the following result is found by the query:

(6) He'll be back as soon as the next one has screwed up Brexit

Superficially, the structure looks like the one intended to be captured by the query. However, to the human analyst it is obvious that the construction *as _ADJ as* does not contribute to a grouping function in this case – the correspondence between the linguistic surface forms is purely incidental. As the grammatical and lexical categories are rather coarse in many positions of the query, it is important to emphasise that the required precision can only be achieved if query development is performed iteratively. Only by regular examination of the concordance lines for each match can the queries gradually become adequately exact.

5. CASE STUDY: SELF-IDENTIFICATION IN THE BREXIT CORPUS

In order to demonstrate the usefulness of our approach to the study of discourse and argumentation, we present a case study of queries designed to extract speakers' self-assignment to a particular group in potentially argumentative contexts. This is a subset of queries matched by formula A, where entity 0 is specified to be the speaker/tweeter themselves. Thus, group membership is the grounds on which a particular claim is made. We focus on self-identification, assuming that these particular statements have special persuasive potential, as the speaker claims expertise or first-hand knowledge of the situation in question.

In order to explore these tweets, the results of three different queries were annotated (*as NP I VP, NP like me VP, {ordinary/common/normal} {people} VP*). While only the first and second query explicitly reference the speaker, we suggest that the third query may also be regarded as a kind of self-identification. Tweets following this pattern will likely often imply that the speaker themselves identifies as belonging to the group "normal" people. Consider example (4) from above – it is unlikely that a speaker would highlight the disadvantages of Brexit for "ordinary people" (whom they understand in contrast to "the wealthy" in an act of othering) and not define themselves as belonging to that group.

The hits of each of the three queries were categorised manually and annotated for stance (leave/ stay/ unclear). Figure 2 shows the empirical distribution of this stance variable for the three argumentational queries.

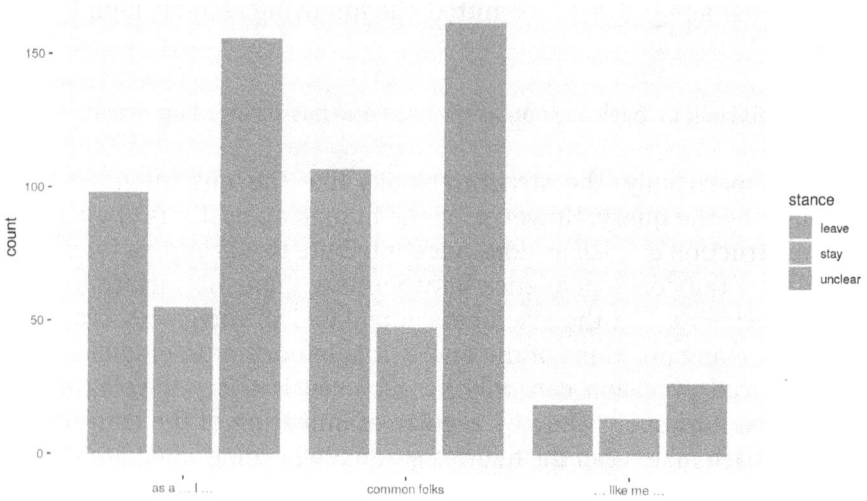

Figure 2: empirical distribution of stance towards Brexit in three annotated queries

Note that this categorisation only applies to the search result itself and not to the overall statement presented in the tweet, which may well be different. While we do aim for a complete analysis of tweets' argumentative content in the future, we take a nested approach where each proposition in the form of a query hit is first evaluated out of context and later may be determined to be a filler in and/ or to contain another proposition. For instance, a statement like X proved that Y is wrong, Y and X are to be captured by separate queries and whatever is claimed in Y is evaluated as false only in the context of evaluating the "outer" proposition X. In this sense, the analysis presented here does not necessarily reflect a subset of Twitter users' opinions on Brexit, as some sequences categorised as "leave" may be embedded in a negative evaluation and vice versa.

A further variable for annotation was the iterative development of group identity types based on the head of the main NP of the sentence. During the examination of concordances, it became clear that most hits could easily be attributed to one of the following categories:

"person": this type includes generic person references as well as kinship terms (*ppl, folks, grandmother, parent*).

"national": this category is reserved for references to nationality, origin or residence (*Brit/Briton, resident, European, Yank, Brummie*)

"voter": rather unsurprisingly, tweeters often specified their position in terms of political or otherwise ideological, and in some cases, spiritual, alignment (*activist, leftie, fan, sceptic, winger*)

"professional": finally, users referred to their profession or educational background (*graduate, banker, fisherman, [a-z-]*worker*).

These categories were used as a basis for wordlists corresponding to particular types of person references and subsequently expanded by systematically searching the corpus for similar words through omitting particular slots within these and other queries developed in the project. Figure 3 shows the distribution of stance towards Brexit across the listed categorization of the NP head.

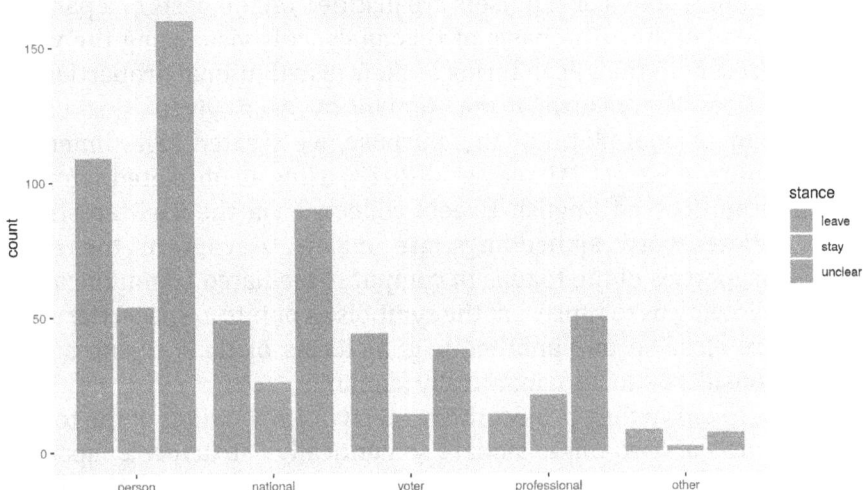

Figure 3: stance across NP categories

6. RESULTS

With regards to the stance towards Brexit, most hits show no clear pattern (cf. figure 2 above). However, "leave" outweighs "stay" in all matches. For the most part, the "unclear" annotation indicates that the statement, though related to the general topic of Brexit, does not in its own suffice to directly deduce a stance. Nevertheless, most matches are clearly used in argumentative contexts and thus promising for further analysis:

(7) the Boris, Gove, Hannan, Farage & Redwood loons in charge Thucydides on experts: "ordinary men usually manage public affairs better than their more gifted fellows"

In cases where stance is present, users ascribing themselves to the same group may use their identification as a ground for opposing arguments:

(8) As an immigrant I find myself closer to the #RemainINEU campaign argument

(9) As an immigrant I find the belief that #UK can integrate into #Europe astonishing

In a subsequent step, we use the manual annotation as a basis for further quantitative analysis to demonstrate the usefulness combining automated and manual approaches. This evaluation concerns the semantics of the VP head. Thus, we address the question of what kinds of actions, states or mindsets are justified on the basis of a particular group membership. The basis of this analysis is visualising the various verbs present in the VPs in terms of their distributional properties.

We visualize the semantics of a given slot using a two-dimensional plot: For this purpose, we created high-dimensional word embeddings (cf. Mikolov et al. 2013) using an unrelated sample of equally preprocessed English tweets collected via the Twitter Streaming API. These word embeddings are meant to capture the distributional properties of the tokens in computer-mediated communication, thereby allowing us to represent the symbols as points in space. Here, vectors that are close to one another (e.g. in terms of their cosine similarity) are considered to be semantically similar.

We then project the respective embeddings of the tokens of a slot onto a two-dimensional semantically structured space using t-distributed stochastic neighbour embedding (van der Maaten and Hinton 2008). This projection is constructed in a way that semantically similar lexical items (i.e. those with similar embeddings in the high-dimensional space) appear in the vicinity of one another in the two-dimensional space. An example can be seen in figure 4. It is obvious that the visualization is reasonable: The more generic items(such as 'people', including the CMC spelling 'ppl', 'folk', 'man', and 'person') cluster together, as do the more specific ones ('worker', 'citizen', 'voter').

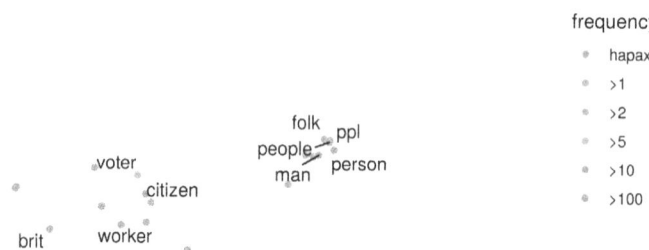

Figure 4: visualizing the NP slot for the {ordinary} {people} VP query

In order to extract more specific subsets of hits, the information from the visualisation can serve as input for more precise queries. For instance, when the NP is specified to express affiliation and the VP slot references a verb of support, the hits are accordingly limited to more narrow types of arguments:

(10) As #Labour member I am voting Brexit

(11) As an anarchist I will vote against Brexit, as it will reinforce the notion of a nation state

Similarly, consider the examples for combining "professional" nouns in the NP and knowledge verbs characteristic for them in the VP:

(12) As a scientist, I know the immense value of #collaboration

(13) As a doctor, I know leaving the EU will be devastating for the NHS

7. CONCLUSION

As has been demonstrated, corpus queries can be a useful tool in extracting numerous argumentative sequences from a large and noisy dataset. In this sense, the queries fulfil the task of handling linguistic variation, partially realised arguments and defeasible logic. Our corpus queries are designed to balance grammatical and semantic flexibility in the argumentation patterns: semantically similar concepts can be grouped into wordlists to fill a lexical slot with multiple options as demonstrated by the $nouns_person_common variable in our query example. Furthermore, grammatical patterns can be queried on different levels; be it parts of speech, phrase chunks or manually written macros to accommodate for structures that the chunker does not specify. All macros and word lists are stored centrally and imported into the query files; making them a reusable resource. Each individual query is one linguistic instantiation of a particular argumentative pattern. Using the CWB's anchor point facilities, it becomes possible to mark sequences of interest and extract them for either logical formalisation or closer linguistic and argumentative analysis. In this sense, we contribute a qualitative approach to handling noisy data in online argumentation.

Current ongoing work includes mapping the developed queries to their logical representations by extracting relevant parts of the query results and filling them into the formulae. We also expect the queries to reversely inform the development of a logical framework equipped to handle noisy data, in that the results will yield insights into which argumentative phenomena are particularly prevalent in the given discourse and others like it.

The visualisation of particular word slots by distributional properties has proven insightful for the case study at hand: similar words were grouped together, thus facilitating the task of drawing quantitatively informed qualitative conclusions. Therefore, future work includes the development of an interactive toolkit integrating visual facilities, allowing researchers to gain a comprehensive overview of the data.

ACKNOWLEDGEMENTS:

This work has been funded by the Deutsche Forschungsgemeinschaft (DFG) within the project "Reconstructing Arguments from Noisy Text", grant number 377333057, as part of the Priority Program "Robust Argumentation Machines (RATIO)" (SPP-1999).

REFERENCES

Al-Hejin, Bandar (2015): Covering Muslim women. Semantic macrostructures in BBC News. In *Discourse & Communication* 9 (1), pp. 19–46. DOI: 10.1177/1750481314555262.

Baker, Paul (2004): 'Unnatural Acts'. Discourses of homosexuality within the House of Lords debates on gay male law reform. In *Journal of Sociolinguistics* 8 (1), pp. 88–106. DOI: 10.1111/j.1467-9841.2004.00252.x.

Bosc, Tom; Cabrio, Elena; Villata, Serena (2016): Tweeties squabbling. Positive and Negative Results in Applying Argument Mining on Social Media. In : Proceedings of COMMA, pp. 21–32.

Cabrio, Elena; Villata, Serena (2013): A natural language bipolar argumentation approach to support users in online debate interactions. In *Argument & Computation* 4 (3), pp. 209–230.

Degano, Chiara (2007): Presupposition and Dissociation in Discourse. A corpus study. In *Argumentation* (21), pp. 361–378. DOI: 10.1007/s10503-007-9058-7.

Dusmanu, Mihai; Cabrio, Elena; Villata, Serena (2017): Argument mining on Twitter. Arguments, facts and sources. In : Proceedings of the 2017 Conference on Empirical Methods in Natural Language Processing.

Evert, Stefan; Hardie, Andrew (2011): Twenty-first century corpus workbench. Updating a query architecture for the new millennium. In : Proceedings of CL.

Goudas, Theodosis; Louizos, Christos; Petasis, Georgios; Karkaletsis, Vangelis (2014): Argument Extraction from News, Blogs, and Social Media. In : SETN 2014.

Kienpointner, Manfred (1992): Alltagslogik. Struktur und Funktion von Argumentationsmustern. Stuttgart: Frommann-Holzboog (Problemata, 126).

Levinson, Stephen (1983): Pragmatics. Cambridge: Cambridge University Press (Cambridge textbooks in linguistics, 8).

Mikolov, Tomas; Chen, Kai; Corrado, Greg; Dean, Jeffrey (2013): Efficient Estimation of Word Representations in Vector Space. In Yoshua Bengio, Yann LeCun (Eds.): 1st International Conference on Learning Representations, ICLR 2013, Scottsdale, Arizona, USA, May 2-4, 2013, Workshop Track Proceedings.

Nigmatullina, Kamilla; Bodrunova, Svetlana (2018): Patterns of emotional argumentation in Twitter discussions. In : Proceedings of the International Conference on Internet Science, pp. 72–79.

O'Halloran, Keiran (2011): Investigating argumentation in reading groups. Combining manual qualitative coding and automated corpus analysis tools. In *Applied Linguistics* 32 (1), pp. 172–196. DOI: 10.1093/applin/amq041.

Owoputi, Olutobi; O'Connor, Brendan; Dyer, Chris; Gimpel, Kevin; Schneider, Nathan; Smith, Noah (2013): Improved Part-of-Speech Tagging for Online Conversational Text with Word Clusters. In : Proceedings of NAACL.

Peldszus, Andreas; Stede, Manfred (2015): An annotated corpus of argumentative microtexts. In : Argumentation and Reasoned Action: Proceedings of the 1st European Conference on Argumentation. Lisbon.

Rayson, Paul (2008): From key words to key semantic domains. In *International Journal of Corpus Linguistics* 13 (4), pp. 519–549. DOI: 10.1075/ijcl.13.4.06ray.

Ritter, Alan; Clark, Sam; Etzioni, Oren (2011): Named entity recognition in tweets: an experimental study. In : Proceedings of EMNLP: Association for Computational Linguistics.

Schäfer, Fabian; Evert, Stefan; Heinrich, Philipp (2017): Japan's 2014 general election. Political bots, right-wing internet activism and PM Abe Shinzō's hidden nationalist agenda 5 (4), pp. 294–309.

van der Maaten, Laurens; Hinton, Geoffrey E. (2008): Visualizing data using t-SNE (Journal of machine learning research), pp. 2579–2605.

van Dijk, Teun (2008): Discourse and Context. A Sociocognitive Approach. Cambridge: Cambridge University Press.

Wachsmuth, Henning; Stede, Manfred; El Baff, Roxanne; Al-Khatib, Khalid; Skeppstedt, Maria; Stein, Benno (2018): Argumentation synthesis following rhetorical strategies. In : Proceedings of the 27th International Conference on Computational Linguistics, pp. 3753–3765.

Walton, Douglas; Reed, Chris; Macagno, Fabrizio (2008): Argumentation Schemes. Cambridge: Cambridge University Press.

Protagoras' principles, disagreement and the possibility of error

MICHEL DUFOUR
University Sorbonne-Nouvelle
<u>michel.dufour@sorbonne-nouvelle.fr</u>

The paper begins with a discussion of what I call Protagoras' principles: the claim that there always are two divergent views on any topic, his famous thesis that man is the measure of all things and his view about error. Then, they are related to contemporary discussions on the nature and management of rational disagreement, especially between experts or epistemic peers. We discuss the case of a possible underlying error, and its limitations.

KEYWORDS: acceptance, *aesthesis*, belief, disagreement, error, mathematics, perception, Protagoras.

1. INTRODUCTION

What should you do when a person, as competent as you, disagrees on a statement about which you think you have convincing evidence? Is a moderation of your epistemic attitude a necessary preliminary requirement to have a rational argument on the point of contention? This kind of question recently raised some interest in social epistemology.[1]

These concerns seem old as Protagoras. Unfortunately we have almost no reliable sources about the old master's thoughts on this topic. Yet, a tension between some of his most famous statements suggests that he was well aware and sensitive to this kind of problem, at the core of his activity of sophist. The first part of the paper is a discussion of Protagoras' views, in connection with the management of disagreement;

---

[1] For an overview see Feldman & Warfield (2010) and Christensen & Lackey(2013).

the second tries to answer our initial questions with a view partly inspired by Protagoras.

2. PROTAGORAS

2.1 A leading sophist

Protagoras (c.-490; c-420) and Gorgias (c.-480; c.-375) are held as the two leading figures of a first generation that De Romilly dubbed the Great Sophists of Pericles' time (De Romilly, 1988). Both were contemporary and less than one generation elder than Socrates.

We have no primary source on Protagoras' philosophy. None of his writings has survived although they were numerous if we trust Diogenes Laertius (1925). Today, scholars consider that hardly more than two or three statements ascribed to Protagoras could be authentic. We should not forget it.

According to Laertius, Protagoras came from Abdera and was a disciple of Democritus. Later on, he made several stays in Athens where he became a star if we trust Plato's eponymous dialogue. The character depicted by Laertius seems verbally more pugnacious than the one in Plato's dialogues. He would have organized dialectical arguments, provided fallacies to quibblers and would then deserve Laertius' comment: "he was the father of the whole tribe of eristical disputants, now so much in evidence" (1925, book 9, 52).

Protagoras is mostly known as a sophist, hence as a travelling teacher.[2] But this should not conceal that besides his project of a renewed model of education, he had an important political involvement. He is said to have been a friend of Pericles who even asked him to write laws for the Panhellenic city of Thuroi. This anecdote, if true, highlights a character able to much more than quibbling or playing to contradict his friends or disciples.

Another famous anecdote illustrates both his taste for vigorous arguments and for political action. One of his books, according to Laertius, would have begun by: "As to the gods, I have no means of knowing either that they exist or that they do not exist. For many are the obstacles that impede knowledge, both the obscurity of the question and the shortness of human life" (1925, book 9, 51). This religious agnosticism would have been the reason why the Athenians would have banned Protagoras and burnt his books.

2.2 Protagoras' first principle.

The term "principles" may not be the right one to qualify the few statements ascribed to Protagoras. "Maxim" could be more appropriate, but we can take them as principles to try to organize the sparse information we have.

---

[2] For a recent brief introduction to Protagoras' life and works, see Silvermintz (2016).

According to Laertius, "He was the first to assert that on any topic there are two contradictory discourses (*logos*) and the first to put them forward in his own speeches ». (1925, book 9, 51) Laertius adds that this principle – that we will hold as the first – was influential. But what did Protagoras mean? We can only venture conjectures.

As such, this saying could appear trivially true. But its use at Protagoras' time suggests that it was also a manifesto for freedom of speech in the context of the young Athenian democracy. One of its consequences is that there is no last and decisive word coming from a political or religious authority. This nicely matches the anecdote on Protagoras' skepticism about the existence of gods. Furthermore, his first principle suggests that after an agreement, you can always restart the game, in a new direction or not: speech is an endless process. In some sense, Protagoras' (first) principle is a principle of instability, quite conducive to investigation and argument.

From a more pragmatic point of view, you can consider this principle as an advertisement for the new job of advocate or counsellor, as illustrated in Aristophanes's *Clouds*. It can also be a more personal advertising: "I, Protagoras, can help you to win your case at any rate, but also at a quite reasonable price!" This reminds the tricky deal he is said to have made with his student Evathlus about the payment of his teaching.

Did Protagoras think that disagreement has a more fundamental role than truth in human communication? Let us then ask an extra question to Protagoras: why only two opposing views? Why not three or more? Is this a proto version of the principle of bivalence? Other tenets of Protagoras' philosophy suggest he could have accepted a larger pluralism, let us say: "On any topic there are several opposing discourses".

2.2 Protagoras' second principle.

This leads to what I will call Protagoras' second principle, the famous "Man is the measure of all things, those that are that they are, and those that are not that they are not » (Quotation from Plato's *Theaetetus* (1961, 152 a)). Let us briefly remind some aspects of the discussion of the status of "Man".

In the *Metaphysics* (1980, K 6 1062b 13-19), Aristotle writes: "He [Protagoras] said that man is the measure of all things, meaning simply that *that which seems to each man also assuredly is* ...". This possible reduction of "man" to "each man", is discussed at length by Socrates in the *Theaetetus*. This individualistic interpretation paved the way to the popular view that Protagoras is the father of the most extreme epistemic relativism. But here again, the principle may have a political flavor leading to an epistemic egalitarianism by holding that any view is as right as the other. However, we shall see that Protagoras did resist this radical view. If he was a relativist, his relativism was mitigated.

Another interpretation takes "man" in a generic sense of which we can distinguish at least two interpretations. The first one is optimistically rationalistic: it grants any man enough cognitive resources to be the measure of all things. The second is less generous but stays optimistic about the global capacity of mankind. It would support something like: "We, humans, are collectively able to measure all things". Yes, but perhaps after an argument.

In harmony with Protagoras' religious agnosticism, the previous interpretations – individualistic or generic – can both have a political side dealing with human freedom: gods do not decide; men do.

A third interpretation is more aristocratic and seems further from Protagoras' rare words. Yet, it can be supported by Protagoras' view on education that I will discuss within a few lines. This last version would amount to: "All things are measured by men, but not by any of them".

2.2 Protagoras' epistemology: the third principle.

This principle is not associated with any particular sentence held to be authentic. Yet, this epistemic principle is ascribed to Protagoras by many commentators and has interesting variations. Let us see three famous versions.

Let us come back to the previous quotation from Aristotle's *Metaphysics*: « ... that which seems to each man also assuredly is ... ». It links the Man-measure-of-all-things principle with the broad epistemic (and ontological) one that what appears is. Other translations rather say: "what is believed is". The important point is that Aristotle's quotation specifies neither the type of this "appearance", nor its origin. So, it does not seem unreasonable to interpret Protagoras' point as "any opinion of each man is true". The path to the impossibility of error is now wide open.

A more limited view if offered in Sextus Empiricus' *Against the Logicians* (1961, VII (1) 389): "One cannot say that *every representation (fantasia) is true*, because this refutes itself as Democritus and Plato taught in opposing Protagoras ». Here, only a 'representation' (*fantasia, aesthesis*) would be always true for any man. This leaves open the possibility that, for Protagoras, not any 'belief' is true, but only if 'representations' are not the only 'thoughts' in the human mind. On this point we can only speculate on how radical Protagoras' empiricism was.

A possible source of Sextus is Protagoras' discourse in Plato's *Theaetetus*: "... it is impossible to think that which is not or to think any other thing than those which one feels (*fantasmata*), and these are always true" (1961, 167a). Two important points are made here. First, you cannot think and say "that which is not", a view close to the Eleats' doctrine. If to make an error amounts to thinking something that is not, i.e a non-being, a false *aesthesis* (representation, sensation or perception) is impossible, since an aesthesis is about 'some-thing'. The second point answers our question about Protagoras' empiricism: any thought is rooted in sensation. Accordingly, since sensations (*aesthesis*) are always true and statements are the expression of sensations, a statement cannot be false. This claim about the impossibility of error is also often associated with the sophistic movement. In any case, it sheds a new light on the first principle: none of two contradictory discourses or statements is false. It is sometimes said that Protagoras held contradiction to be impossible; but, in virtue of the first principle, it also seems that he considered it possible. So, a charitable interpretation suggests that what he denied was not that two arguers oppose each other, but the necessary falsity of at least one of the conflicting points of view.

This does not mean that everybody is right, as claimed by the individualistic interpretation of the Man-measure-of-all-things principle. For, if Plato is a reliable source, Protagoras thought that some appearances are better than others and that you can make somebody's mind change. So, he would not have been an epistemic egalitarian, even if, in a weak sense, everybody's is right. Listen to him through Socrates' mouth:

I do not by any means say that wisdom (*sophia*) and the wise man do not exist. On the contrary, I say that if bad things appear and are to any one of us, precisely that man is wise who causes a change and makes good things appear and be to him (166 d) ....
...in education a change has to be made from a worse to a better condition; but the physician cause the change by means of drugs, and the professor of wisdom (*sophist*) by means of words. And yet, in fact, no one ever made anyone think truly of who previously thought falsely, since it is impossible to think that which is not or to think any of the things than those which one feels (*fantasmata*) and these are always true.... I call some appearances better than the others, but in no wise truer. (167a)

Protagoras likely thought he was a wise man, able to change the *fantasmata* of people less gifted than him. This aristocratic attitude probably protected him from Socrates' argument that he should acknowledge that his own opinion is false because some people think it is and nobody is wrong (171 d). But Socrates also stresses that Protagoras' opinions are often badly presented, so that you should be very careful when you criticize ideas ascribed to him. And Socrates also grants that there may be some truth in the sophist's view on the infallibility of *fantasmata*. Socrates is now speaking for himself:

... not every opinion of every person is true ... but it is much more difficult to prove that opinions are not true in regard to the momentary states of feeling of every person, from which our perceptions and the opinions concerning them arise. But perhaps I am quite wrong; for it may be impossible to prove that they are not true, and those who say that they are manifest and are forms of knowledge may perhaps be right ... (179 c)

But this is not relevant to identify *episteme* that is somewhere else, beyond sensorial *aesthesis*:

Knowledge is not in the sensations, but in the process of reasoning about them; for it is possible, apparently, to apprehend being and truth by reasoning, but not by sensation. (186d)

This clearly goes against Theaetetus' proposal to identify knowledge and *aesthesis*. But this may also be not so far from Protagoras whose principles show a tension between the passive stability of individual opinion and the endless possibility to use discourse to move people's opinions. Unfortunately textual evidence about Protagoras' epistemology is too rare and sparse to clarify this point.

3. THE MANAGEMENT OF DISAGREEMENT

3.1 From Protagoras to social epistemology.

In harmony with his first principle, Protagoras claimed to be able to argue for or against any thesis. However, we may wonder whether he thought he could be a good advocate of a view he disagreed with? Today, to be able to argue against one's own convictions seems a common requirement in contemporary legal practice, perhaps in politics and certainly in classroom debates organized to train students to argue.

Protagoras' claim that some views are better than others, suggests that he also acknowledged that a defense can be better than another and a fair evaluative comparison between rival arguments is always possible. But would he have granted that he should agree with his opponents, as suggested by Socrates' ironical conclusion based on the individualistic interpretation of his second principle?

If Protagoras really was a proponent of a strict epistemic egalitarianism, the answer should have been 'yes'. But this has weird consequences, for he should not only have granted that his opponents are right when they say he is wrong, but also that his proponents are right when they say he is right. Then, would he have been ready to acknowledge anybody's authority? Believe that he is stupid because an epistemic peer says so? To grant anything anybody asserts leads to fairly uncomfortable positions. This is especially dreadful for an orator who claimed to be expert at persuasion and ready to assume a political leadership.

I have used the term 'epistemic peer' to stress an analogy with a contemporary debate in social epistemology about disagreement. The concept of disagreement is not that clear, even limited to the paradigmatic case of two parties, A and B, who disagree about a single proposition[3]. A much discussed question bears on the epistemic attitude one should adopt when you disagree with an 'epistemic peer'. An epistemic peer of A in domain D can be roughly defined as someone who has more or less the same competence as A in domain D and is approximately as reliable as A about the truth of a proposition relevant to D. Some people could complain that this definition remains a bit vague – at least because of the "approximately". Furthermore, the concept of epistemic peer seems highly relative and it is often in front of non-expert people that you get the status of experts and epistemic peers. Like authority and expertise, epistemic equivalence appears more salient when seen from far away.

You may require that A and B have exactly the same competence (in D) and are exactly as reliable as the other; but, in practice, you can doubt the plausibility of such a situation. Strict epistemic equality seems a very rare bird, if it ever happened. By the way, this suggests that the embarrassment we had with Protagoras' epistemology depends on how strictly is interpreted the individualistic version of his second principle making of each man an epistemic peer of any other. In any case, the concept of 'epistemic peer' reminds familiar situations, for instance physicians can easily have the status of epistemic peers

---

[3] MacFarlane (2007) has a general discussion on disagreement and relativism.

in front of their patients, or professors (in the same field) in front of their students.

The ongoing discussion in epistemology of disagreement is centered on the kind of questions we have just asked about Protagoras' attitude about his opponents' views and his ability to argue against his own beliefs. What should you do when you disagree with an epistemic peer when you share the same evidence and reasoning abilities? Should you stick to your view and your own reasons, as suggested by Socrates advising not to go and listen to Protagoras if it is true that he claims that your opinion is a good as his (161e)? Should you suspend your judgment? Should you adopt your peer's position, as suggested again by Socrates with the very example of Protagoras who should agree with his opponents? Should you shift only to the « equal weight view », i.e. the view that you should give no epistemic privilege to one or the other, when you disagree with an epistemic peer? Should you opt for another solution?

I doubt that there is a single general normative solution to the management of one's own attitude in case of disagreement, even with an epistemic peer. Here are two main objections to an *a priori* systematic attitude. First, the disagreement may be spurious. To this, you may object that you are only interested in genuine disagreements. Second, one or several errors or misunderstandings are possible on both sides and you may not know where they lie. Furthermore, to solve a disagreement, changing the initial positions and motives is not a necessary condition. To decide who is right or not and to explain the emergence of the disagreement are two different issues; but the understanding of its origin is often an epistemic benefit to make this decision. Hence, you should at least keep score of the original situation. I hold this benefit to be an epistemic intrinsic virtue of Protagoras' first principle – if you don't disagree only for fun –, because a disagreement (and/or a doubt) stimulates a revision of the truth of any statement. Even if you and your epistemic peer now agree, you may still be wrong. On the contrary, a disagreement – old or new – challenges your positions and their reasons and, strictly speaking, calls to think twice about the whole process that led to your point. To illustrate this, let us examine two cases.

3.2 Aesthetic disagreement

It is no surprise that in the *Theatetus*, Protagoras' view about disagreement and error is discussed about *aesthetic* judgements. They are often considered as 'subjective', in the sense that the subject who utters them is presumed to have a privileged and authoritative access to their truth. This privilege makes this case extremely supportive to the individualistic interpretation of Protagoras' principle that each man is the measure of all things. In such a case, when a disagreement occurs each subject is supposed to be the most competent to decide for oneself. Accordingly, it suggests that one should stick to one's own opinion and stay unmoved by diverging opinions.

Yet, you can doubt that the disagreement is genuine, because of a possible equivocation. Suppose that, speaking of the ambient temperature in the room where we are, I say "It's cold" and you reply "No, it's not cold". We usually think that people are competent and reliable about ambient temperature, so that you and I can consider each other as an epistemic peer. But do we disagree? The answer is not obvious, because in some languages statements about taste and perception are indifferently expressed by first or third-person expressions, then held as synonymous. In practice, "It's cold" often means "I'm cold". So, if my "It's cold" only means "I am cold", or your "It's not cold" only means "I am not cold", we may not disagree but only speak at cross purposes. The disagreement is spurious.

You could say that this is just a consequence of equivocal expressions and does not answer the question of the epistemic attitude to adopt in face of an authentic disagreement. A genuine one would occur if I would say "I am cold" and you would reply "You are not cold". But this can hardly seem an argument between epistemic peers, for even if I may be victim of sensory illusions, I am generally held in a better epistemic position than anybody else on these personal matters. So, even if I seem to shift towards your position by granting that I might be misled in my feeling of coldness, I may still stick to my initial statement. 'Belief' being an equivocal term, you can both stick to a first belief but think that something goes wrong somewhere and so, start an inquiry fed by the dissatisfaction induced by your opponent's opinion. The options offered when you disagree with epistemic peers are not exclusive.

In some sense, this kind of ambivalent or pluralistic position in a disagreement seems more plausible than a deliberate management of a single belief about the point at issue, either by sticking to an initial position or shifting towards the opponent's view. It also has the advantage to take into account our lack of command on our beliefs or, at least, on most of them. Catherine Elgin (2010) discusses this point by calling to Jonathan Cohen's distinction between belief and acceptance (Cohen 1989, 1992). Contrary to authors who consider that you can "shift", "move" or "abandon" your beliefs or your confidence (see, among others, Kelly 2013, Elga 2010) – formulations that are common in normative approaches of disagreement –, Elgin stresses this lack of direct command on our beliefs. This reminds us of Protagoras' thesis about the infallibility of aesthesis and the impossibility to err in this field. Following Cohen, Elgin makes a distinction between involuntary beliefs and propositions we accept, that is propositions we are ready to use in a reasoning. According to me, a reasoning can sometimes be stimulated by a disagreement, beyond and independently of our belief about the issue at stake.

Notice a similar ambivalent epistemic attitude about doubt. A nonexclusive distinction can be made between a passive doubt, a doubt that affects us, and an active doubt, like the Cartesian methodological doubt. The first amounts to an acknowledgement of ignorance or powerlessness, while the second chooses to wait for reasons or better reasons.

3.2 Disagreement in Mathematics

To support the view that in face of a contradiction expressed by an epistemic peer I should lower my confidence in my initial position, David Christensen (2007) designed the following case. I go out to dinner with a friend (let us call him Protagoras). We have decided to share the check. When the time to pay has come, each of us makes his own calculation. We are not experts at math, but we have reached the same level at school, so that we consider ourselves as epistemic peers. According to me, our shares amount to 43€; according to Protagoras they amount to 45€. Christensen suggests that in such a case you should lower your confidence into your own position.

The punch of this example relies on the implicit idea that there must be a mistake somewhere. Arithmetic has the reputation not to be governed by subjective feelings or preferences: in this field truth is commonly held to be exclusive. Therefore, one of us is likely to be wrong and should weaken his confidence in his calculation or his result. Since we are epistemic peers, there is no *a priori* reason to think that you, rather than I, have made the mistake. Hence Christensen's point: each of us should weaken his confidence into his result. Unfortunately, both of our calculations may be right. The discrepancy of our results may come from a non-mathematical mistake. We may wrongly assume that we share the same data: one of us may have misread a price or forgotten an item. And none of us may be responsible for our diverging calculations: for instance, the waiter may have given to each of us different tickets. Even expert epistemic peers may reach diverging results.

Finally, this example is not very telling, because we consider that our diverging results are incompatible because we grant as "normal" that a restaurant check has a single value. So, (at least) one mistake must have been made. At least for practical reasons, it has to be discovered. Does this requires that each of us drops or moderate his faith into his own result or in the correctness of his own calculation? I do not think so, even if it would be better for sake of politeness or friendship. Here is another example to stress that in front of a disagreement a rational agent does not always have to moderate her first conclusion.

Our friend Zeno wrote the equation $X^4 = 1$ and asked Protagoras and me to find the solution. After one second I say: "X = 1, I can prove it". A few seconds later, Protagoras, who loves contradiction, shouts: "No, X = -1, I can prove it". Protagoras is wrong to say "No", since both solutions are true. You could however say that there is no real disagreement here: Protagoras was perhaps badly inspired by Zeno's tricky expression about "the" solution of the equation. In any case, from a mathematical point of view, Protagoras and I should not lower our credence in our own results. Yet, each of us should also agree with the other answer. Our results are not exclusive.

We are epistemic peers, therefore we are as reliable as each other. But at least one of us may be wrong. In a new version of the same story I say "X = 1 and I can prove it" and Protagoras, who knows that there can be more than two solutions to this kind of equation says "No, X=-1 and X=2 are the two solutions". What should each of us do? Again, it seems to me that both of us should take into account that the other has reached a different result and the value of this result. But I still doubt that although we are epistemic peers, we should weaken our belief in our own result. This is not a rational requirement to safely proceed into our investigation. In any case, it is not the authority of the other that should move our opinions, but mathematical rules.

Now, imagine that after putting forwards two couples of false results, we finally agree that the solution of the equation is the couple X=1 and X=-1. We are quite certain of our result and announce to Zeno that we have the solution. He disagrees: "No, you don't have the solution yet". What should we do? He kindly goes on: "Have you ever heard of imaginary numbers?" He explains that there is a strange number "j", that mathematicians more expert than us call an imaginary number, such that $j^2=-1$. Thus, $X^4 = 1$ has four solutions (1, -1, j, -j). Zeno likes the idea that a fourth order equation has four solutions, whereas Protagoras and I find all this rather puzzling. But we finally accept the quasi-existence of that mysterious *j*. Finally, Zeno is right: the equation has four solutions, even if we do not really believe that *j* really exists.

Many examples in the literature on disagreement with epistemic peers are based on more familiar situations where it is commonly assumed that the core question has a correct answer and even a single one. Incompatible opinions then suggest that at least one mistake or confusion has been made. But as shown by the disagreements between Protagoras and me (and Zeno), this may be too hasty a conclusion. If the initial question is still on the agenda, an option that could be fruitful is to take our disagreement and its reasons as data to be explained. Unless we and our opinions are part of the problem, their fate becomes a secondary issue. When Protagoras stated his first principle, he perhaps intended to make a similar point by stressing the normative possibility to prolong or reopen a debate, even if it challenges the most stubborn of our beliefs.

REFERENCES

Aristotle. (1980). *Metaphysics.* Trans. Hugh Tredennick. Cambridge (Mass.): Harvard University Press.
Christensen D. (2007). Epistemology of disagreement: The good news. *Philosophical Review,* 116(2), 187-217.
Christensen D. & Lackey J. (2013). *The epistemology of Disagreement.* Oxford: Oxford University Press.
Cohen L. J. (1989). Belief and acceptance. *Mind,* 98(91), 367-389.
Cohen L. J. (1992). *An essay on belief and acceptance.* Oxford: Oxford University Press.
Diogenes Laertius. (1925). *Lives of eminent philosophers. Vol II. Books 6-10.* Trans. R. D. Hicks. Harvard : Harvard University Press.
Elga A. (2010). How to disagree about how to disagree. In R. Feldman & T.A. Warfield (Eds), Disagreement (pp 175-186). Oxford: Oxford University Press.
Elgin, C. Z. (2010). Persistent disagreement. In R. Feldman & T. A. Warfield (Eds.), *Disagreement* (pp. 53-68). Oxford: Oxford University Press.
Feldman R. & Warfield T. A. (2010). *Disagreement.* Oxford: Oxford University Press.
Kelly, T. (2013). Disagreement and the burdens of judgement. In D. Christensen & J. Lackey (Eds), *The epistemology of Disagreement* (pp 31-53). Oxford: Oxford University Press.
MacFarlane, J. (2007). Relativism and disagreement. *Philosophical Studies,* 132, 17-31.
Plato. (1961). *Theaetetus.* Trans. Harold North Fowler. Cambridge (Mass.): Harvard University Press.

Romilly J. (1988). *Les grands sophistes dans l'Athènes de Périclés*. Paris : Editions de Fallois. Janet Lloyd (trans.). (1992). *The great sophists of Periclean Athens*. Oxford: Clarendon.
Sextus Empiricus. (1961). Trans. Robert Gregg Bury. *Against the Logicians*. Cambridge (Mass.): Harvard University Press.
Silvermintz D. (2016). *Protagoras.* London: Bloomsbury.

Changing minds through argumentation: Black Pete as a case study

CATARINA DUTILH NOVAES[1]
VU Amsterdam – Arché, St. Andrews
c.dutilhnovaes@vu.nl

EMILY SULLIVAN
Eindhoven University of Technology
eesullivan29@gmail.com

THIRZA LAGEWAARD
VU Amsterdam
t.j.lagewaard@vu.nl

MARK ALFANO
Macquarie University
mark.alfano@gmail.com

Deep disagreements are often thought to be unresolvable. In this paper, we discuss a specific case of apparent deep disagreement, namely the public debate on the polemic figure of Black Pete in the Netherlands, where a noticeable change in public opinion has occurred in recent years. We present the preliminary findings of a study on Twitter interactions on the topic, focusing in particular on how arguments spread outside 'epistemic bubbles' and 'echo chambers'.

KEYWORDS: Black Pete, Deep disagreement, Public debate, Racism, Twitter

---

[1] CDN came up with the idea for the study, coded Twitter accounts, and wrote most of the article; ES advised on study design, collected and analyzed the data, and wrote sections 4 and 5.1-5.2; TL coded Twitter accounts and contributed to writing process; MA advised on study design and contributed to writing process. All authors contributed valuable ideas at various stages. CDN acknowledges the support of ERC-Consolidator grant 771074 for the project 'The Social Epistemology of Argumentation'.

1. INTRODUCTION

Views on the efficacy of argumentation to change minds in public discourse vary widely. On the one hand, there is a long-standing tradition that emphasizes the significance of argumentation and deliberation for public life (Mill, Habermas etc.), in particular to resolve societal disagreements. On this view, what is specific to argumentation as opposed to some other (non-rational) means to change minds (e.g. propaganda) is that, ideally at least, through argumentation people may change their minds by means of *reasons*, which they reflect upon and come to embrace consciously.[2] Thus understood, argumentation promotes and supports epistemic autonomy. However, the well-documented phenomena of group polarization and confirmation bias suggest that attempts to change minds through argumentation in public discourse are often futile. When presented with information that contradicts their well-entrenched beliefs, rather than examining the reasons and evidence offered objectively, people tend to seek ways to discredit them so as to maintain their original beliefs intact.

One challenge to argumentation as a means to manage disagreement in societies is the phenomenon of *deep disagreement*, a concept introduced in (Fogelin, 1985). As (Kappel, 2012) (p. 7) describes it: "We sometimes disagree not only about facts, but also about how best to acquire evidence or justified beliefs within the domain of facts that we disagree about. And sometimes we have no dispute-independent ways of settling what the best ways of acquiring evidence in these domains are." In situations of deep disagreement, often there does not seem to be enough common ground for a fruitful exchange of arguments to occur, as there is insufficient background agreement on what counts as evidence or as correct argumentation. Reasons given by one side of the disagreement are not accepted as such by the other side, and vice-versa. In such cases, it would seem that argumentation cannot change minds.

However, in some real-life situations that qualify as deep disagreements, exchange of reasons does seem to lead to changes of opinion at least for some of those involved. These cases suggest that deep disagreements may not be insurmountable after all (which would be good news for argumentation in public discourse), at least if they are deep but not *too deep*; arguably, disagreement depth is a gradable, comparative notion (Aikin, 2018). In this paper, we discuss a specific case of apparent deep disagreement, namely the public debate on the polemic figure of Black Pete in the Netherlands, where a noticeable

---

[2] Of course, there may well be other rational ways to change minds beside argumentation.

change in public opinion has occurred in recent years. In particular, we present the preliminary findings of a study on Twitter interactions on the topic, focusing in particular on how arguments spread outside 'epistemic bubbles' and 'echo chambers'.

2. BLACK PETE

Black Pete is a popular folk character in the Netherlands. He is presented as the servant of St. Nicholas, and is a crucial figure in the massively popular St. Nicholas festivities of early December. The festivities are meant in particular for children, who enjoy the gifts they receive but also the playful rituals involved. Black Pete, the servant, is traditionally represented with stereotypical racialized features associated with sub-Saharan Africans and their descendants: black face, curly hair, thick red lips. Moreover, he has features such as golden earrings, a servant costume, goofy behavior and (sometimes) a 'funny' accent. (The character is typically played by white people in blackface.)

There have been expressions of concern with what many see as racist aspects of the character for decades, but in recent years the polemic has intensified: critics are vocal in the press and on social media; protests are now regularly organized demanding that the tradition be significantly changed. However, at first sight it may seem that these protests have only led to further group polarization, with much pushback from those who want to maintain the tradition as is. This has included counter-protesters blocking a highway so as to prevent protesters (who had been issued a legal permit to protest) from reaching the main site of the festivities in 2017, and physical attacks on protesters perpetrated by organized groups of football supporters in 2018.

Prima facie, the controversy on the Black Pete character appears to be a clear instance of deep disagreement. In particular, the question of whether it is a racist tradition seems intractable, as the different parties disagree on what counts as evidence of racism, especially as they seem to disagree on what counts as racism in the first place. Typically, those who support the tradition associate the phenomenon of racism with explicit attributions of inferiority to a certain group of people vis-à-vis other groups, often accompanied by acts of violence against the group seen as inferior. On this narrow conceptualization of racism, the Black Pete figure is not obviously racist, since he is presented as very likeable and friendly.

However, there are at least two other senses of racism that seem relevant here: historical/structural racism, and implicit racism. Historical/structural racism is a consequence of European colonization, with millions of Africans brought as slaves to the Americas. These historical events of tremendous implications still now entail racist institutions as well as overall attributions of inferiority to people of color (Mills, 2015). Implicit racism, in turn, pertains to the internalization of these perceived hierarchies such that even those who consciously embrace egalitarian values may harbor implicit negative associations with members of certain groups (people of African descent in this case) (Levy, 2017). From a historical perspective, Black Pete is arguably a colonial figure, the black servant reminiscent of African slaves (even if he is no longer a slave himself), and thus may plausibly be seen as reaffirming racist hierarchies. Similarly, by reinforcing the association between servitude and people of color, the figure of Black Pete perpetuates a perception of people of African descent as inferior, which becomes internalized by children from early on.

Now, if different segments of the population adopt different conceptions of racism, the debate over whether Black Pete is a racist figure, and thus whether it should be modified or remain as is, seems intractable. However, there have been some noticeable changes over the last years, both in public opinion and in how the festivities occur. For example, in a number of larger cities (Amsterdam, The Hague, Utrecht), associations of primary schools decided to exclude the racialized representation of Pete from their celebrations (opting for example for Petes whose faces are covered with 'soot' from the chimneys that they allegedly climb to bring presents). In past years, roughly 5% of people per year changed their minds on the acceptability of the tradition and joined the critical camp (which however remains a minority). While in 2013, 89% were against changes, in 2017 this number went down to 68% (see tables below).

Percentage of people interviewed supporting changing the Black Pete tradition[3]

2014	12%
2015	17%
2016	21%

2016

Population of Surinamese or Caribbean origin	43%
Others	18%

---

[3]https://www.nrc.nl/nieuws/2016/11/02/heimelijk-onderzoek-eenvijfde-wil-andere-zwarte-piet-5100360-a1529881

Percentage of people interviewed supporting changing the Black Pete tradition[4]

2013	11%
2017	32%

Thus, it does seem that arguments by critics are having uptake and changing at least some people's minds on the (non-)acceptability of the traditional figure of Black Pete (though again, it may well be that non-argumentative factors also play a role). Perhaps a number of people have come to think that the phenomenon of racism and its negative consequences go beyond what was described above as 'explicit racism', thus recognizing the relevance of more 'subtle' manifestations of racism. Perhaps some people came to appreciate the discomfort experienced by children of African descent during the festivities, as registered in a report by the Children's Ombudsman of the Netherlands in 2016.[5] In sum, while the majority of the Dutch population continues to support the tradition, there have been significant changes in public opinion in a short period of time, which suggests that this controversy is a (deep?) disagreement that is not entirely intractable

3. DEBATES ON TWITTER

But how do switchers come to change their minds about the (non-)acceptability of the traditional Black Pete figure? Given the (presumed) phenomena of epistemic bubbles and echo chambers in social media and elsewhere (Nguyen, forth.), it is not immediately obvious how they get exposure to arguments supporting changes to the tradition. In order to study potential networks of propagation for these arguments, we conducted a pilot study on Twitter. To our knowledge, the Black Pete controversy specifically has never been studied on Twitter, but a number of other prominent controversies have been studied recently with corpora of Twitter interactions, including by some of the present authors (Sullivan, et al., forth.).

The motivating idea for our study was the following observation: activist accounts (both pro- and anti-Pete) are likely followed and interacted with only by people who already have a firm opinion on the controversy (either people who follow them because they already agree with the position being defended, or people who

---
[4]https://eenvandaag.avrotros.nl/panels/opiniepanel/alle-uitslagen/item/draagvlak-voor-traditionele-zwarte-piet-loopt-terug/
[5]https://www.nrc.nl/nieuws/2016/09/30/kinderombudsman-zwarte-piet-in-strijd-met-kinderrechtenverdrag-a1524070

vehemently disagree and follow them to engage in overt confrontation). By contrast, accounts whose profiles are not strongly associated with a specific position in the controversy (and thus are followed for unrelated reasons) are likely to have followers with less firm opinions on Black Pete, and thus more susceptible to change their minds. Such accounts would arguably have uptake also outside of the relevant bubbles and echo chambers. Our hypothesis is that accounts that are *verified* by Twitter, which are presumed to be of general public interest, might (among others) be playing the role of broadcasters of messages supporting changes to the Black Pete tradition. They not only have wider reach across bubbles and echo chambers, but their followers presumably attribute a certain degree of epistemic trust to them for reasons unrelated to this specific controversy. These include accounts for news organizations such as newspapers and accounts of public figures such as journalists, celebrities, and artists. More generally, in a cacophony of messages being broadcast and competing for the receivers' limited attention (what has been described as the 'economy of attention' (Franck, 2019)), there are gigantic disparities in how much each of the 'voices' in the conversation is heard.[6]

The role of celebrities in politics has been a topic of interest for decades, but in recent years interest has intensified in view of the pervasiveness of social media. For example, a recent study (Archer, Cawston, Matheson, & Geuskens, forthcoming) presents an analysis of the role of celebrities in politics from the perspective of social epistemology. In particular, the authors describe celebrities as having the core feature of capturing attention, and attribute to celebrities a high degree of epistemic power: "A person has epistemic power to the extent she is able to influence what people think, believe, and know, *and* to the extent she is able to enable and disable others from exerting epistemic influence." (Archer, Cawston, Matheson, & Geuskens, forthcoming)

Beside simply having a wider following on social media—which translates in what is described as 'attention capital' (Franck, 2019)—the concept of epistemic power thus understood suggests that celebrities may also inspire a high level of epistemic trust given their artistic or otherwise achievements. In other words, a fan is likely to be open to considering carefully the views professed by their favorite celebrities also on matters that do not pertain to the achievements they are famous for. Imagine a person with a certain political leaning, who will typically dismiss outright views that clash with their political convictions (Taber

---

[6] More generally, the role of social factors and social influence in the spread of beliefs and information is now increasingly recognized as crucial (O'Connor & Weatherall, 2019).

& Lodge, 2006). If these views are defended by their favorite artist, this may have the upshot of disabling the otherwise default response of outright rejecting views clashing with one's own original convictions. It is in this sense that arguments put forward by celebrities and people with significant social influence may be able to change minds more readily than when the source of an argument is perceived either negatively or neutrally by the receiver.

4. STUDY DESIGN

To investigate the role of public figures in societal debates, we conducted a pilot study on the Black Pete discussion on Twitter.[7] The main theoretical hypothesis we sought to explore was whether social power predicts content uptake, in particular given that those with greater social power both off- and online are likely to have a wider reach than 'regular' Twitter users, and to inspire an overall sense of epistemic trust. Our study focused on two aspects of this thesis: do public figures get higher engagement with their tweets about Black Pete, compared to 'regular' accounts tweeting about Black Pete? Do public figures get higher engagement with their tweets about Black Pete, compared to their other, non-Pete-related content? More concretely, we considered the following initial hypotheses:

(H1) Tweets about Black Pete from **verified** accounts will have more engagement (i.e. more retweets and more likes) than tweets about Black Pete from **non-verified** accounts.
(H2) Tweets about Black Pete from **verified** accounts will have more engagement than tweets from **verified** accounts *not* about Black Pete.
(H3) Tweets about Black Pete from **non-verified** accounts will have more engagement than tweets from **non-verified** accounts *not* about Black Pete (as a control group).

4.1 Collecting users

From October 10, 2018 to October 29, 2018, using the free developer version of the Twitter Stream API we collected tweets that contained

---

[7] Of course, it may be objected that Twitter debates are not an accurate representation of public debates at large. While this is possible, it is now widely (though not unanimously) thought that social media significantly influences public opinions, so we assume that the results presented here reveal at least something significant about the debates on Black Pete at large.

the string 'zwarte piet', 'black piet', and their variations.[8] By collecting the target users for our analysis this way, we hoped to avoid researcher bias of hand picking particular accounts. Our search resulted in 16,384 distinct users who tweeted about Black Pete at least once. Of these users, 116 were from verified accounts, which (as mentioned) tend to be news organizations and public figures. Thus, 16,286 of the users identified were from non-verified accounts, with 2,690 users tweeting about Black Pete at least 5 times during the collection period.

We included all 116 of the verified accounts in the main study, and took a random sample of non-verified users who tweeted 5 times or more about Black Pete during the initial collection, resulting in 114 accounts (in order to have a similar sample size between verified and non-verified accounts). Since the collection window from October is slightly outside the peak discussion season (which ranges from early November until December 5th, the day of the festivities), this suggests that the users we identified have strong interest in the controversy.

4.2 Following identified users

From November 5th to December 31st, 2018, using the free developer version of the Twitter Stream API, we collected all the tweets (on Black Pete or otherwise) that each user in our identified user list (verified and non-verified) tweeted during this window. We used Twitter's follow function that allows us to collect tweets, retweets, and replies created by the user during the requested time period.

4.3 Getting engagement statistics

Collecting tweets through the Twitter stream API collects tweets as they happen, thus there is no retweet or like count provided with the tweet in real time. On March 16, 2019 we made another call to the Twitter rest API that received the updated information for each tweet based on each specific tweet-id. There were several tweets for which we were unable to get the engagement data because the tweets were no longer available. This can be because these tweets were deleted by the user, removed by the platform, or the user set their account to private.

---

[8] The full search query contained the following terms: 'zwarte piet', 'black piet', 'zwartepiet', 'zwartepieten', '#zwartepiet', '#zwartepieten', '#blackpiet', 'zwarte', 'black', 'piet', 'pieten'.

4.4 Data pre-processing

We engaged in data pre-processing with the data collected from November 5th to December 31st, 2018. We were specifically interested in adding particular labels to the data:

- **Verified versus non-verified:** This is a built-in Twitter category that is directly taken from the data provided by the Twitter API.
- **Black Pete tweets versus non-Black Pete tweets:** Using the same criteria as our initial search criteria from October, we labeled particular tweets as being about Black Pete or not.
- **News organization versus non-news organization:** Within the verified accounts there exist public figures in addition to news and journalistic outlets. We labeled specific accounts as news organization. It is possible that the uptake of journalistic accounts display different patterns, and that users share them for different reasons. Therefore, we wanted to have this information for exploratory purposes.[9] (The labeling of news versus non-news was done by someone with thorough familiarity with the Dutch media landscape.)
- **Deleted tweets versus non-deleted tweets:** The tweet-ids that were not found as of March 15th, 2019, when we collected the engagement metrics, were labeled as a deleted tweet.
- **Anti-Black Pete leaning users versus Pro-Black Pete leaning users:** For each user we had two independent Dutch speakers hand-label whether particular Twitter accounts are pro-Black Pete or anti-Black Pete, neutral or irrelevant. Evaluation was done by each evaluator looking at the user's tweet history and profile description to determine whether the user was likely to be Anti-Black Pete (i.e. believing the tradition should be ended or changed significantly) or Pro-Black Pete (i.e. believing the tradition should be maintained as is). Interrater reliability was 74% (Fleiss's Kappa of .64), indicating adequate agreement between the raters. The most common points of disagreement were between labeling an account as neutral versus irrelevant and labeling an account as neutral or irrelevant versus anti or

---

[9] In our sample, for the original Black Pete tweets there was not much difference in engagement between news and non-news accounts, at an average of 16 retweet count versus 14, respectively. For this reason, we will not discuss this distinction further, though for our purposes news and non-news verified accounts are treated differently (for example, all news accounts are labeled as neutral).

pro. Each disagreement was resolved by taking the more extreme position. For example, if a user was evaluated as irrelevant by one rater and pro by the other rater, we gave the user a pro label. If a user was evaluated as both irrelevant and neutral we labeled the user as neutral. (All news organizations were labeled neutral, despite the fact that some news organizations express a particular ideological slant, e.g *De Telegraaf* for conservative positions.) The results were as follows:

	Anti-Pete	Pro-Pete	Neutral	Irrelevant
Verified	42%	13%	35% (half of them news)	10%
Non-verified	11%	71%		16%

We were somewhat surprised by such a high preponderance of pro-Black Pete accounts among our sample for non-verified accounts (71%), which was selected randomly. This suggests that there is a high preponderance of pro-Black Pete users among the non-verified accounts that tweet about Black Pete as a whole. By contrast, among verified accounts, anti-Pete accounts were the largest group, and this already partially confirms our initial hypothesis that celebrities are among the disseminators of anti-Pete arguments.

5. RESULTS

Our dataset from November and December resulted in a total of 438.610 tweets, with only 2,3% of those tweets about Black Pete, as shown in Table 1. 8,4% of the tweets about Black Pete were deleted or removed by March. This resulted a filtered dataset of 402.782 tweets for further analysis. Table 2 shows the number of tweets broken down by account type for our final dataset. The first interesting observation is that non-verified accounts tweet more often about Black Pete compared with verified accounts, both in terms of the raw number of tweets and the ratio between Black Pete tweets and other tweets. We also see that the percentage of tweets about Black Pete is small. This suggests that the identified accounts are largely not single-issue accounts, but rather focus on several topics.

	Black Pete tweets	Non-Black Pete	Total	Percentage of Black Pete
Tweets from non-verified accounts	8.109	267.432	275.541	2,94%
Tweets from verified accounts	2.001	161.068	163.069	1,23%
Total	10.110	428.500	438.610	2,31%

Table 1 - Number of tweets collected from Nov. 5 – Dec. 31, 2018

	Black Pete tweets	Non-Black Pete	Total	Percentage of Black Pete
Tweets from non-verified accounts	6.160	245.412	251.572	2,40%
Tweets from verified accounts	1.753	149.457	151.210	1,16%
Total	7.913	402.782	402.786	1,94%

Table 2 - Number of tweets remaining after filtering for engagement

5.1 Hypothesis testing

(H1) Tweets about Black Pete from verified accounts will have more engagement (i.e. more retweets and more likes) than tweets about Black Pete from non-verified accounts.

Table 3 shows the summary statistics for tweets about Black Pete that originated from verified and non-verified accounts. Figure 1 shows the density distribution of favorite and retweet count. We excluded retweets in our analysis because our central interest is in the engagement of the tweets that originated from our identified users. Our results show that verified accounts do indeed get more engagement for their Black Pete tweets compared to the Black Pete tweets from non-verified accounts. A Wilcoxon rank-sum test shows these results to be significant, with w = 607650 and a p value of < .0001 for favorite count, and w = 681140 and a p value < .0001 for retweet count. Of course, it should not be surprising that verified accounts get greater engagement, since they have a greater number of followers compared with non-verified accounts.

	Favorite Count		Retweet Count	
	Non-Verified	Verified	Non-Verified	Verified
Total Tweets	1746	1277	1746	1277
Mean	2.572	38.888	1.426	15.425
Std. Deviation	11.902	141.863	7.848	49.855
Variance	141.656	20125.159	61.589	2485.566
Minimum	0.000	0.000	0.000	0.000
Maximum	206.000	2701.000	183.000	714.000

Table 3 - Summary Results for H1: Black Pete tweets

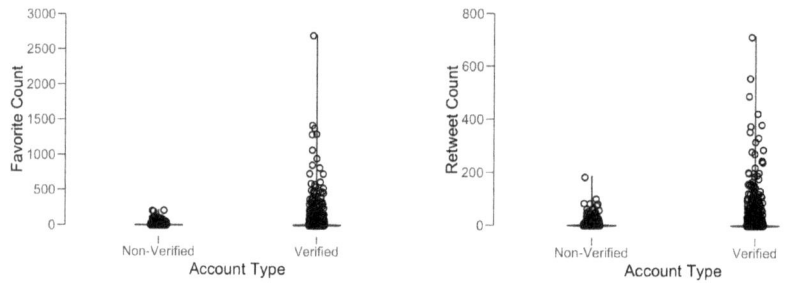

Figure 1 - Summary Results for H1: Black Pete tweets

H2) Tweets about Black Pete from verified accounts will have more engagement than tweets from verified accounts not about Black Pete.

Table 4 shows the summary statistics for original tweets that were created by verified accounts, broken down by tweets about Black Pete and all other tweets. Figure 2 shows the density distribution of favorite and retweet count. We see that on average original tweets about Black Pete get more than double the engagement compared to tweets not about Black Pete from verified accounts, which indicates in particular uptake of anti-Pete arguments (recall that 42% of the verified accounts were labelled anti-Pete, as opposed to 13% pro-Pete verified accounts). However, it is important to notice that the most engaged with tweets are not about Black Pete; these are so-called 'viral' tweets that get through-the-roof levels of engagement. But the Black Pete tweets taken as a whole show consistent patterns of higher engagement than most other topics. A Wilcoxon signed-rank test shows these results to be significant with w = 241348 and a p value of < .001 for favorite count, and w =164782 and a p value of < .001 for retweet count.

	Favorite Count		Retweet Count	
	Black Pete	Not BP	Black Pete	Not BP
Total Tweets	1277	117891	1277	117891
Mean	38.888	16.671	15.425	6.636
Std. Deviation	141.863	160.627	49.855	57.182
Variance	20125.159	25800.905	2485.566	3269.760
Minimum	0.000	0.000	0.000	0.000
Maximum	2701.000	36992.000	714.000	11920.000

Table 4 - Summary Results for H2: Verified accounts

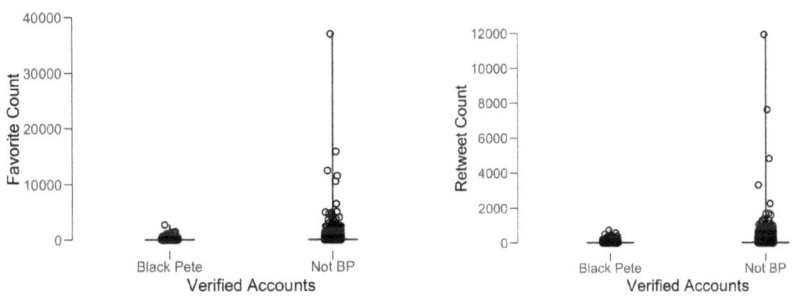

Figure 2 - Summary Results for H2: Verified accounts

(H3) Tweets about Black Pete from non-verified accounts will have more engagement than tweets from non-verified accounts not about Black Pete.

Table 5 shows the summary statistics for original tweets that were created by non-verified accounts, broken down by tweets about Black Pete and all other tweets. Figure 3 shows the density distribution of favorite and retweet count. The same trend appears: on average, tweets about Black Pete get more engagement. In the case of retweets there is nearly three times as much engagement with tweets about Black Pete compared to the other tweets created by the same users. However, again, the highest engaged-with tweets are not about Black Pete. This suggests that there is a limited reach that Black Pete tweets get compared to other tweets. A Wilcoxon signed-rank test shows these results to be significant with w = 254127and a p value of < .001 for favorite count, and w =83461and a p value of < .001 for retweet count.

	Favorite Count		Retweet Count	
	Black Pete	Not BP	Black Pete	Not BP
Total Tweets	1746	84076	1746	84076
Mean	2.572	1.403	1.426	0.594
Std. Deviation	11.902	10.108	7.848	5.303
Variance	141.656	102.172	61.589	28.121
Minimum	0.000	0.000	0.000	0.000
Maximum	206.000	1499.000	183.000	672.000

Table 5 - Summary Results for H3: Non-verified accounts

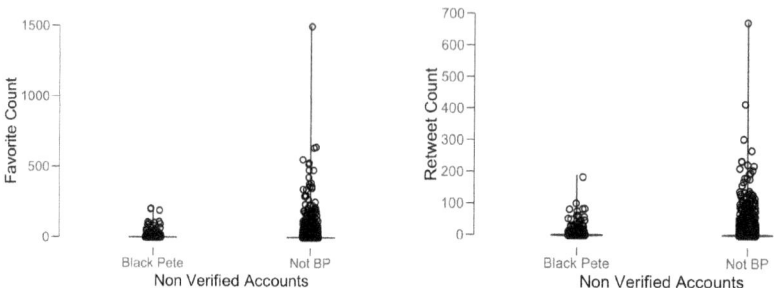

Figure 3 - Summary Results for H3: Non-verified accounts

5.3 Word clouds

On the basis of the corpus of collected tweets about Black Pete, we generated word clouds that give us clues as to the specific contents being discussed (so far, we have only considered engagement statistics without looking 'inside' the tweets). The word clouds indicate the concepts and themes that are viewed as significant by the different groups. Let us first consider the word clouds for pro-Pete accounts (both verified and non-verified) and for non-verified accounts. These two groups largely overlap, as the 81 pro-Pete non-verified accounts dominate. (The pro-Pete accounts are: 81 non-verified, 15 verified. The non-verified accounts are: 81 pro-Pete, 33 for the rest).

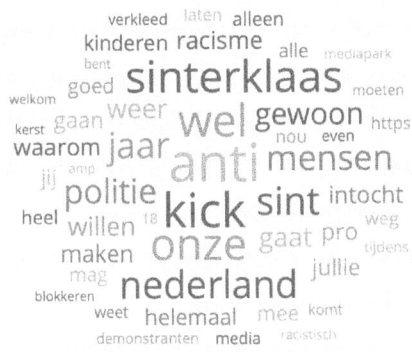

Figure 4 - Word cloud for Black Pete tweets from non-verified accounts

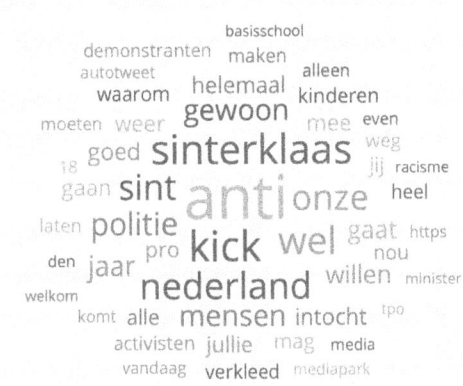

Figure 5 - Word cloud for Black Pete tweets from pro-Pete accounts

Some of the words that stand out here are very telling about the pro-Pete mindset: 'Nederland' (Netherlands), 'onze' (our, used for tradition, culture etc.), 'Sinterklaas'. These word clouds thus reflect the main worry that motivates defenders of the tradition: it is 'our' traditional Dutch culture that is under attack, being rejected by these 'intruders' who do not belong here (i.e. people with a migrant background, in particular people of color). Notice also that, while they appear, words such as 'racisme' (racism) and 'kinderen' (children) are comparatively much less prominent.

By contrast, a word cloud for the Black Pete tweets by critics of the tradition (48 verified accounts, 13 non-verified) gives a very different picture of what they take to be at stake.

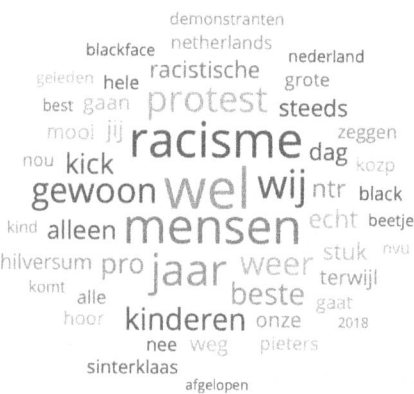

Figure 6 - Word cloud for Black Pete tweets from anti-Pete accounts

Here we see words for 'Netherlands' and variations as much smaller, and no occurrence of 'our'. By contrast, 'racisme' and 'kinderen' figure very prominently; 'racisme' for obvious reasons, and 'kinderen' because one of the main arguments of critics is that the festivities must be inclusive and enjoyable for children of all races, which reportedly is not the case for children of color with the traditional, racialized Black Pete.[10]

These word clouds do not offer particularly surprising information, but they provide objective evidence for what those who follow the debate closely already suspect unsystematically: the debate reveals a clash of values. On one side, proponents of Black Pete praise tradition and cultural identity highly; on the other side, critics highlight the negative effects of the racialized character for people of color, especially children. It appears thus to illustrate R. Talisse's description of political polarization as primarily related to identities rather than to facts (Talisse, 2019).

6. CONCLUSIONS

Our results by and large confirm the initial hypothesis that celebrities may be important broadcasters of anti-Black Pete arguments. It is noteworthy that a high percentage of verified accounts (42%) were classified as anti-Pete, as opposed to 13% of pro-Pete accounts among verified accounts (though this may also reflect Twitter's bias towards left-leaning accounts receiving the verified seal more frequently). We've

---

[10] Rossana Kluivert, the wife of football player Patrick Kluivert, has been vocal about her decision to let her children stay home on December 5th instead of going to school so as to spare them of the festivities with the traditional, racialized Black Pete figure. https://www.libelle.nl/mensen/rossana-kluivert-5-december/

also confirmed that there is significant engagement (likes and retweets) with the Black Pete tweets by the verified accounts, indeed on average more than for their non-Black Pete tweets (the same holds for the non-verified accounts), which reflects the importance of the debate in Dutch society. Moreover, as expected, tweets by verified accounts tended to have more reach and uptake than those by non-verified accounts, and this was the case especially of Black Pete tweets.

For this study we did not investigate the structure of networks connecting these different accounts, that is in terms of who follows who and who reacts to whom. A natural continuation would be to investigate these structural factors. This would allow us to further probe the extent to which celebrities do have epistemic power, as conjectured in (Archer, Cawston, Matheson, & Geuskens, forthcoming), especially across putative epistemic bubbles and echo chambers. Moreover, we did not consider patterns of retweets and instead only looked at the original tweets of the accounts in our sample. This too is a distinction worth investigating in future work.

Our aims with this study were modest; indeed we view it as a pilot study, and intend to repeat the data collection in October-December 2019 (with some refinements motivated by what we have learned so far). Nevertheless, our results already lend support to the conceptually motivated idea that celebrities and public figures have some degree of epistemic power when it comes to changing people's minds on societal matters where there is substantive disagreement, as they mitigate to some extent the phenomena of epistemic bubbles and echo chambers.

REFERENCES

Aikin, S. (2018). Deep Disagreement, the Dark Enlightenment, and the Rhetoric of the Red Pill. *Journal of Applied Philosophy*, forth., 1-16.
Archer, A., Cawston, A., Matheson, B., & Geuskens, M. P. (forthcoming). Celebrity, Democracy, and Epistemic Power. *Perspectives on Politics*, 1-45.
Fogelin, R. (1985). The logic of deep disagreements. *Informal Logic*, 7, 3-11.
Franck, G. (2019). The economy of attention. *Journal of Sociology*, 55, 8-19.
Kappel, K. (2012). The Problem of Deep Disagreement. *Discipline Filosofiche*, 22, 7-25.
Levy, N. (2017). Am I a Racist? Implicit Bias and the Ascription of Racism. *Philosophical Quarterly*, 67, 534-551.
Mills, C. (2015). Global White Ignorance. In M. Gross, & L. McGoey, *Routledge International Handbook of Ignorance Studies* (pp. 217-227). London: Routledge.

Nguyen, C. T. (forth.). Echo chambers and epistemic bubbles. *Episteme*, 1-21.
O'Connor, C., & Weatherall, J. (2019). *The Misinformation Age.* New Yaven, CO: Yale University Press.
Sullivan, E., Sondag, M., Rutter, I., Meulemans, W., Cunningham, S., Speckmann, B., et al. (forth.). Can Real Social Epistemic Networks Deliver the Wisdom of Crowds? In T. Lombrozo, J. Knobe, & S. Nichols, *Oxford Studies in Experimental Philosophy, Volume 1.* Oxford: Oxford University Press.
Taber, C. S., & Lodge, M. (2006). Motivated Skepticism in the Evaluation of Political Beliefs. *American Journal of Political Science*, *50*, 755-69.
Talisse, R. (2019). *Overdoing Democracy.* Oxford: Oxford University Press.

The pragma-dialectical view of comparison argumentation

FRANS H. VAN EEMEREN
ILIAS and University of Amsterdam
f.h.vaneemeren@uva.nl

BART GARSSEN
ILIAS and University of Amsterdam
b.j.garssen@uva.nl

The pragma-dialectical typology of argument schemes consists of three main categories: symptomatic argumentation, comparison argumentation and causal argumentation. The subcategories and variants of these main categories have not yet been distinguished systematically. In this contribution the authors start doing so by distinguishing the most important subtypes and variants of comparison argumentation. They argue that two distinct subtypes of comparison argumentation need to be distinguished: descriptive comparison argumentation and normative comparison argumentation.

KEYWORDS: argument scheme, comparison argumentation, descriptive comparison argumentation, normative comparison argumentation pragma-dialectical typology of argument schemes

1. INTRODUCTION

Comparison argumentation, also called *argumentation by analogy*, belongs with symptomatic argumentation and causal argumentation to the main categories of the pragma-dialectical typology of argument schemes. In our present research we are out to specify the general category of argumentation by comparison into relevant subcategories. This research is part of a more comprehensive research project aimed at extending the pragma-dialectical theory of argument schemes by specifying the typology that was earlier developed (van Eemeren and Grootendorst, 1992: 93-102) and relating it to the various macro-contexts in which argumentative discourse takes place.

In the pragma-dialectical theory of argumentation the rationale for distinguishing between the various categories of argument schemes has a pragmatic as well as a dialectical dimension (van Eemeren, 2018:

45-39; van Eemeren and Garssen, 2019: Section 3). The pragmatic dimension relates to the kind of justificatory principle that legitimizes the transfer of acceptance from the argumentation that is advanced to the standpoint that is defended. Unlike logical validity, this justificatory principle has a pragmatic instead of a formal basis, because it is grounded in the arguers' practical experiences in justifying standpoints in argumentative discourse. The dialectical dimension relates to the evaluation procedure associated with the argument scheme that is used, i.e. to the critical questions that need to be answered in legitimizing the use of an argument scheme. When taken together, these two dimensions constitute the *principium divisionis* underlying the pragma-dialectical typology of argument schemes. The various types and subtypes of argumentation included in this typology and the critical questions instrumental in their evaluation are presumed to be part of the intersubjectively accepted starting points for conducting a critical discussion.

Since each type or subtype of argumentation characterized by the use of a particular argument scheme included in the typology calls out its own set of critical questions, the three categories of argument schemes that are distinguished are all associated with specific dialectical routes for resolving a difference of opinion on the merits. The differences between the dialectical routes instigated by the use of symptomatic argumentation, comparison argumentation and causal argumentation are in the first place determined by the "basic" critical question relating to the bridging premise (usually implicitly) used in the argument schemes concerned (van Eemeren and Garssen, 2019: Section 3). This basic critical question is in principle the same for all subtypes of a certain type of argumentation, but in the various subtypes may take a more specific shape and it may be accompanied by other (general or specific) critical questions. In addition, the macro-context in which the argumentative discourse takes place may have an influence on the way in which in a particular case the critical questions will be shaped.

In this contribution we aim to provide a characterization of two prominent subtypes of argumentation by comparison based on their pragmatic rationale and the critical questions that need to be answered when they are used. In addition we will indicate some kinds of "variants" of each of the subtypes of comparison argumentation coming about in certain kinds of argumentative practices. These variants consist of different manifestations of the subtype that are as a rule connected with the kind of macro-context in which the argumentation is used. In the empirical component of a fully-fledged research programme for examining argumentative discourse describing such different manifestation of the various subtypes of argumentation in argumentative reality is an important task.

Argumentation by comparison creates a relation of comparability or analogy between something that is already accepted and something still to be accepted. According to van Eemeren and Grootendorst, in analogy argumentation the argumentation is presented "as if there were a resemblance, an agreement, a likeness a parallel, a correspondence or some other kind of similarity between that which is stated in the argument and that which is stated in the standpoint" (1992: 97). The pragmatic principle exploited in argumentation by comparison is that something which is comparable to something else is to be treated or dealt with in the same or a similar way.

In our view, based on pragmatic distinctions between them and differences in the dialectical procedures that are to be followed in their evaluation, two vital subtypes of comparison argumentation are to be distinguished: "descriptive analogy argumentation" and "normative analogy argumentation" (Garssen, 2009: 134). In descriptive analogy argumentation the standpoint defended is descriptive and refers to a certain state of affairs, while in normative analogy argumentation the standpoint is evaluative or prescriptive and involves a value judgment. Because, as a consequence, the evaluation procedure for normative analogy argumentation requires the inclusion of an extra critical question, descriptive analogy argumentation and normative analogy argumentation are in the pragma-dialectical typology seen as two separate subtypes of comparison argumentation. Each of these two subtypes has its own dialectical profile and its own variants.

2. DESCRIPTIVE ANALOGY ARGUMENTATION

It is characteristic of descriptive analogy argumentation that the standpoint defended and the argumentation advanced in its support are both descriptive: each of them expresses a certain state of affairs. The standpoint defended is invariably a prediction ("Y will be the case"), a hypothetical prediction ("If X will be the case, then Y will be the case") or a quasi-prediction claiming something about the past or the present ("At time t, Y was to be expected").

In descriptive analogy argumentation a comparison is made between the actual characteristics of a thing, person, group, institution, event or situation and the actual characteristics of another thing, person, group, institution, event or situation. This type of argumentation can be characterized in the following way:

               Y is true of X
because: Y is true of Z
and: Z is comparable to X
(van Eemeren and Snoeck Henkemans, 2017: 87)

This argument scheme is, for example, displayed in (1):

(1)   Camera surveillance in the Amsterdam metro will be effective, because it is also effective in the London underground [and the situation in Amsterdam is comparable to the situation in London].

In (1) the protagonist argues that something will be the case in Amsterdam because it is already the case in London (and Amsterdam is in the relevant respects comparable to London). In this argumentation a comparison is made in which it is assumed that there are a number of directly relevant similarities between the situation in Amsterdam and that in London. Because of these similarities we may take it that the camera surveillance effectiveness applying to London mentioned in the argument will also apply to Amsterdam. The presence of the property at issue is as it were extrapolated from properties the two cities already share.

It is important to realize that the similarities relied upon in comparison argumentation are in argumentative practice often not mentioned explicitly in the argumentation; the relevant similarities are only tacitly assumed. As we shall see, in the evaluation of the analogy argumentation these similarities play a crucial role.

In descriptive analogy argumentation the justificatory principle of analogy is used in extrapolating a property from a list of commonalities between what is conveyed in the argumentation and what is conveyed in the standpoint. Because the two things, persons, groups, institutions, events or situations that are compared have a series of properties in common, they are assumed to share also another property claimed in the standpoint. In this sense this subtype of analogy argumentation, which is sometimes called *case by case reasoning*, involves an inductive process. A general characterization of this subtype is provided in Figure 1.[1]

Situation (implicitly) referred to in premise	Situation (implicitly) referred to in standpoint
Relevant similarity 1	Relevant similarity 1
Relevant similarity 2	Relevant similarity 2
Similarity 3	Similarity 3 extrapolated

Figure 1. *Descriptive analogy argumentation as extrapolation of properties*

---

[1] See Fearnside and Holther (1959: 23) for a similar representation of the internal organization of this subtype of analogy argumentation.

The first step to be taken in the testing procedure for descriptive analogy argumentation is to ask whether the elements that are compared in the argumentation are, in principle, comparable. If asked to answer this question, the protagonist is obliged to show that they do indeed belong to the same category or class. Next the testing procedure can take different routes. Since the extrapolation is based on similarities which remain implicit, the antagonist may ask the protagonist to add the relevant similarities explicitly. The protagonist is then forced to provide additional argumentation in which relevant similarities are mentioned. These similarities are the properties that allow for the extrapolating step to be made. The mentioning of similarities can in its turn lead to further criticism from the antagonist, who may not recognize them as real similarities or see them as not relevant to the issue at hand.

The antagonist may also criticize the argumentation by pointing at differences between the two elements that are compared. The protagonist is then forced to show that the differences mentioned by the antagonist are not relevant or that the existing similarities outweigh the differences. In this way it is established by the testing procedure whether or not the intended extrapolation of characteristics is allowed. The more relevant similarities are mentioned, the more likely it is that an extrapolation is successful. The crucial question in the testing procedure for descriptive analogy argumentation is whether the step of extrapolating properties mentioned in the argumentation in line with the standpoint is indeed justified and acceptable to both parties.

It is important to emphasize that in descriptive analogy argumentation: the similarities and differences pointed at must also be relevant to the standpoint defended. Decisive in this regard is the claim that is made in the standpoint. As we have pointed out, in the case of descriptive analogy argumentation the standpoint is some kind of prediction, but underlying this prediction is always a causal claim. In Example (1), for instance, the (descriptive) standpoint is that in Amsterdam camera surveillance in the metro will lead to greater safety. All conditions that are necessary for being able to uphold this causal claim are in this case relevant factors. The commonality between Amsterdam and London that they are both capitals of a country, for instance, is not relevant here, since it is not directly related to the safety of transport. Instead, it may be an important factor that Amsterdam and London are both rather big cities. It is clear that in cases in which the implicit causal claim cannot be so easily reconstructed it will generally be harder to determine exactly which factors are relevant and which factors are not. This predicament may prompt the antagonist to ask the protagonist for a clarification of the standpoint.

The dialectical profile for the testing procedure of descriptive analogy argumentation is as follows:

1 P: Standpoint Y
2 A: Doubt Y?
3 P: Comparison argumentation X
4 A: Basic critical question: is Y comparable with X?
5 P: Answer to basic critical question: Y is indeed comparable with X.
6 A: Additional critical question 1a: Are there relevant similarities between Y and X?
7 P: Answer to additional critical question 1a: There are relevant similarities between Y and X.
8 A: Additional critical question 1b: Are there relevant differences between Y and X?
9 P: Answer to additional critical question 1b: There are differences between Y and X but they are not relevant in this case.
10 A: Additional critical question 1a-b: Do the relevant similarities between Y and X outweigh the relevant differences between Y and X?
11 P: Answer to additional critical question 1a-b: Yes, the relevant similarities between Y and X outweigh the relevant differences between Y and X.

3. VARIANTS OF DESCRIPTIVE ANALOGY ARGUMENTATION

The different ways in which a certain subtype of argumentation manifests itself in various argumentative practices can be viewed as variants of the subtype concerned. In describing these manifestations of argumentative reality in the empirical component of the research program distinguishing between these different variants is an important task. In this endeavour more precise distinctions can for instance be made between variants that differ primarily in the kind of selection from the topical potential (relating to differences in subject-matter), variants that differ first of all in the way they appeal to the audience (relating to different ways of associating with the listeners or readers), and variants that differ first of all in the choice of presentational devices (relating to differences in the means of expression) (van Eemeren and Garssen, 2019: Section 5). In descriptive analogy argumentation variants can also be distinguished based on the kind of standpoint that is defended: a straight prediction, a hypothetical predication or a quasi-prediction.

In this contribution we start making an inventory modestly by distinguishing between two variants of descriptive analogy argumentation. The first variant that can be found regularly in political argumentation and is connected with the use of pragmatic argumentation. Pragmatic argumentation is based on the causal claim that following the course of action proposed will lead to desirable (or – in the case of negative pragmatic argumentation – undesirable) consequences. In argumentative practice this causal claim, which can be seen as a hypothetical prediction,[2] is in communicative activity types in the political domain prototypically defended by descriptive analogy argumentation. Although we have not conducted any quantitative research to check this claim, we expect that in the political domain the use of this variant of descriptive analogy argumentation will even be stereotypical.[3]

Take the defence of the prescriptive standpoint that the United States should adopt a policy of gun control by means of pragmatic argumentation involving the causal claim that doing so leads to a safer social environment. This causal claim can be readily defended by means of descriptive analogy argumentation in which the situation in the United States is compared to that in Canada, where the introduction of gun control has indeed led to fewer casualties. Carrying out this particular defence results in the creation of a prototypical argumentative pattern consisting of pragmatic argumentation defended by the use of descriptive analogy argumentation in a subordinative argument structure.[4]

A second context-determined variant of descriptive analogy argumentation prototypically occurs in the academic domain. In evolutionary biology, to take a case that illustrates our point, the standpoint that a certain extinct animal has certain characteristics can be defended by comparing this animal to much better-known animals that lived much later. We quote in (3) an example from a scientific

---

[2] If the proposed course of action is chosen, desirable consequences will follow.
[3] For the notion of "stereotypical" argumentative pattern see van Eemeren (2018: 165-167).
[4] For the notions of "argumentative pattern" and "prototypical argumentative pattern" see van Eemeren (2018: 149-154).

discussion about the anatomy of Archaeopteryx (a forerunner of the birds). The palaeontologist Heilmann argues that Archaeopteryx probably had a patagium – a fold of featherless skin – connecting the inner wing to the side of the body, analogous to flying mammals:

(3)   Such a fold of skin is the first to appear when the evolution of a "flying" mammal sets in, and therefore it does not seem unlikely that an incipient patagium was present in some forerunner of birds, in due time giving place to the fully developed wing of feathers (Shelly, 2003: 43).

In other words: since we know that some flying mammals are equipped with a featherless skin connecting their wings to the body it is not unlikely that forerunners of the birds also had such a flying skin.

4. NORMATIVE ANALOGY ARGUMENTATION

In the second subtype, normative analogy, the *principle of consistency* plays a central role. In this kind of analogy argumentation it is claimed that a something mentioned in the standpoint belongs to the same category as something mentioned in the argumentation advanced in its support, and that the former should be treated in the same or a similar way as the latter.

Just like in the descriptive subtype, the arguer claims in this kind of analogy argumentation that what is mentioned in the standpoint is comparable to what is mentioned in the premise. This happens, for instance, in the normative analogy argumentation advanced in Example (4).

(4)   The employees in the administration department should get a salary raise because the salespersons in our firm also get a salary raise.

The argument scheme for normative analogy argumentation is as follows:

	Y is appropriate for X
because:	Y is appropriate for Z
and:	Z is comparable to X

Normative analogy argumentation differs from descriptive analogy argumentation because the use of the principle of consistency does not involve an extrapolation of characteristics. Instead, the central issue is whether the two elements that are compared really belong to the same category. Another difference is that in normative analogy argumentation the standpoint is by definition normative whereas in descriptive analogy argumentation it is descriptive. Normative analogy argumentation invariably involves a call for consistent behaviour in the sense that the standpoint always claims that for the sake of consistency something should be treated in a certain way.

Just like in descriptive analogy argumentation, it is in normative analogy argumentation always presumed that there are relevant similarities between what is mentioned in the standpoint defended and what is mentioned in the argumentation advanced in its defence. Again, these similarities are not so much mentioned in the premises that make up the analogy argumentation, but come to the fore in the dialectical testing procedure. Although the critical questions going with the two subtypes of analogy argumentation are to a large extent identical, the dialectical testing procedure proceeds in a different manner.

A necessary preliminarily step in testing normative analogy argumentation is, again, to establish whether what is mentioned in the standpoint defended is really comparable to what is mentioned in the argumentation. After it has been established that this is indeed the case, the antagonist can ask the protagonist of a normative analogy to mention the similarities that justify the claim that the two elements compared really belong to the same category. The protagonist is then forced to present additional argumentation in response. In case the antagonist has pointed at differences which show that the elements that are compared do not belong to the same category, the protagonist's response has to make clear that the differences are either not relevant or are outweighed by the similarities. Again, like in testing descriptive analogy argumentation, it has to be determined in evaluating normative analogy argumentation whether the similarities and differences that have been observed in the argumentation are relevant to the claim that is made in the standpoint.

An additional critical question to be answered in evaluating normative analogy argumentation relates to the application of the principle of consistency. There are sometimes reasons why this principle is not pertinent in the case at hand so that the arguer may refuse to apply it. This is so when the normative argumentation refers to circumstances where requiring consistency would be unproductive or even absurd. This could, for instance, be the case if it would be argued that the same raise in salary should be given to the people serving in the canteen as was given to the employees that designed the product that determined the company's commercial success. Hence an extra critical question must be added that pertains to the need to apply the principle of consistency: is there in this case a special reason why the principle of consistency should not be applied?

Normative analogy argumentation has some crucial characteristics in common with the form of reasoning that Govier (1987) calls *"a priori* reasoning". In both cases the persons, groups or institutions responsible may, for instance, be prompted to act consistently by being confronted with a critical question based on the following observation: "You should do X, because in a similar situation you would also do X". According to Govier, it is an essential property of *a priori* reasoning that the case advanced in making the comparison can be fictitious in the sense that in reality the similar situation mentioned in the comparison does not necessarily exist.

The consistency issue is crucial to all normative analogy argumentation. The basic idea is that you *should do x now* because you *acted in the same way in a very similar case*. Doing otherwise would make your attitude automatically inconsistent. Of course, in argumentation – and perhaps in human communicative interaction in general – the principle of consistency always plays a role. However, when it comes to normative analogy argumentation this principle is applied in a special way because it is not the consistency of the propositions that are part of the argumentation that is at issue, but the consistency of these propositions with other acts and judgments which are not part of what is mentioned in the argumentation. The decisive role of this specific consistency requirement (and the different kinds of critical questions ensuing from it) constitutes the biggest difference between normative and descriptive analogy argumentation.

In normative analogy argumentation the problem of the relevance of similarities and differences between what is stated in the argumentation and what is stated in the standpoint is more complex than in descriptive analogy argumentation. The standpoint defended by descriptive analogy argumentation always involves some kind of prediction about a certain state of affairs or event. Since the standpoint defended by normative analogy argumentation does not involve any such prediction, there is in normative analogy argumentation, unlike in descriptive analogy argumentation, no assumption of a causal relation. In the case of normative analogy argumentation the question of the relevance of similarities and differences between what is stated in the argumentation and what is stated in the standpoint is directly related to the notion of consistency and the need for being consistent. In order to determine whether certain similarities and differences are relevant, one has to check whether the similarities or differences observed are indeed pertinent to the normative judgment conveyed in the standpoint at issue.

We may take it that in example (4) the fact that both groups of employees work in the same firm constitutes a relevant kind of similarity. Since these differences are directly related to the question whether a salary raise is appropriate, observations like that the one group of employees is more productive than the other and has, unlike the other group, already for some time not been given a salary raise would refer to relevant differences. A difference however such as that most employees in the administration department are locals whereas those in the sales department are from out of town would probably not be relevant because being a local is not a factor likely to be of any importance in making decisions about salary raises.

The dialectical profile for the testing procedure of normative analogy argumentation is as follows:

1   P: Standpoint Y

2   A: Doubt Y?

3   P: Comparison argumentation X

4   A: Basic critical question: is Y comparable with X?

5   P: Answer to basic critical question: Y is indeed comparable with X.

6   A: Additional critical question 1a: Are there relevant similarities between Y and X?

7   P: Answer to additional critical question 1a: There are relevant similarities between Y and X.

8   A: Additional critical question 1b: Are there relevant differences between Y and X?

9   P: Answer to additional critical question 1b: There are differences between Y and X but they are not relevant in this case.

10   A: Additional critical question 1a-b: Do the relevant similarities between Y and X outweigh the relevant differences between Y and X?

11   P: Answer to additional critical question 1a-b: Yes, the relevant similarities between Y and X outweigh the relevant differences between Y and X.

12   A: Additional critical question 2: Should the principle of consistency be applied in this case?

13   P: Answer to additional critical question 2: Yes, the principle of consistency should be applied in this case.

5. VARIANTS OF NORMATIVE ANALOGY ARGUMENTATION

Just like in the case of descriptive analogy argumentation, within normative analogy argumentation a distinction of variants can be made that is based on its specific uses, forms and appearances. The variants that can be distinguished depend primarily on the specific way in which the principle of consistency plays a role.

In a well-known variant of normative analogy argumentation the way in which the principle of consistency is used boils down to employing it as "the rule of justice". According to this rule, persons, groups and institutions which belong to the same category should be treated in the same way or at least similarly. This results in a legal variant of normative analogy argumentation which plays an important role in de macro-context of law when a judicial decision is defended by reference to a precedent that belongs to the same category as the case at issue.

This variant of normative analogy argumentation was for instance used in 2018 when the then 69 years old Dutch celebrity Emil Ratelband asked the court to adjust his date of birth to his emotional age, which he claimed to be 49. Because neither law nor jurisprudence provided any grounds for having this request granted, the counsel for Ratelband supported it by means of analogy argumentation. The lawyer asked the court to apply the existing legal regulations for a name change (Article 1: 4 and 1: 7 BW[5]) and for gender change (Article 1:28 and further BW) analogically to the desired adjustment of the date of birth of his client. The lawyer argued that name, gender and age are all part of a person's identity and if name and gender can be changed, this should also be possible for age. *A fortiori*, because a change of age is less drastic than a change of gender, if the more is permitted, the less is also allowed.

The rule of justice also plays a role in another variant of normative analogy argumentation, which is based on a manifestation of the principle of consistency known as the "principle of reciprocity", which is prototypically used in the interpersonal domain. See, for instance, Example (6).

(6)   Since I just helped you with your homework, you should now help me doing the dishes.

---

[5] BW refers to the *Burgerlijk Wetboek*, the Dutch Civil Code.

We conclude this brief introduction of some variants of normative argumentation by noting that normative analogy argumentation can also be used as a technique of refuting arguments that is sometimes called *parallel reasoning* (Juthe, 2009). Parallel reasoning boils down to refuting in the case of two arguments which are structurally similar one of the arguments by showing that the other one is flawed. This variant of normative analogy argumentation occurs prototypically in discussions about social issues. A clear instance is Example (7), which stems from Prince Philip, the Duke of Edinburgh, who reacted to the demand for banning handguns in Great Britain after the shooting of 28 school children in March 1996 in Dunblane, Scotland:

(7)   If a cricketer, for instance, suddenly decided to go into a school and batter a lot of people to death with a cricket bat, which he could do very easily, are you going to ban cricket bats?[6]

6. CONCLUSION

In this contribution we have made a distinction between descriptive and normative analogy argumentation and we have also mentioned some variants of each of these two subtypes that are to a large extent context-specific. We do not claim that our description of subtypes of analogy argumentation is exhaustive, but we do believe that they cover most occurrences of analogy argumentation.

A distinctive feature of our overview is that so-called *figurative analogy* has been left out. As we have argued elsewhere (Garssen, 2009; van Eemeren and Garssen, 2014) figurative analogy should not be seen as subtype of comparison argumentation. In a figurative analogy things are compared that stem from completely different spheres of life. In such a case a metaphorical relation is established between what is mentioned in the standpoint and what it is compared with in the argumentation. This is a form of indirectness which calls for a reconstruction in the analysis that makes clear that the relationship between the argumentation and the standpoint is not analogical but either causal or symptomatic.

Although our discussion of variants of the two subtypes has remained limited to a few contextually-determined variants, we think to have indicated what kind of variants may be expected to be found. In extending the overview many more variants should be described, and in much more detail. In this endeavour the institutional constraints are to

---

[6] This example is taken from Shelly (2002: 1-2).

be taken into account that go with the specific types of argumentative interaction in particular communicative activity types in the specific domains in which the argumentative discourse takes place.

REFERENCES

Eemeren, F. H. van (2018). Argumentative patterns viewed from a pragma-dialectical perspective. In F. H. van Eemeren (Ed.), *Prototypical argumentative patterns: Exploring the relationship between argumentative discourse and institutional context* (pp. 7-29). Amsterdam/Philadelphia: John Benjamins.
Eemeren, F. H. van, & Garssen, B. (2014). Argumentation by analogy in stereotypical argumentative patterns. In H. J. Ribeiro (Ed.), Systematic approaches to argument by analogy (pp. 41-56). Cham etc.: Springer.
Eemeren, F. H. van, & Garssen, B. (2019). Argument schemes. Extending the pragma-dialectical approach. In F. H. van Eemeren & B. Garssen (Eds.), From argument schemes to argumentative relations in the wild (Chapter 2). Cham etc.: Springer.
Eemeren, F. H. van, & Grootendorst, R. (1992). Argumentation, communication, and Fallacies: A pragma-dialectical perspective. Hillsdale NJ etc.: Lawrence Erlbaum.
Eemeren, F. H. van, & Snoeck Henkemans , A. F. (2017). Argumentation: Analysis and evaluation. New York/London: Routledge.
Garssen, B. (2009). Comparing the incomparable: Figurative analogies in a dialectical testing procedure. In F. H. van Eemeren & B. Garssen (Eds.), Pondering on problems of argumentation: Twenty essays on theoretical issues (pp. 133-140). Cham etc.: Springer.
Govier, T. (1987). Problems in argument analysis and evaluation. Dordrecht/Providence: Foris.
Juthe, A. (2009). Refutation by parallel argument. Argumentation, 23(2), 133–169.
Shelly C. (2002). Multiple analogies in science and philosophy. Amsterdam: John Benjamins.

An indexical characterization of disagreement based on possible worlds semantics

LÉA FARINE
Institute of Cognition and Communication Sciences, University of Neuchâtel, Switzerland
lea.farine@unine.ch

In this paper, I discuss Robert Fogelin's definition of "deep disagreement" as a "clash of propositional frameworks", challenging the view that there are "deep disagreements" as opposed to "normal disagreement" (Fogelin, 1985). For this purpose, I use a possible worlds semantics (Lewis, 1986) to characterize the notion of "propositional framework" as a set of possible worlds and I explain why this perspective leads to question the opposition between "deep disagreement" and "normal disagreement".

KEYWORDS: [belief; belief network; David Lewis; deep disagreement; disagreement; indexical; modal realism; possible worlds semantics; propositional framework; Robert Fogelin]

1. INTRODUCTION

Robert Fogelin, in his canonical paper *The logic of deep disagreement* (1985), defines deep disagreement as resulting from a clash between two propositional frameworks. In his view, in certain rare circumstances, the sources of the propositions constituting the disagreement are such that its rational resolution is impossible. For example, in the context of the debate on the right to abortion, "parties on opposite sides of the abortion debate can agree on a wide range of biological facts-when the heartbeat begins in the fetus, when brain waves first appear, when viability occurs, etc., yet continue to disagree on the moral issue." (Fogelin, p. 6) In this case, still according to Fogelin, the primitive sources of the conflict are located in an opposition between religion and secularism. Indeed, if, in a religious context, one believes that "shortly after conception, an immortal soul enters into the fertilized egg" (Fogelin, p. 6) and if, in a secular context, one does not believe in the immortal soul, then the parties stand in conflicting frameworks. In other words, the source of the disagreement is "a whole system of mutually supporting propositions (and paradigms, models, styles of acting and thinking) that constitute [...] a form of life." (Fogelin, p. 7) In contrast to cases of deep disagreement, Fogelin also explains, "an argument, or better, an argumentative exchange is normal when it takes place within a context of broadly shared beliefs and preferences", that is when the frameworks of the participants are globally shared.

Nevertheless, the distinction between deep and normal disagreement is not always clear. It seems to me that this difficulty is due to the ambiguity of the notion of "propositional framework". What is it? What is the nature of this "system of mutually supporting propositions"(Fogelin, p. 7)? According to Fogelin's paper, one might think that the notion refers to a set of propositions that base some given systems, for example, propositions of the Bible, of the Declaration of the Rights of Man and of the Citizen or propositions contained in constitutions.

However, these sets of propositions are not always precisely determined and localized, do not always constitute a clear source and are not always written. For example, the dream culture of the Australian Aborigines or the symbolic thought of the Middle Ages could be considered as propositional systems in the sense of Fogelin, without however it being possible to define precisely their outlines. They exist, in fact, as shared belief systems, in other words, as belief networks shared by many individuals.

Consequently, if a propositional framework is a network of shared beliefs, then any individual constitutes a particular form of life, because any individual has a particular belief network. Hence, any case of disagreement results from a clash between two propositional frameworks, that is, from a clash between two or more individual belief networks, necessarily dissimilar in some respects and necessarily similar in others. It is therefore difficult to distinguish between deep and normal cases of disagreement. Indeed, an apparent case of deep disagreement may in fact be normal if the belief networks of the disagreeing parties coincide relevantly with respect to that disagreement. On the other hand, an apparent case of normal disagreement may in fact be deep, if the belief networks of the disagreeing parties diverge relevantly with respect to that disagreement.

To illustrate this second possibility, assume that Mia and Laura disagree about the proposition "The speed must be limited to 30km/hour in the city". Mia, because of a whole series of experiences, learning, etc. since her birth, believes and expresses the opinion that speed must be limited. The reasons she gives to support this standpoint is that speed limit statistically minimizes the risk of accidents and that this minimization of risk is worth more than the freedom of the driver. Laura expresses the opinion that speed must not be limited, because unlike Mia, even if she acknowledges that limitation minimizes the risk of accidents, she favours the freedom of the driver. Here, it seems that the disagreement described cannot be rationally resolved, because Laura and Mia stand in two incompatible frameworks (their individual framework). So, "dialogues of the deaf" (Angenot, 2008) are perhaps more frequent than Fogelin claims.

In this paper, I attempt to propose a more inclusive approach to disagreement than the one suggested by Fogelin. In my opinion, if disagreement is always the result of a clash between two propositional frameworks, these frameworks, relevantly to the disagreement in question, may be similar or dissimilar. In other words, there would not be two opposite categories of disagreement, deep disagreement and normal disagreement, but a continuum.

To illustrate this idea, I suggest using a semantics of possible worlds. This semantics has the merit, in my opinion, of being adapted to an accessible and user-friendly characterization of the notion of "propositional framework". From this semantics, I will first show that the propositional framework specific to an agent, i.e. the set of his or her beliefs, can be characterized as a set of possible worlds in which the propositional content of these beliefs is true. Then, I will define disagreement as a difference between two sets of possible worlds, before concluding by saying why such a definition allows us to adopt an indexical point of view on disagreement.

2. DISAGREEMENT SEEN AS A DIFFERENCE OF BELIEFS

Before explaining how the semantics of possible worlds could be useful in defining the notion of propositional framework, I would like to explain why it is relevant to consider these frameworks as belief networks and disagreement as a difference of beliefs, more broadly, as a difference between two parts of belief networks.

It is difficult to deny that disagreement implies ontologically the presence of mental states. Indeed, the very nature of disagreement is to be a difference of beliefs between two subjects. For example, Jean-Blaise Grize writes:

> "The problem [of argumentation] is not through [the] discourse to preserve a supposed truth, but to give to see - more exactly to give to look - plausible representations, i.e. to manipulate belief values." [translation my own] (1996, p. 48)

However, recognizing the triviality of the link between mental states and argumentation does not imply, for the analyst, to consider these mental states at a methodological and metatheoretical level. On the contrary, according to David M. Godden:

> " Those theories of argumentation which hold the goal of persuasive argumentation to be the settling of a difference of opinion by rational means [...], not wanting to get bogged down in a quagmire of psychological considerations, hold that argumentation ends when there is some change in the commitments – rather than the beliefs – of the disputants." (2010, p. 1)

One of those theories is the pragma-dialectics, which adopts as a metatheoretical premise a principle of externalisation. In short, pragma-dialectics analyses the argumentative discourse at the level of commitments attributed to the parties, which must be:

> "(1) externalized by the parties themselves in the discourse,
> (2) externalizable from what has been said in the discourse, or
> (3) on other grounds regarded as understood in the discourse." (Van Eemeren et al., 2014, p. 526)

In other words, adopting an externalization principle for the analysis of argumentation makes it possible to consider the phenomenon in a pragmatic way: argumentation is an observable communicative act aimed at resolving a disagreement considered as a difference of standpoints and not as a difference of beliefs, because the standpoint is external, whereas the belief is not.

Nevertheless, if externalism serves its purpose adequately, the principle is limiting when it comes to explain the nature of disagreement and to understand specific argumentative phenomena, such as deep disagreement. Indeed, if one argues to resolve a difference of opinion, it is also a question, for the agents involved in an argumentation process, of acting both on their own belief network and on that of the other (see e.g. Sperber, 2001). In other words, deep disagreements are deep because the roots of a difference of opinion are cognitive. Therefore, while it is sometimes possible to identify one or other external propositional framework within which this disagreement falls, this is not always the case.

For example, suppose that S1 and S2 disagree about the proposition "Between the Earth and Mars there is a china teapot revolving around the sun in an elliptical orbit" (Van Inwagen, 2012). In this case, the propositional framework in which the disagreement falls is very specific and not describable other than referring to the belief networks of S1 and S2. Thus, if the propositional framework as described by Robert Fogelin is often a system or a form of life globally identifiable and roughly circumscribed (the aboriginal dream culture, the liberal-democratic framework, Christianity), its most elementary unit is a network of shared or unshared beliefs. As well, the most elementary units of disagreement are beliefs and therefore, its most elementary definition is in terms of a difference of beliefs.

3. DISAGREEMENT AND POSSIBLE WORLDS

David Lewis publishes *On the Plurality of Worlds* in 1986. In this book, he proposes a strange thesis: there is an infinity of possible worlds, which realizes all complete and consistent conceptualizations of logical domains alternative to reality. In other words, according to Lewis, "There are so many other worlds, in fact, that absolutely *every* way that a world could possibly be is a way that some world *is*." (Lewis, p. 2)

The possible worlds semantics was developed from the middle of the twentieth century in order to provide the modal logic with an extension principle (for more explanations, see Copeland, 2002). In this view, for example, the meaning of a modal proposition such as "Possibly, all crows are blue" is "In some worlds, all crows are blue", and the meaning of a modal proposition such as "Necessarily, P or non-P" is "In all possible worlds, P or non-P." This way, the definition of possibility and necessity is not circular.

From this point on, the question arises of the ontology of possible worlds. Roughly, there are three opposing conceptions. First, supporters of non-representational abstractionism consider that possible worlds are abstract entities and that this definition is sufficient (see e.g. Van Inwagen, 1986). Second, supporters of representational abstractionism consider that possible worlds are abstract constructions and specify the nature of these constructions, which can be sets of propositions; sets of properties, etc. (see e.g. Adams, 1974 or Heller, 1998). Third, David Lewis proposes the thesis that possible worlds are concrete entities, exactly in the same way that our world is concrete. According to him, possible worlds are strictly separated from each other spatially and temporally, but linked by "counterpart" relations (see Lewis, 1968).

For example, a world W1 and in this world W1, S1 having the property of being Bertrand Russel (and all properties related to the property of being Betrand Russell, such as that of being British, being born on May 18, 1872, etc.). Likewise, a world W2 and in this world, S2 also having the property of being Betrand Russell. In the Lewisian approach, S1 and S2 are not identical (there are numerically two objects having the property of being Betrand Russell), but they are counterparts of each other. Indeed, Bertrand Russell in W2 is the closest individual, in terms of properties, to Bertrand Russell in W1.

Now, a world W1 containing Bertrand Russell (a living being) and a W2 world containing only one goat (a living being) and stones. In W2, the counterpart of Bertrand Russell is a goat, because what most closely resembles Bertrand Russell in W2 is a goat. Moreover, if W1 contains other living beings in addition to Bertrand Russell, the goat is also their counterpart in W2 (the counterpart relation is not an equivalence relation). To simplify, we can say that, in general, worlds have relations of similarity and dissimilarity.

Eventually, a world W, which is the world where we live. This world is "actual" from our perspective. However, from the perspective of the inhabitants of a world W2, the world W2 is the actual world. In other words, the actuality is indexical. Each world is actual for itself and all possible propositions are true relatively to at least one possible world. (for complements and an overview of numerous objections to modal realism, see Menzel, 2017)

3.1 A possible worlds semantics to characterize belief

In *On the Plurality of Worlds* (1986), David Lewis suggests an application of modal realism to doxastic and epistemic modalities, by approaching this second type of modality in a superficial way, because they pose additional difficulties. Like him, I focus here only on doxastic modalities. In addition, I leave aside technicalities and objections and I present only the foundations of the semantics, thus described:

> Like other modalities, [doxastic modalities] may be explained as restricted quantifications over possible worlds. To do so, we may use possible worlds to characterise the content of thought. [...] The content of someone's system of belief about the world (encompassing both belief that qualifies as knowledge and belief that fails to qualify) his given by his class of doxastically accessible worlds. World W is one those iff he believes nothing, either explicitly or implicitly, to rule out the hypothesis that W is the world where he lives.
> Whatever is true at some epistemically or doxastically accessible world is epistemically or doxastically possible for him. It might be true, for all he knows and for all he believes. He does not know or believe it to be false.
> [...] If he is mistaken about anything, that is enough to prevent his own world from conforming perfectly to his system of belief (Lewis, 1986, p.27).

For example, assuming that Laura believes that "Speed should not be limited to 30 km/h" (P), that "Drivers' freedom is worth more than minimizing accident risk" (Q) and that "P justifies Q", there is a set of worlds W where P is true, where Q is true and where P justifies Q[1].

This perspective may seem bizarre, because it forces us to adopt an absolute and almost frantic realism, even about evaluative and deontic judgments. However, it makes sense if we consider the direction of fit of belief, which aims to truth (see e.g. Anscombe, 1957, Searle, 2002, Humberstone, 1992). In this perspective, if Laura sincerely believes that P, she believes that P is true, regardless of the nature of P (epistemic, deontic, evaluative, etc.). In the same way, if Laura supports the standpoint that P, she supports the standpoint that P is true. She could, of course, practice zealous *zététique* (scepticism) and suspend its judgment when the truth of a proposition is not positively ascertainable. Nonetheless, in practice, it seems that subjects tend to believe and optimistically support the truth of all kinds of propositions. In addition, they justify these propositions with other propositions in which they also believe, in the space of their own rationality[2].

Thus, modal realism applied to doxastic modalities preserve the bivalence of belief and explain at the same time the diversity of standpoints, without taking a position on the epistemic nature of justification, since the truth or falsity of propositions in the actual world is left aside. Therefore, the approach is meta-dialectic and minimally normative: as long as an inferential process does not contain a logical contradiction, it is true with respect to at least one possible world.

---

[1] There would be much more to say (and to criticize) about the inference relation between P and Q in W but I leave this thorny question aside for now.

[2] For example, suppose a child goes to his parents and asks them: "I want to kill someone, do I have the right?" or "I want to kill someone, is that wrong?". To this question, it is doubtful, even if the parent is a specialist of argumentation or a philosopher, that he answers: "It depends on the point of view", "It is not right or wrong, but in the current context most people consider it wrong", "Do not do it because it is forbidden by law and you risk going to prison" or "Let me explain why it is not reasonable to be a moral (or deontic) realist". Instead, parents will answer, "Yes, it's wrong" (it's *really* wrong) or "You have no right" (and then they will take their child to the psychiatrist). In other words, if we are forced to recognize that it is difficult to determine the truth-value of certain statements and what bases it, in practice we tend to act as if propositions were true or false.

3.2 A possible worlds semantics to characterize disagreement

From the possible worlds semantics proposed by Lewis for doxastic modality, I suggest characterizing disagreement in the following way. In a world W, S1 and S2 sincerely disagree about P if and only if:

1. In some set of possible worlds W1, P is true.
2. In some set of possible worlds W2, P is not true.
3. S1 expresses P and S2 expresses non-P.
4. S1 has doxastic access to W1 and S2 has doxastic access to W2.
5. S1 and S2 are individuals of the same world.

Note that, on this basis, since the propositions are expressed, S1 has modal access to W2, S2 has modal access to W1 and the observer has modal access to W1 and W2 (in addition, if he/she believes P or not-P, he also has doxastic access to W1 or W2). As well, S1 has epistemic access to some of the content of the belief network of S2, S2 has epistemic access to some of the content of the belief network of S1 and the observer has epistemic access to some of the content of the belief network of S1 and S2.

4. AN INDEXICAL PERSPECTIVE ON DISAGREEMENT

At the beginning of this paper, I explained that, in my opinion, the propositional framework, in its most elementary form, corresponds to the belief network of an individual and that disagreement is a difference of belief. Further, I pointed out that the definition of disagreement as a difference of belief does not seem sufficient, insofar as the content of a belief, from the point of view of the subject, is not only plausible, it is true. Moreover, this is not only the case for factual judgments, but also for evaluative judgements, deontic judgements, etc. Therefore, from the point of view of subjects disagreeing about a proposition such as "shortly after conception, an immortal soul enters into the fertilized egg" (Fogelin. 1985, p. 6), there is no difference in nature between the proposition and its negation. Indeed, in the set of worlds W1 doxastically accessible to S1, the proposition is the case and is therefore a judgment of fact from the point of view of W1 where this judgement is true. As well, in the set of worlds W2 doxastically accessible to S2, the proposition is not the case and is therefore, in its negative form, a judgment of fact from the point of view of W2.

In addition, since there is no essential difference between P and non-P, there is no essential difference neither between the architecture of the justification of P and non-P in W1 and W2. For example, in the set of worlds W1 where it is case that "shortly after conception, an immortal soul enters into the fertilized egg", then it is also the case that God exists, that he has revealed a message through the Bible writers, etc. In the set of W2 worlds where this proposition is not the case, then it is not the case, for example, that God exists, or that he has revealed a message, etc.

Consequently, the difference between the set of worlds W1 and the set of worlds W2 that characterize a sincere disagreement (i.e. when expressed propositions are contained in the beliefs of the subjects) is a difference of properties, since in one of these worlds, P is the case and in the other, P is not the case. From this difference, S1 and S2, by explicitly expressing propositions that justify P or non-P, exhibit other properties (other facts) of W1 and W2 that they consider causally related to the propositions they support. This causal relation is true in W1 and W2, assuming that the subjects believe that there is a causal relation between the propositions they defend and their justifications. Therefore, from the difference between W1 and W2 expressed in the disagreement, subjects exhibit properties of W1 and W2 relevant to P and non-P. Note that relevance, in this definition, is also indexical, i.e. relative to W1 and W2. For example, if a subject sincerely believes that "shortly after conception, an immortal soul enters into the fertilized egg" because "God exists" and "God sends an immortal soul into the egg", these justifications are relevant to W1, since in W1, God sends an immortal soul into the fertilized egg.

In this perspective, since the worlds W1 and W2 are discernible worlds, any case of disagreement is the result of a "clash" of worlds or the result of something else than a "clash", depending on how one considers a difference of properties. This approach is indexical, because the use of the possible worlds semantics allows to index the truth of the propositional content of a belief not to a certain context, a form of subjectivity, a norm, etc. in the actual world (which is logically problematic), but to a set of possible worlds in which this content is true. The depth or superficiality of the disagreement would therefore be connected to the extent rather than the kind of differences between the worlds in which the propositions constituting the disagreement are true.

5. CONCLUSION

The conception of disagreement as presented in this paper may seem strange. It is because it is. Moreover, the problems it raises are many and perhaps insurmountable; when people argue, do not they talk about their world and not about another one? How can we reasonably believe in an infinity of possible worlds that physically exist? How to distinguish a reasonable argument from a fallacious one? How is the disagreement resolved?

However, it has the merit, in my opinion, of reflecting the idea of ordinary sense that disagreements arise when and because individuals "do not live in the same world". In a certain way, this is true. They live physically in the actual world, but doxastically in worlds that are sometimes very far from it and very far from each other. Moreover, from a methodological point of view, the indexical approach helps to distinguish the question of epistemic truth from that of doxastic truth. Of course, the analyst of argumentation knows that the Earth is ellipsoid, that the existence of an immortal soul is not provable in the same way as the biological processes involved in the development of the foetus or that Donald Trump often lies. Yet some subjects with an operational cognitive system believe that the Earth is flat, that the existence of an immortal soul is provable in the same way as the biological processes involved in the development of the foetus or that Donald Trump is trustworthy, with reasons and reason (even though they are wrong).

Furthermore, it seems to me that individual belief systems are rarely homogeneous. Indeed, it often happens that two systems contradict each other even though the argumentative exchange "takes place within a context of broadly shared beliefs and preferences" (Fogelin, 1985, p. 4). For example, a Christian may believe that God created the world, but not in seven days, a biologist may believe in the reality of biological processes while believing that the love he/she has for his/her friends is not reducible to chemical attachment processes or an evolutionary function, etc.

In conclusion, I would like to echo Marc Angenot's incipit of *Dialogues de sourds* (quoting Saint-Jérôme speaking of the controversies between Christians and pagans): "We judge each other the same way: we seem crazy to each other" (Angenot, 2008, p. 7). However, I do not think we are all crazy. Why, how are we not crazy? Because our reason is so broad, that it embraces, to a certain extent, the diversity of *possibilia*.

REFERENCES

Adams, R. M. (1974). Theories of actuality. *Noûs*, 8, 211-231.
Angenot, M. (2008). *Dialogues de sourds: traité de rhétorique antilogique*. Paris: Mille et une nuits.
Anscombe, G. E. M (1957). *Intention*. Blackwell: Oxford.
Copeland, B. J. (2002). *The genesis of possible worlds semantics*. Journal of Philosophical logic, 31(2), 99-137.
Fogelin, R. (1985). *The logic of deep disagreements*. Informal Logic, 7(1), 3-11.

Godden, D. M. (2010). *The importance of belief in argumentation: belief, commitment and the effective resolution of a difference of opinion.* Synthese, 172(3), 397-414.
Grize, J.-B. (1996). *Logique naturelle et communication.* Paris: Presses Universitaires de France.
Heller, M. (1998). *Five layers of interpretation for possible worlds.* Philosophical Studies, 90(2), 205-214.
Humberstone, I. L. (1992). *Direction of fit.* Mind, 101(401), 59-83.
Lewis, D. K. (1986). *On the Plurality of Worlds.* Blackwell: Oxford.
Lewis, D. K. (1968). *Counterpart theory and quantified modal logic.* The Journal of Philosophy, 65, 113-126.
Menzel, C. (2017). "Possible Worlds". *The Stanford Encyclopedia of Philosophy,* Edward N. Zalta (Ed.).
URL: <https://plato.stanford.edu/archives/win2017/entries/possible-worlds/>.
Searle, J. R., & Willis, S. (2002). *Consciousness and language.* Cambridge: Cambridge University Press.
Sperber, D. (2001). *An Evolutionary Perspective on Testimony and Argumentation.* Philosophical Topics, 29(1/2), 401-413.
Van Eemeren, F. H., Krabbe, E. C. W., & Henkemans, A. F. S. (2014). *Handbook of Argumentation Theory.* Dordrecht etc.: Springer.
Van Inwagen, P. (2012). Russell's China Teapot. In D. Łukasiewicz & R. Pouivet (Eds.), *The Right to Believe. Perspectives in Religious Epistemology* (pp. 11-26.). Ontos verlag: Heusenstamm.
Van Inwagen, P. (1986). *Two concepts of possible worlds.* Midwest Studies in Philosophy, 11(1), 185-213.

The visual rhetoric of iconic photographs as *topoi* in editorial cartoons: an argumentative analysis

EVELINE FETERIS
*Department of Speech Communication,
Argumentation Theory and Rhetoric
University of Amsterdam*
e.t.feteris@uva.nl

This paper describes the visual rhetoric of the use of iconic photographs as *topoi* in editorial cartoons and indicate how the audience can reconstruct the argumentative message underlying the cartoon. With the aid of a model that is developed for the reconstruction of the argumentation underlying cartoons in which an iconic photograph is used as *topos* an exemplary analysis is given of the visual rhetoric and the underlying argumentation of an editorial cartoon.

KEYWORDS: argumentation, argumentative activity, argumentative pattern, editorial cartoon, iconic photograph, *topos*, visual metaphor, visual communication, visual rhetoric

1. INTRODUCTION

In public discourse, editorial cartoons can be considered as an argument criticizing a current event. It is claimed that a particular situation or behaviour should be evaluated negatively because it is not in accordance with certain values of a society.[1] To represent these values, a cartoonist may refer to a certain image, or *topos*, that is considered as a visual symbol for certain commonly shared values. To express his critique in a visual way, the cartoonist uses visual rhetorical techniques. These techniques consist in performing certain changes in the symbolic image of the *topos*.

An example of a reference to a *topos* and the changes that are made to convey the critique can be found in the cartoon in Figure 1. This cartoon by the Dutch political cartoonist Joep Bertrams criticizes the behaviour of Donald Trump during his meeting with the communist Chinese leader Xi on November 9, 2017. The *topos* is the famous 'iconic' photograph of the socialist fraternal kiss made by Régis Bossu on

---

[1] See Feteris (2013, 2019) and Feteris, Groarke, Plug (2011).

October 4, 1979 of the meeting of Soviet leader Brezhnev and GDR leader Erich Honecker (Figure 2).[2] The way in which the elements of the original photograph are changed provide a clue for the interpretation of the negative message. To reconstruct the argument, the audience will have to translate the visual rhetoric of these changes in terms of components of the claim and supporting arguments.

Figure 1. Joep Bertrams editorial cartoon '#Xi Too' https://www.cagle.com/joep-bertrams

---

[2] For a discussion of the concept of iconic photographs see Hariman & Lucaites (2007), Kjeldsen [2017], Melching [2019], Paul [2011], Perlmutter [1998]. For a discussion of the photo of the socialist fraternal kiss see Kleppe [2013].

Figure 2. Régis Bossu, *Paris Match* October 4, 1979, 'The socialist fraternal kiss'

In research of multi-modal and visual argumentation in editorial cartoons, authors such as Edwards and Winkler (1997) and Schilperoord (2013) have addressed the visual rhetoric of the use of the iconic photograph of Iwo Jima in editorial cartoons to convey critique with regard to current events. However, they have not addressed the question of how the visual rhetoric is used to present aspects of a particular claim and supporting arguments.

In this paper I will describe the visual rhetoric of the use of iconic photographs as *topoi* in editorial cartoons and indicate how the audience can reconstruct the argumentative message underlying the cartoon. To this end, in section 2 I discuss the use of iconic photographs as *topoi* in editorial cartoons and the characteristics of the visual rhetoric in these cartoons. In section 3 I present a model for the reconstruction of the argumentation underlying cartoons in which an iconic photograph is used as *topos*. With the aid of this model, in section 4, I given an exemplary analysis of the visual rhetoric and the underlying argumentation.

2. THE VISUAL RHETORIC OF REFERENCES TO ICONIC PHOTOGRAPHS AS *TOPOI* IN EDITORIAL CARTOONS

Iconic photographs can be used as part of the visual rhetoric of editorial cartoons. Like other well-known images (such as biblical images) which have a symbolic value, they can be used as a *topos*. This means that the cartoonist can re-use and exploit the image because it represents a well-known commonplace that is shared within a particular community. Edwards and Winkler (1997) and Schilperoord (2013) explain how iconic photographs can be re-used in an afterlife as *topoi* in editorial cartoons to evaluate contemporary events.[3]

The possibility to re-use iconic photographs as commonplace is based on certain characteristics with regard to the symbolic form and content of the original photo. Iconic photos have a recognizable symbolic form, like certain well-known biblical images, consisting of a combination of visual features and a narrative structure. These features and narrative structure make use of symbolic attributions. These symbolic attributions constitute the symbolic content that represents the essence of certain cultural beliefs and ideals. These characteristics make it possible to use the photo as a moral standard to judge contemporary events in the *afterlife* of the photo in other forms of visual communication.

The photo as moral standard is used as a *topos* in the visual rhetoric in editorial cartoons. This visual rhetoric makes use of certain graphic changes. These changes concern the insertion of new elements, the substitution of existing elements, the removal of existing elements and the distortion of existing elements. The aim of the changes is to express critique on the behaviour or event in light of certain common values embodied by the *topos* of the photo.

3. THE RECONSTRUCTION OF THE ARGUMENTATION IN EDITORIAL CARTOONS WITH AN ICONIC PHOTOGRAPH AS *TOPOS*

The question has to be answered how the visual rhetoric of the changes in the symbolic form of the *topos* of an iconic photo can be interpreted in terms of elements of an argumentative message supporting a standpoint that criticizes the behaviour of a public official. To this end it has to be established how the message conveyed by the visual rhetoric

---

[3] For a more detailed discussion and analysis of the visual rhetoric of the iconic Iwo Jima photograph by Joe Rosenthal and its re-use as *topos* in editorial cartoons see Edwards & Winkler (1997) and Schilperoord (2013). For other discussions of this iconic photograph see Bertelsen (1989) and Hariman & Lucaites (2002).

that refers to an iconic photograph as *topos* can be translated in terms of certain parts of the argumentation.

As is indicated in Feteris (2019), editorial cartoons are a specific argumentative activity type that is governed by certain conventions and constraints. In an editorial cartoon the behaviour of a politician, public official, institution or situation is criticized in light of certain common norms that are shared by the cartoonist and his audience. The critique is presented in a specific way, that is by means of a particular image containing certain forms of visual rhetoric such as a visual metaphor, a hyperbole, etcetera, often in combination with verbal elements. In an editorial cartoon the cartoonist can use various graphic techniques that create possibilities to formulate the critique in an indirect and often humorist way and to leave the interpretation to the audience.

In Feteris (2019) it is explained how the complete argumentation can be reconstructed on the basis of an analysis of the visual rhetoric.[4] The argumentative pattern of Figure 3 represents the commitments of a cartoonist who gives a negative evaluation of the behaviour or event X on the basis of certain characteristics Y that conflict with value W.[5]

1   The behaviour/event X must be evaluated negatively
1.1a   The behaviour/event X has characteristics Y1, Y2 etc.
1.1b   The characteristics Y1, Y2 etc. of the behaviour/event X must be evaluated negatively
    1.1b.1a   Characteristics Y1, Y2 etc. conflict with value W
    1.1b.1b   Value W is a generally accepted value

Figure 3. Argumentative pattern based on a symptomatic relation in an editorial cartoon

In an editorial cartoon, the largest part of the complex argumentation remains implicit. The cartoon contains certain visual and verbal information about the behaviour/situation X, as well as a characterization of X with characteristics Y and the negative evaluation of these characteristics Y. The value W that is considered as an accepted value is left implicit.

---

[4] Feteris (2019) also describes a similar argumentative pattern for causal argumentation in which a particular behaviour or policy is evaluated negatively because it will lead to certain negative consequences.

[5] For a discussion of the concept of an argumentative pattern and the discussion of various prototypical argumentative patterns in different institutional contexts see van Eemeren (2017).

The content of the characteristics Y, the negative evaluation of Y as well as the content of the values W must be reconstructed by the reader on the basis of an analysis of the visual rhetoric, in combination with certain visual and verbal information provided in the cartoon. Often the cartoon does not contain information about the negative evaluation and the values or goals because the audience and the cartoonist are supposed to share certain common values. Because these values are tacitly shared, the expression of the criteria for evaluating Y in a negative way on the basis of W can be considered as superfluous because they concern the tacit common values of a particular audience or culture. Such a reconstruction by the reader is possible on the basis of the shared knowledge of the conventions of the genre of the cartoon.

In light of the characteristics of the visual rhetoric of editorial cartoons that refer to an iconic photo as topos (section 2), for editorial cartoons with an iconic photograph as *topos* the following characteristics of the sub-genre apply, where (1) concerns the *symbolic form* and the changes performed in the original image, and (2) the *symbolic content*:[6]

(1) The negative evaluation of the characteristics Y of X (in argument 1.1b) is represented by deviation of the *symbolic form* of the image of the iconic photo performed by means of the visual changes by the cartoonist in the original image of the *topos* of the photo.

(2) The value W on the basis of which the characteristics Y are evaluated as negative because they conflict with this value (as stated in argument 1.1b.1a) is the *symbolic content*, represented by the *topos* of the original iconic photograph (that may change over time).

With regard to (1), the negative evaluation of the characteristics Y of X that form part of the symbolic form, this evaluation must be reconstructed on the basis of the way in which the cartoonist has presented the symbolic form and the original narrative structure with the elements of the photo. Characteristic for the genre of cartoons based on an iconic photo is that certain changes are made. The operations

---

[6] For a discussion of the role of *topoi* in editorial cartoons as argumentative message see Feteris, Groarke & Plug (2011) who give a reconstruction of the function of the *topos* as part of the complex argumentation. See also Schilperoord (2013:2007) who characterizes the role of an iconic photo as *topos* as the minor premise in an enthymematic argument. See Feteris (2013) for an argumentative analysis of allusions to cultural sources in general in editorial cartoons.

described by Schilperoord (2013) such as insert, substitute, remove and distort that perform certain changes in the original image indicate the way in which the reader is supposed to interpret the message from the cartoonist.

When the elements X, Y1, Y2 etc., and W have been reconstructed on the basis of the analysis of the visual rhetoric (in combination with certain verbal elements), the complete complex argumentation can be reconstructed on the basis of the argumentative pattern of Figure 1 for editorial cartoons. The way in which the arguments can be formulated depends on the shared background knowledge of the audience and the cartoonist with regard to the historical and factual knowledge, the knowledge of the symbolic content of the value attached to the *topos* of the photo as well as the values W that are referred to by the cartoonist. As Edwards & Winkler (1997) indicate, these values are often presented on a high level of abstraction so that the audience can give its own interpretation of these values in the concrete case. The argumentative commitments reconstructed on the basis of the model define the interpretation space for reconstructing the content of the argumentation in light of the interpretation of the visual rhetoric of references to iconic photos as *topos*.

4. EXAMPLARY ARGUMENTATIVE ANALYSIS OF AN EDITORIAL CARTOON WITH AN ICONIC PHOTOGRAPH AS *TOPOS*

By using the model for the reconstruction, I give an exemplary analysis of an editorial cartoon with an iconic photograph as *topos*. For my exemplary analysis I have chosen the iconic photograph by Régis Bossu of October 4, 1979, also known as 'The socialist fraternal kiss', and its afterlife in an editorial cartoon of Joep Bertrams referred to in earlier parts of this contribution.

4.1 The iconic photograph of 'The socialist fraternal kiss' and its afterlife

The iconic photograph of 'The socialist fraternal kiss' was made by the French photographer Régis Bossu at a festive meeting after the 30th annual celebration of the German Democratic Republic's foundation as a Communist republic on October 4, 1979. The photo shows Leonid Brezhnev, General Secretary of the Central Committee of the Communist Party of the Soviet Union (left) and Erich Honecker, General Secretary of the Socialist Unity Party of the German Democratic Republic (right) engaged in a kiss. The kiss is known as the Socialist fraternal kiss which was a form of greeting between statesman of Communist countries, demonstrating the special connection that exists between Socialist states. The photograph was first published in the *Paris Match* and later in *Stern, Bunte* and *Time*. It was reproduced as an image in print and became widespread.

The photograph was re-used by Dmitri Vrubel as a painted representation in the form of a mural in the East Side Gallery in Berlin on the remains of the Berlin Wall. Vrubel entitled his painting 'My God, Help Me to Survive This Deadly Love'. The mural was not made as a direct reaction to a particular event, but was a re-popularization of the image in 1990, 11 years after its publication. It was after the re-use of the photo in the East Side Gallery that the image became an iconic image that was given new re-interpretations. The symbolic value of the original photograph was 'neutral' in the sense that it was published as a news picture of a ritual meeting of two Communist leaders embraced in a greeting in the form of the traditional Socialist Fraternal Kiss.[7] However, since the GDR was dependent on the Soviet Union, the kiss was not a kiss between equals but between the GDR as a dependent 'satellite' state of the Soviet Union. In light of later historical developments the kiss can be seen as a 'Judas kiss'.

The cartoon '#Xitoo' made by the Dutch cartoonist Joep Bertrams comments on the behaviour of the US president Donald Trump during the meeting with the Chinese leader Xi Jinping on November 9, 2017. The cartoon was sent as a twitter message with the commenting text 'Sudden Big Friend #Trump # 'XiJinping #MeToo' ('Plotselinge Grote Vriend #Trump # 'XiJinping #MeToo') on November 11, 2017.[8] The cartoon shows the US president Donald Trump (on the left) and the Chinese leader Xi Jinping (on the right).

The meeting of Trump and Xi followed a period of tension between the US and China. Trump had criticized the economic policies of China because they would damage the interests of the US. The meeting with Xi concerned trade negotiations. Apart from the trade issue, for Trump an important issue was to get Xi's help in stopping the nuclear threat of North Korea.

During the short meeting of November 9, Trump gave an eight-minute address in the Great Hall of the People (the ceremonial heart of Communist Party Rule) in which he praised Xi and asked him to 'act

---

[7] See Kleppe (2013) for a discussion of the reactions in the West where such a kiss was seen as something unusual as a way of greeting of leaders, whereas in Socialist countries this was a normal way of greeting between statesmen.

[8] The primary audience of the twitter message were Dutch followers of Joep Bertrams (who can understand the Dutch text 'Plotselinge grote vriend'). Later the cartoon was published on Bertrams site https://www.cagle.com/joep-bertrams that is also accessible for an international audience. The image and the caption of the verbal text '#Xi Too' are accessible for a wider international audience.

faster and more effectively' to extinguish North Korea's nuclear 'menace' and said that he was convinced of Xi's capacities to rein in North Korea's weapons programmes.

The cartoon by Joep Bertrams makes use of the symbolic form and content of the original photo and uses these aspects of the *topos* as a vehicle to convey the critical message. The meaning of the symbolic form of the image and the changes Bertrams makes in the original form constitute the source domain of the visual metaphor that he uses to present his critique. The arguments in support of his critique can be reconstructed on the basis of the changes that have been made in the original image.

The message of Bertrams is that the behaviour of Trump in his address is improper. The US and China are rivals from a geopolitical, economical and ideological perspective. Trumps exuberant speech is not in line with the conventions of the reserved way in which American leaders used to behave towards Chinese leaders. The behaviour of Trump is also inconsistent with Trumps' earlier critical comments on China's economic policy. Furthermore it is considered as exaggerated from the perspective of these conventions and Trumps earlier behaviour.

4.2 Analysis of the visual rhetoric of the re-use of the iconic photograph 'The socialist fraternal kiss' as topos in the '#Xi too' cartoon

In the analysis of the visual rhetoric it must be established how the original image of the *topos* and the changes that are performed can be interpreted in terms of the source domain of the visual metaphor and how the target domain can be interpreted in light of the knowledge of the actual event.

The reference to the original photo of the 'Socialist fraternal kiss' is the basis for the way in which the source domain of the *topos* of the iconic photo must be interpreted. As is indicated in section 4.1, the *topos* of the 'Socialist fraternal kiss' represents the way in which leaders of the former communist countries used to greet each other at official meetings. The extra meaning of the *topos* in its afterlife is that the message conveyed by the *topos* is also a visual representation of 'Deadly love', love out of hate, as a Judas kiss. The cartoon also refers to other forms of re-use of the photo as *topos* in which Trump is depicted, mouth-kissing with other world leaders.

In light of the symbolic value of the *topos* of the photo as source domain, the reference to the photo can, in general terms, be interpreted as a negative characterization of the improper behaviour of a political leader, given the fact that the other leader should be considered as an enemy. The narrative structure of the photo characterizes the leader on the left side as the more powerful one in relation to the one on the right side. The less powerful leader on the right side puts his arm around the more powerful on the left side. The facial expression of the two leaders while kissing is neutral, both have their eyes closed.

For a further analysis of the source domain, the graphic techniques used to perform the changes must be interpreted in terms of indications of differences between the photo and the cartoon. In the cartoon Bertrams has made the following changes:

(1) The negative evaluation of the characteristics Y of X is represented by deviation of the *symbolic form* of the image of the iconic photo, performed by means of the following visual operations in the original image of the *topos* of the photo:
Changes:
- The actors (Trump and Xi)
- The nature of the embrace: the form of the kiss
- The position of the arm (in the cartoon the arm is put by Trump on the left around the Xi on the right, in the iconic photo the arm is put by the less powerful Honecker on the right around the more powerful Brezhnev on the left)
- The facial expressions

Addition:
- The saliva of Trump
- The text '#Xitoo'

Removal:
- The faces in the background

(2) The value W (the convention that an American president should be reserved towards a Chines communist leader who is a traditional 'enemy' of the US) on the basis of which the characteristics Y are evaluated as negative because they conflict with this value (as stated in argument 1.1b.1a) is the *symbolic content*, represented by the *topos* of the original iconic photograph.

These changes can be interpreted as a visual expression of a violation of the behaviour of the conventions for the way in which US presidents behave towards communist leaders (and more in general towards other world leaders). By presenting the behaviour of Trump as that of a fellow communist leader, it is made clear that his behaviour is not in line with leaders of western democratic countries who do not behave according to traditions of communist leaders. The distortion of the embrace and the addition of the spit represent the exaggeration of the way in which Trump addresses Xi.

On the basis of this analysis of the source domain of the cartoon, the target domain of the cartoon can be interpreted as a critique on the behaviour of Trump by viewing his behaviour in terms of a representation of 'Deadly love', love out of hate for the Chinese communist leader. The changes can be interpreted as a further specification of this behaviour in terms of a violation of the conventions of the behaviour of US presidents towards communist leaders that are traditional 'enemies' of the US.

By presenting Trump as a male who wants to kiss the other against his will (which is clear from the facial expression of Xi) and by using the reference to the #MeToo movement Bertrams adds an element to the original photo by presenting Xi as an 'object of desire' for Trump and the intrusive behaviour of Trump as a form behaviour that constitutes a critique in terms of a violation of certain conventions among human beings. The reference to the #MeToo movement and Xi's facial expression make clear that part of the message is that Xi does not appreciate Trump's behaviour as a fellow communist.

4.3 The analysis of the argumentation underlying the '#Xi too' Cartoon that uses the 'Socialist fraternal kiss' as a topos

In the analysis of the argumentation, it must be established how the results of the rhetorical analysis can be translated in terms of elements of the argumentative pattern of an editorial cartoon with an iconic photo as *topos*. To this end I will explain how the negative evaluation of the characteristics of the behaviour of Trump can be analysed as characteristics (Y) that must be evaluated negatively. Furthermore I will explain how the standards on the basis of which Y must be evaluated negatively can be analysed as values (W). The complete reconstruction is represented in Figure 4.

1   The behaviour of Donald Trump in his exuberant praise of Xi (X) must be evaluated negatively (is improper)

1.1a The behaviour of Donald Trump in his exuberant praise has as characteristics that it is not in line with the behaviour of former US presidents (Y1) and that it is inconsistent with Trump's earlier behaviour towards Xi (Y2)

1.1b Behaviour that has as characteristics that it is not in line with the behaviour of former US presidents (Y1) and that it is inconsistent with Trump's earlier behaviour towards Xi (Y2) must be evaluated negatively

1.1b.1a   Behaviour that is not in line with the behaviour of former US presidents (Y1) conflicts with the convention that an American president should be reserved towards a Chines communist leader who is a traditional 'enemy' of the US (value W1) and should be evaluated negatively
    1.1b.1a.1   An US president should be reserved towards Chinese communist leaders since China represents the opposite of the values of the US nation as a capitalist country and constitutional democracy that respects fundamental human rights.
1.1b.1b   The convention that an US president should be reserved towards a Chinese communist leader who is a traditional 'enemy' of the US (Value W1) is a generally accepted value in the US foreign policy towards communist leaders
1.1b.1c   Behaviour that is inconsistent with earlier behaviour towards another world leader (Y2) conflicts with the convention that an American president should be consistent in his/her foreign policy (W2) and should be evaluated negatively
    1.1b.1c.1   An US president should have a consistent foreign policy towards Chinese communist leaders since China is an aspiring world power and should be handled with care.
1.1b.1d   The convention that an US president should be consistent in his foreign policy (value W2) is a generally accepted value in US foreign policy

Figure 4. Reconstruction of the argumentation underlying the cartoon '#Xitoo' by Joep Bertrams

In this reconstruction the evaluation of the behaviour of Trump (X) is represented by the characteristics Y1 and Y2 and their negative evaluation. The characteristics Y1 and Y2 are a translation of elements of the target domain of the visual metaphor. This target domain, as was indicated in section 4.2, is based on an interpretation of the source domain formed by the *topos* of the 'Socialist fraternal kiss' and the graphic changes in the symbolic form of this *topos*. The negative evaluation of Y1 and Y2 is based on the values W1 and W2 that are a translation of elements of the target domain in terms of the values that form part of the argumentation. These elements of the target domain are the values that are to be upheld in US foreign policy in general, and toward communist leaders of China in particular. In their turn, the interpretation of these elements of the target domain is based on an interpretation of the source domain of the symbolic values of the original image and its afterlife referred to by the cartoon.

Given the reconstruction of the elements of X, Y1 and Y2, W1 and W2, the further content of the arguments that form part of the argumentative pattern can be reconstructed as is indicated in Figure 3. The complex argumentation represents the commitments of the cartoonist on the basis of the interpretation of the visual rhetoric and the argumentative commitments of a cartoonist. As has been indicated earlier in section 2, the values referred to in a cartoon are of a general nature and are represented on a high level of abstraction. For this reason the audience has a certain interpretation space to formulate the values for the case at hand. Also the way in which the changes can be interpreted and the meaning to be attached to these changes leaves the audience ample space for its own reconstruction of the message. This interpretation space is part of the nature of the argumentative activity of editorial cartoons and their suitability to criticize and mock the behaviour of public officials and current events in a creative and humorous way. This gives images and the values they refer to a powerful tool for contributing to discussions about the behaviour of politicians in the public arena.

5. CONCLUSION

In this contribution it has been explained how iconic photographs can be used in editorial cartoons as *topoi* in the context of an argument in which a certain behaviour of situation is criticized. They have been characterized as an argumentative activity type in which it is one of the characteristics that there is 'reason to dissent' about the behaviour of a public official or situation. Starting from the conception of an editorial cartoon as an argumentative activity, I have implemented the general model of the argumentative pattern in editorial cartoons developed in Feteris (2019) for the specific genre of editorial cartoons with an iconic photograph as *topos*. This implementation makes it possible to clarify the argumentative commitments of the cartoonist that are specific for this genre. In this way the commitments can be made explicit in the interpretation of the visual rhetoric and the translation of the message conveyed by the visual rhetoric that forms part of the argumentation.

Using this implementation of the model, I have given a demonstration of the way in which the visual rhetoric that refers to an iconic photo as *topos* can be analyzed as a means for conveying part of the argumentative message. I have done this by indicating how changes with respect to the original iconic photograph represent a key to the interpretation of the critique that is conveyed indirectly by the cartoonist. I have shown how the visual rhetoric can be analyzed and how the critique can be reconstructed as part of the argumentative message underlying the cartoon. On the basis of this analysis I have given a reconstruction of the complex argumentation underlying the cartoon.

REFERENCES

Bertelsen, L. [1989]. Icons of Iwo. *Journal of Popular Culture*, 22(4), 79-95.
Bertrams, Joep [2017]. Plotselinge grote vriend. # Xi too. https://politicalcartoons.com/cartoonist/joep_bertrams; https://twitter.com/joepbertrams
Bredekamp, H. [2007]. *Theorie des Bildakts: Frankfurter Adorno-Vorlesungen*. Berlin: Suhrkamp.
Edwards, J.L. & Winkler, C.K. [1997]. Representative form and the visual ideograph: The Iwo Jima image in editorial cartoons. *Quarterly Journal of Speech*, 83, 289-310.
Eemeren, F.H. van (Ed.) [2017]. *Prototypical argumentative patterns. Exploring the relationship betwee argumentative discourse and institutional context*. Amsterdam: John Benjamins.
Feteris, E. [2013]. The use of allusions to literary and cultural sources in argumentation in political cartoons. In H. van Belle, P. Gillaerts, B. van Gorp, D. van de Mieroop, & K. Rutten (Eds.), *Verbal and visual rhetoric in a media world*. (pp. 415-428). (Rhetoric in Society). Amsterdam: Leiden University Press.
Feteris, E.T. [2019] The reconstruction of visual argumentation in editorial cartoons with a visual metaphor. In: B. Garssen, D. Godden, G.R. Mitchell, J.H.M. Wagemans (Eds.), *Proceedings of the Ninth Conference of the International Society for the Study of Argumentation* (pp. 370-378). Amsterdam: Sicsat.
Feteris, E.T., L. Groarke, H.J. Plug [2011]. Strategic maneuvering with visual arguments in political cartoons: A pragma-dialectical analysis of the use of *topoi* that are based on common cultural heritage. In: E.T. Feteris, B.J. Garssen, A.F. Snoeck Henkemans (Eds.), *Keeping in touch with pragma-dialectics* (pp. 59-75). Amsterdam: Benjamins.
Hariman, R. & Lucaites, J.L. [2002]. Performing civic identity: The iconic photograph of the flag raising on Iwo Jima. *Quarterly Journal of Speech*, 88, 363-392.
Hariman, R. & Lucaites, J.L. [2007]. *No caption needed. Iconic photographs, public culture, and liberal democracy*. Chicago: The University of Chicago Press.
Kjeldsen, J.E. [2015]. The study of visual and multimodal argumentation. Argumentation, 29(2), 115-132.
Kjeldsen, J.E. [2017]. The rhetorical and argumentative potentials of press photography. In: A. Tseronis & Ch. Forceville (Eds.), *Multimodal*

Argumentation and Rhetoric in Media Genres (pp. 51-79). Amsterdam: John Benjamins.
Kleppe, M. [2013]. Canonieke foto's. De rol van (pers)foto's in de Nederlandse geschiedschrijving. Delft: Eburon.
Medhurst, M.J. & Desousa, M.A. [1981]. Political cartoons as rhetorical form: A taxonomy of graphic discourse. *Communication Monographs* 48, 197-236.
Melching, W.F.B. [2019]. Clash of the icons. The iconoclasm of the image of the United States. Notre Dame University Press (forthcoming).
Paul, G. [2011]. *Bilder die Geschichte schrieben. 1900 bis heute*. Bonn: Bundeszentrale fur politische Bildung.
Perlmutter, D.D. [1998]. *Photojournalism and foreign policy: icons of outrage in international crises*. Westport, CO: Praeger.
Schilperoord, J. [2013]. Raising the issue: A mental-space approach to Iwo Jima-inspired editorial cartoons. *Metaphor and Symbol* 28(3), 185-212.

Conceptual analysis of an argumentation: using argumentation schemes and the Toulmin model

MARIE GARNIER
Cultures Anglo-Saxonnes, Université Toulouse 2 Jean Jaurès, France
mgarnier@univ-tlse2.fr

PATRICK SAINT-DIZIER
CNRS-IRIT, University Toulouse 3, France
stdizier@irit.fr

This investigation explores the correlations that exist between argumentation schemes and the Toulmin model, and how these can be used conjointly to develop a more accurate conceptual representation of an argumentation structure. It is based on a natural language corpus of claims and attacks or supports developed in the social domain. Claims and justifications are annotated using a system of XML-based frames focusing on linguistic and conceptual features.

KEYWORDS: argumentation schemes, comparison, Toulmin model, warrant

1. INTRODUCTION AND CHALLENGES

In this contribution, we explore the relations and the potential cooperation between the Toulmin model on the one hand and the model of argumentation schemes on the other hand. The goal is to deepen our understanding of the interactions between these two models and to develop a more accurate semantic representation of arguments in context.
    The investigation which is reported in this paper is based on the elaboration of schemes from Warrants/Backings as given in (Walton et ali. 2012), (Walton 2015) via (Eemeren et al. 1996). These authors have shown the general defeasible character of the reasoning that is involved in schemes.
    With respect to these two approaches, our goal is to:
- investigate precise forms of cooperation between the Toulmin system and argument schemes to have a precise and concrete analysis of the validity of an argument in context;

- consider corpora and work on selected case studies as a preliminary analysis level to validate our analysis;
- suggest elements of a formal model that develops the cooperation of the two models with related logical devices.

In this article, we first offer a reminder of the main aspects of the two models considered, the Toulmin model and argumentation schemes, then we introduce the corpus of authentic texts used for this exploratory study, from which three case studies are extracted and analyzed in the following sections. The last sections of this paper focus on offering a formal representation of the possible integration of warrants and argument schemes, and an exploration of the use of warrants to answer critical questions.

2. OVERVIEW OF THE TWO PARADIGMS

2.1 The Toulmin Model

The goal of this model is to organize the structure of an argument. Following is a representation of a typical argument cell as described in the Toulmin model (Toulmin 1958), (Toulmin 2001) (Freeman 2005a):

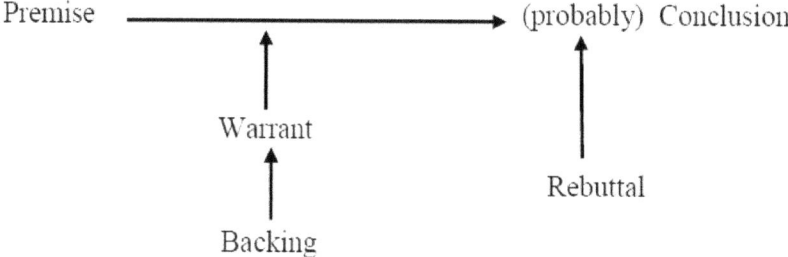

Figure 1 –Toulmin's model (Toulmin, 1958)

This representation can be illustrated with the following example:
- **Premise**: Temperatures are below 0°C this morning.
- **Conclusion**: The garden will suffer.
- **Warrant**: Plants in general are damaged by temperatures below 0°C.
- **Backing**: It is a physical law in botanic.
- **Rebuttal**: Unless they got early sun or were well protected, etc.

The different elements of the model can be defined as follows:

- **Conclusion** (also called "claim"): the conclusion being drawn from the premise(s);
- **Premises** (which can be called Grounds/Facts/Evidence): the data and facts offered to support the claim (not represented above);
- **Warrant**: the element that logically connects the grounds to the conclusion;
- **Backing**: the element that supports or explains the warrant. While the warrant may be a precise and contextual element, the backing is usually a general fact, such as the physical law in the example above;
- **Qualifiers**: statements about the strength of various elements, to be understood in context (not represented above);
- **Rebuttal**: exceptions to the claim.

Some versions of this model include additional, more peripheral, structures which will not be considered here.

2.2 The role of argument schemes

Argument schemes are specific modes of arguing which are not logical forms of reasoning and occur in everyday arguments. Several classes of argument schemes have been identified, covering most argumentative situations. For example, one of the most common argument schemes is "Argument from expert opinion" (Walton et ali. 2008), which states that:

Major premise: Source E is an expert in subject domain S containing proposition A.
Minor premise: E asserts that proposition A is true (false).
Conclusion: A is true (false).

One of the specificities of argument schemes is the inclusion of critical questions that enable the conclusion of the argumentation to be questioned by a respondent, and can serve to sound the quality of an argument. For the argument scheme introduced above, one of the critical questions is: *Is A consistent with what other experts assert?* If it is not the case, doubts on E's judgement on the veracity of proposition A may arise, as well as doubts on the credibility of E.

Given an argument, one of the main difficulties is to evaluate the underlying (defeasible) reasoning mechanisms at stake, their scope and validity. It is in general quite challenging to identify the argument scheme(s) that are used in a specific argument, linking premises to conclusions. Our aim is to identify schemes from mined resources, then, to evaluate the scope, the strength and the validity of schemes in precise contexts, We then attempt to offer complements to the schemes by identifying the role the system of warrants and backings described by the Toulmin model of argument could play in a given argument scheme. Our objective is to investigate how these two frameworks might cooperate to produce a more accurate analysis of the validity of an argument in context.

Our experimental protocol is as follows:
- consider a claim or an argument;
- use our argument mining tools to extract supports, attacks and warrants from these corpora (Saint-Dizier 2018), as well as induce warrants if possible. These tools are based on grammatical clues and are implemented on our TextCoop platform. Other systems include for example (Nguyen and Litman, 2015);
- manually identify a scheme or a family of schemes which can be candidates, show how the variables of the schemes get instantiated via warrants, following, in particular (Garcia et ali. 2007);
- provide some evaluation of the validity of a scheme in context;
- consider the critical questions associated with the argument schemes identified and show how these can be (partly) answered.

3. CONSTRUCTION OF A CORPUS

The first step of our work is to construct a valid corpus from which case studies are taken. For that purpose, we consider six topics related to different domains linked to social issues. The corpus is constructed from a "seed" which is a controversial statement. This technique is also known as bootstrapping. This seed is submitted to the web, from which statements that support or attack it are then extracted. Duplicates are frequent and are eliminated to prevent the overrepresentation of some statements.

The seeds considered for our experiment are the following:

(1) Ebola vaccination is necessary,
(2) Women's conditions have improved in India,
(3) The development of nuclear plants is necessary,
(4) Brexit is good for the UK,
(5a) Affirmative Action is good for the economy,
(5b) Gender parity is reachable.

Table 1 presents the size of the text extracts integrated into the corpus; the texts considered are those which contain a support or an attack, including the context for the controversial statement at stake. The last column indicates the rough number of arguments identified, given that arguments often present themselves in clusters and are notoriously difficult to separate and count. It is interesting to note that the number of arguments is not large. Indeed this means that the linguistic resources and knowledge resources needed for a certain topic may not be so large, and therefore their development and implementation possibly from already existing resources may be possible.

Topic	Corpus size	Nb. of arguments identified
(1)	16 extracts, 8300 words	50
(2)	10 extracts, 4800 words	27
(3)	7 extracts, 5800 words	31
(4)	23 extracts, 6200 words	22
(5)	5 extracts, 2200 words	26
Total	59 extracts, 27300 words	156

Table 1. Presentation of the exploration corpus

From this corpus, we consider several case studies of different degrees of complexity. They are presented and analysed in the next section.

4. ANALYSIS OF CASE STUDIES

Let us consider in this section three case studies taken from different domains.

4.1 Case 1: Argument scheme based on practical inference

Let us consider the following argument, where the seed is the first part of the statement. It is followed by a justification which may be supported or attacked:

Brexit is a good step forward because citizens want a healthy economy.

This argument is based on several supports, attacks or warrants extracted or induced from our corpora via a grammar dedicated to argument analysis implemented on our TextCoop platform. Let us consider here two of these warrants:
 W1: Citizens want to live in optimal conditions,
 W'1: Isolating a country is the best way to get a healthy economy.

We analyse this argument as being an argument cluster, part of which can be interpreted as fitting the scheme "Practical inference" (Walton et al., 2008, p. 323).

Practical inference (subset of Practical reasoning, slightly adapted):

Major premise: A group of people have a goal G.

Minor premise: Carrying this action A is a means to realize G.

Conclusion: Therefore, they ought (practically speaking) to carry out this action A.

Let us note that this conclusion is not expressed in the claim above, but constitutes an intermediary conclusion leading to the identification of action A as positive, possibly through the scheme "Argumentation from ends and means" (Perelman and Olbrechts-Tyteca, 1969, p. 273-278, via Walton et al., 2008, p. 325).

Then, merging schemes and the Toulmin model can be expressed as follows:
- a general version of G is expressed in W1,
- a general version of A is expressed in W'1 (note that there are several actions A_i which can have the same effect).

Then, given these elements, if G = W1 and A = W'1, and the action is considered generally positive in virtue of allowing an end to be reached, then follows:

Brexit is a good step forward.

This example shows that argument schemes are more generic than the Toulmin model, in a certain sense, since the parameter values extracted from the linguistic data which instantiates the Toulmin model, can, in a next stage, instantiate the variables in the argument scheme considered. The analysis of these values and their relevance with respect to the scheme allows to evaluate how relevant and valid the selected scheme is. Then a certain semantic representation of the argument can be developed.

4.2 Case 2: Going deeper into the scheme: deliberation based on several warrants

In this section, we illustrate how linguistic material collected via argument mining, within the framework of the Toulmin model, can be reused to motivate or help determine which argument scheme is the most appropriate. The complexity of this problem is developed in (Kock 2003). Let us consider now the following argument:

> Vaccination against Ebola is necessary because it prevents disease dissemination.

Let's assume a deliberation based on several factors Di, with supports or attacks collected from an argument mining process, such as the following examples:

Examples of supports:
- Vaccine protection is very good;
- Ebola is a dangerous disease;
- There are high contamination risks, etc.

Examples of attacks:
- There is a limited number of cases and deaths compared to other diseases;
- Seven vaccinated people died in Monrovia;
- Vaccine may have dangerous side-effects, etc.

It is possible to induce some warrants from the supports and attacks extracted using our method. For example, the following warrants could be induced by our system:
W2: it is necessary to protect a population against major diseases.
W'2 it is important to care about side effects of medicines.

Then, the deliberation being illustrated in this section consists in comparing and weighing the different warrants which were induced from the attacks and supports taken from the corpus.

It is also possible, from the examples above, to manually elaborate a synthetic warrant W"2 such as the one introduced below. This could be viewed as reconstructing an enthymeme (Jackson et al. 1980). It would summarize the deliberation and give its argumentative direction or polarity, which is useful for the scheme to be validated:

W"2: (generalization) it is important to have good management of medical situations to make good decisions, for example regarding possible side effects, incurred costs, in order to effectively protect a population.

Then, the scheme that is the most appropriate depends on (1) the propositional content but also on (2) the sources of attacks or supports, in particular whether they come from simple bloggers or political commentators, expert reports, the general population, etc. For example a source identified as coming from experts would trigger a scheme related to expertise.

In the case of our example, the source is identified has not being from an expert. Given these considerations, a potential argument scheme could be "Argument from deliberation *Ad Populum*", assuming there has been a deliberation of a sample of a standard population (in contrast with a group of experts, as mentioned above):

Argument scheme from deliberation *Ad Populum*:

> *Premise 1:* Everybody in group G accepts A.
> *Premise 2:* Group G has deliberated intelligently and extensively on whether to accept proposition A or not using the considerations Di.
> *Conclusion:* Therefore, A is (plausibly) true.

In this context, the variable A, which represents the opinion being deliberated, can be associated with several, possibly weighted, warrants which are instances of a rule R, a hypothesis H or a situation A.

The warrants W and the other elements of the deliberation, because of the additional content they provide, give more validity and context to the scheme being considered. In the case of a group of experts being involved in the deliberation, a slightly different argument scheme would have been proposed, namely the scheme "Argument from expert opinion".

4.3 Toward a formal account of the cooperation between argument schemes and the Toulmin model

Let us now introduce a first level of formalization of the cooperation between the two models. In this investigation, it is possible to use a compositional and monotonic approach. Let us consider in this section a few well-known schemes and the integration of elements of the Toulmin model.

For the abductive reasoning scheme, a global formal expression of the entailment proposed in this scheme can be summarized as follows:

$$\text{explanation}(E, F, C) \Rightarrow A.$$

This formula simply paraphrases the language formulation.

Let SW be the set of warrants W that are relevant to this scheme:

> SW={W1..., Wn}.

Then, the integration of explanations considered as warrants can be realized compositionally via a lambda expression which scopes over warrants W as follows:

> $\lambda E \, (explanation(E, F, C) \Rightarrow A)(SW)$.

Let us now consider the argument scheme "Argument from deliberation *Ad Populum*" as illustrated in the previous section. Several elements related to the deliberation are considered in this formal model. The global formal expression is:

> $\forall i \in [1, m], D=\{d_i, deliberation(G, A, D) \Rightarrow A$.

where the set of induced warrants is computed by the function: induce(D, SW).
The integration of these elements in the initial formal expression above yields:

> deliberation based on induced warrants:
> $\lambda D \, (deliberation(G, A, D) \Rightarrow A) \, (SW)$.

Although this first formal level is compositional and seems to capture the main intuitions behind the scheme, it lacks the weights on the Wi which are essential elements, as some warrants may have higher weight or priority than others.

4.4 Argument scheme with conflicting Warrants

Let us now consider a case where a scheme can be based on conflicting, i.e. a priori incoherent, warrants as presented in (Saint-Dizier, 2018). To illustrate this aspect, we consider a few motivational examples. From the statement:

> *This film is good because it is politically correct.*

A number of warrants can be mined, as explained above. For example:
W3: (only) politically correct ideas are appreciated by the public,
But opposite warrants/attacks were also mined, for example:
W'3: it is good to criticize standard education via political incorrectness to promote the evolution of minds.

If, for example, we consider the scheme "Argument from popular opinion":

> *General acceptance premise:* A is generally accepted as true,
> *Presumption:* If A is generally accepted as true, this gives a reason in favor of A,
> *Conclusion:* There is a reason in favor of A.

In this scheme, if the opinion A remains valid, it is nevertheless weakened by W'3 and related elements since W'3 introduces a kind of attack. The weight of the attack is not considered here. From this example, it can be noted that the cooperation between schemes and warrants is therefore quite tricky, in particular when there is the inclusion of authentic data collected on the basis of argument mining techniques.

5. EXPLORING THE USE OF WARRANTS TO ANSWER CRITICAL QUESTIONS

The critical questions associated with argument schemes are a means to show that schemes are not logical deduction, but may be subject to contradictory evaluations. A problem that is frequently encountered is the identification of relevant critical questions. It is particularly interesting to base this search on corpora, where real case situations can be identified and weighted.

To illustrate this feature, let us consider again the "Deliberation *Ad Populum*" argument scheme, with the group of people G presented above. The critical questions proposed are, among others:
(1) how competent and representative is the group G?
(2) what are the elements considered during the deliberation?
(3) are they sufficient to allow A to be `inferred'?
Let us now consider one of the claims that we have investigated: *Vaccine protection is very good* which has been debated by the group G. Let us assume that this deliberation has originated a number of supports and attacks which constitute the d_i, which are repeated from above for an easier reading:

Example of supports:
- Ebola is a dangerous disease;
- there are high contamination risks;

Examples of attacks:
- there is a limited number of cases and deaths compared to other diseases;
- seven vaccinated people died in Monrovia;

- vaccine may have high side-effects.

Given these mined supports and attacks, our approach allows to go deeper and in a more concrete way in the debate represented in the argument scheme. In particular, our approach allows to answer critical question (2) above via the d_i defined above, possibly weighted.

Then critical question (3) can be partially answered via the induction of SW (section 4.3) where its scope and generality can be analyzed. Finally, critical question (1) is more delicate to answer, however the scope and quality of the d_i (relevant, insightful, etc.) can contribute to answer this question.

6. RESULTS AND EPILOGUE

In this article, we have investigated precise forms of cooperation between the Toulmin model and argument schemes with the goal of being able to offer a precise and concrete analysis of the validity of an argument in context, using argument mining tools. The analysis presented in this article is clearly preliminary and exploratory. It is based on a few test cases taken from real life situations, with all their complexity and contextual effects. In spite of its exploratory character, we feel that our analysis raises interesting questions and some elements of solutions.

This analysis needs to be extended in at least the following directions:

a larger corpus needs to be compiled and also analyzed, possibly with a higher diversity of sources, or on the contrary with a focus on a specific type of source in order to streamline the type of claims and argument schemes found,

the impact of context in general and on the validity of argument schemes needs to be more thoroughly investigated,

the related aspects of argument mining need to be developed in order to be able to access to a large amount of relatively reliable data.

ACKNOWLEDGEMENTS: This research was partially supported by a grant from the *Maison des Sciences de l'Homme et de la Société de Toulouse*.

REFERENCES

van Eemeren, F., Grootensdorf, F., Hoekemens, F., (1996). *Fundamentals of Argumentation Theory*. Routledge.
Freeman, J.B. (2005a). Systematizing Toulmin's warrants: an epistemic approach. *Argumentation* 19(3).

Freeman, J.B. (2005b). *Acceptable Premises: An Epistemic Approach to an Informal Logic Problem*. Cambridge University Press.

Garcia, A. J., Chesnevar, C. I., Rotstein, N. D., Simari, G. R. (2007). An Abstract Presentation of Dialectical Explanations in Defeasible Argumentation. *Proceedings of the International Workshop on Argumentation and Nonmonotonic Reasoning* (ArgNMR) 2007. Arizona, USA.

Kock, C. (2003.) Multidimensionality and non-deductiveness in deliberative argumentation. In V. Eemeren et ali. (Eds), *Anyone who has a view: Theoretical contributions to the study of argumentation*. Kluwer Academic.

Nguyen, H., Litman, D. (2015). Extracting argument and domain words for identifying argument components in texts. In *Proceedings of the 2nd Workshop on Argumentation Mining*. Denver, Colorado.

Perelman, C., Olbrechts-Tyteca, L. (1969). *The New Rhetoric: A Treatise on Argumentation*. University of Notre-Dame Press.

Saint-Dizier, P. (2018). A knowledge-based approach to warrant induction. In *Proceeding of Comma 2018*. Warsaw.

Toulmin, S.E. (1958). *The Uses of Argument*, Cambridge: Cambridge University Press.

Toulmin, S.E. (2001). *Return to Reason*. Harvard University Press.

Walton, D., Reed, C., Macagno, F. (2008). *Argumentation Schemes*. Cambridge: Cambridge University Press.

Walton, D. (2015). *Goal-based Reasoning for Argumentation*. Cambridge: Cambridge University Press.

Representing second-order arguments with Adpositional Argumentation (AdArg)

FEDERICO GOBBO
University of Amsterdam
f.gobbo@uva.nl

JEAN H.M. WAGEMANS
University of Amsterdam
j.h.m.wagemans@uva.nl

This paper extends a high-precision method for representing 'first-order' arguments to the linguistically and pragmatically more complex 'second-order' arguments (such as the argument from authority). It thereby contributes to the further development of *Adpositional Argumentation* (AdArg), an approach to representing argumentative discourse with applications in corpus linguistics and computational argumentation that combines Gobbo and Benini's linguistic representation framework of *Constructive Adpositional Grammars* (CxAdGrams) and Wagemans' argument categorisation framework of the *Periodic Table of Arguments* (PTA).

KEYWORDS: *Adpositional Argumentation*, argument from authority, argument type, argument scheme, argumentative adpositional trees, *Constructive Adpositional Grammars*, constructive pragmatics, *Periodic Table of Arguments*, second-order arguments

1. INTRODUCTION

In response to the need for high-precision tools for analysing and evaluating arguments, Gobbo and Benini's (2011) linguistic representation framework of *Constructive Adpositional Grammars* (CxAdGrams) has recently been combined with Wagemans' (2016, 2019) argument categorisation framework of the *Periodic Table of Arguments* (PTA). The resulting approach of *Adpositional Argumentation* (AdArg) (Gobbo & Wagemans, 2019a, 2019b, 2019c; Gobbo, Benini & Wagemans, to appear) enables the analyst of argumentative discourse to represent arguments expressed in natural language by means of so-called 'argumentative adpositional trees' (or 'arg-adtrees'). Such trees contain not only very detailed linguistic information about the statements that make up the argument, but they also include pragmatic information concerning the order of presentation of these statements, the type of argument they substantiate, and the argumentative function of their constituents.[1] At the same time, an arg-adtree is flexible in that the analyst can show, hide, and highlight any piece of information according to her needs.

So far, this method for representing arguments has been successfully applied to so-called 'first-order' arguments such as the 'argument from sign' and the 'argument from analogy'. In the process of identifying their type, the statements that function as the premise and the conclusion are analysed on the level of the proposition, i.e., the specific constellation of their subjects and predicates is determined (see Wagemans, 2019).

Argumentative discourse, however, also contains so-called 'second-order' arguments such as the 'argument from authority' and the 'argument from disjuncts'. These arguments differ from first-order ones in that the analyst, in the process of identifying their type, has to shift from the level of propositions to that of assertions. This means that the statement functioning as the conclusion (and sometimes also that functioning as the premise) should be complemented with a predicate expressing the arguer's epistemic commitment regarding its truth or acceptability, thereby changing the nature of the statement from a proposition to an assertion (see Wagemans, 2019). The addition of 'is true' to one or both of the statements poses a challenge to the method for representing arguments just described. As yet, it is unclear how this additional pragmatic information about the statements that make up the argument should be included in the corresponding arg-adtree.

In this paper, we make a proposal for constructing arg-adtrees of second-order arguments by examining the consequences of the abovementioned shift in the level of the analysis for the representation of the linguistic and pragmatic information contained in the argument. We start with a short exposition of our representation method as applied to first-order arguments (Section 2). Then, we explain the nature and constituents of second-order arguments, emphasising how they differ from first-order ones, and describe the extra steps the analyst should take in order to identify their type (Section 3). Next, we consider how to represent the additional linguistic and pragmatic information in an arg-adtree and illustrate our solution by providing the arg-adtrees of two examples of second-order arguments (Section 4). We

---

[1] For an explanation of the very possibility of representing pragmatic information in adtrees see Gobbo and Benini (2011, chapter 6).

conclude with a summary of our findings and a brief discussion of newly arisen challenges that should be addressed in further research (Section 5).

2. BUILDING ARGUMENTATIVE ADPOSITIONAL TREES

Our high-precision method for representing arguments expressed in natural language is the result of combining the linguistic representation framework of *Constructive Adpositional Grammars* (CxAdGrams) with the argument categorisation framework of the *Periodic Table of Arguments* (PTA). We have explained the theoretical background of both frameworks and their combination into an approach we named *Adpositional Argumentation* (AdArg) elsewhere (see Gobbo & Wagemans, 2019a, 2019b, 2019c; Gobbo, Benini & Wagemans, to appear). For the present purposes, we shall briefly elucidate the characteristics of the argumentative adpositional tree (arg-adtree) of the example of a first-order argument pictured in Figure 1, *The suspect was driving fast, because he left a long trace of rubber on the road*.

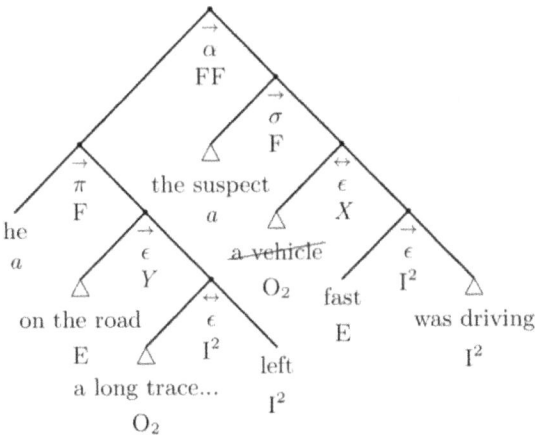

Figure 1 – The arg-adtree of a first-order argument (Gobbo & Wagemans, 2019c, p. 417)

The arg-adree consists of two main branches: the right one conventionally representing the statement that functions as the conclusion (σ) of the argument, *The suspect was driving fast*, and the left one representing the statement that functions as its premise (π), *He left a long trace of rubber on the road*. Each of the two branches contains linguistic information on the word level expressed by means of five different grammar characters (A, E, I, O, U), which are taken from the

317

linguistic representation framework of CxAdGrams. Table 1 explains their meaning – adapted from Gobbo and Benini (2011, p. 41).[2]

gc	name	function	examples
A	adjunctive	modifier of O	adjectives, articles, determiners
E	circumstantial	modifier of I	adverbs, adverbial expressions
I	verbant	valency ruler	verbs, interjections
O	stative	actants	nouns, pronouns, name-entities
U	underspecified	transferer	prepositions, derivational morphemes

Table 1 – The meaning of grammar characters in adtrees

Under the top hook of the arg-adtree, where the two main branches connect, one finds first of all pragmatic information about the order of presentation in the discourse. In this case, the order is retrogressive (conclusion, *because* premise), which is represented by a right arrow (→). Under this arrow, the analyst places information about the type of argument in terms of the theoretical framework of the PTA. The example can be identified as a first-order predicate argument combining a statement of fact with another statement of fact, which is represented in the arg-adtree in abbreviated form (α FF).

Moving down to the branches representing the statements themselves, one finds information about their argumentative function as conclusion (σ) or premise (π). This level of the arg-adtree also reiterates the information about the type of statement in terms of the tripartite typology of statements distinguished within the PTA: statements of fact (F), statements of value (V) and statements of policy (P). In this case, both the conclusion and the premise of the argument are classified as a statement of fact (F).

Finally, the arg-adtree contains information about the predicates and subjects of the propositions expressed in the conclusion and the premise of the argument. Following logical conventions, the predicates are indicated with a, b, etc., and the subjects with X, Y, etc. In this case, *the suspect* (a) / *he* (a) is the shared subject of these propositions, while *was driving fast* (X) and *left a long trace of rubber on the road* (Y) are their respective predicates. The argument thus has the form 'a is X, because a is Y', which is why it is identified as a first-order predicate argument.

---

[2] The apexes and pedices serve to indicate the valency of the verbants, to identify their actants, and to indicate their level of saturation. For a more detailed explanation of the linguistic information represented in this argumentative adtree, see for instance Gobbo and Wagemans (2019c, pp. 414-419).

3. WHAT ARE SECOND-ORDER ARGUMENTS?

As we have illustrated in the previous section, first-order arguments can be identified on the basis of an analysis of the components of the propositions that express their conclusion and premise. More precisely, the form of the argument is determined by the constellation of the linguistic subjects and predicates of these propositions. The theoretical framework of the PTA distinguishes two possible constellations, which we will now describe in more detail.

If the propositions share a common subject, they have the form '*a* is *X*, because *a* is *Y*' and are classified as 'predicate' arguments. In technical terms, the common subject (*a*) functions as the 'fulcrum' of the argument and the relationship between the different predicates (*Y* and *X*) as its 'lever', i.e., as its underlying argumentative mechanism (see Wagemans, 2019). The example just mentioned, for instance, has *the suspect* (*a*) / *he* (*a*) as its fulcrum and the relationship between *leaving a long trace of rubber on the road* (*Y*) and *driving fast* (*X*) as its lever.

> Example 1
> *The suspect* (*a*) *was driving fast* (*X*), *because he* (*a*) *left a long trace of rubber on the road* (*Y*)

A subsequent determination of the types of statement gives the analyst the systematic name of the argument under scrutiny (in this case, '1 pre FF' or 'α FF'), while a determination of the nature of the relationship between the predicates provides its traditional name (in this case, 'argument from effect').

The other possible constellation is when the propositions expressed in the conclusion and the premise share a common predicate. In this case, the argument has the form '*a* is *X*, because *b* is *X*' and is classified as a 'subject' argument. An example is *Cycling on the grass is forbidden, because walking on the grass is forbidden*, which has *is forbidden* (*X*) as its fulcrum and the relationship between *cycling on the grass* (*a*) and *walking on the grass* (*b*) as its lever.

> Example 2
> *Cycling on the grass* (*a*) *is forbidden* (*X*), *because walking on the grass* (*a*) *is forbidden* (*X*)

Like with first-order predicate arguments, the systematic name of first-order subject arguments indicates their argument form as well as the specific combination of the types of statement (in this case, '1 sub VV' or 'β VV'). The determination of the nature of the relationship between the subjects provides their traditional name (in this case, 'argument from analogy').

How do second-order arguments deviate from first-order ones? One way to explain the difference is to assume that in order to identify an argument, the analyst has to determine the 'fulcrum', i.e., the common term of the propositions involved (see Wagemans, 2019). As illustrated by means of Example 3 and Example 4, this sometimes poses a problem.

The first problematic case is when the conclusion does not have anything in common with the premise and the search for the fulcrum thus yields a negative result.

> Example 3
> *He must have gone to the pub, because the interview is cancelled*

From analysing the conclusion and the premise in terms of the constituents of the propositions, the only thing to report is that the argument has the form '*a* is *X*, because *b* is *Y*'. As a result, it also remains unclear how the premise supports the conclusion or, in other words, how to formulate the 'lever' or underlying mechanism of the argument.

In other cases, as illustrated by means of Example 4, the propositions do share a common element, but it cannot unambiguously be identified as their common term (subject or predicate).

> Example 4
> *We only use 10% of our brain, because that was said by Einstein*

In analysing the conclusion, one may take the proposition *we only use 10% of our brain* to consist of the subject *we* and the predicate *only use 10% of our brain*. But neither of these terms functions as such in the premise. Since instead, it is the proposition as a whole that functions as the subject of the premise, the only thing the analyst can say is that the argument has the form '*a* is *X*, because *a* is *X* is *Z*'.

As Wagemans (2019) explains, the problems illustrated through these two examples can be solved by adding the predicate 'is true' to the conclusion or to both the conclusion and the premise of the argument. This epistemic predicate expresses the commitment of the arguer to the truth or acceptability of the statements, which means that the level of analysis changes from that of the 'proposition' to that of the 'assertion'.

If we revisit the examples and perform this shift in the level of the analysis, Example 3 now has the predicate 'is true' (T) as its fulcrum and the relationship between the propositions *he must have gone to the pub* (*q*) and *the interview is cancelled* (*r*) as its lever.

> Example 3 – revisited
> *He must have gone to the pub* (*q*) [*is true*] (T), *because the interview is cancelled* (*r*) [*is true*] (T)

The addition of the epistemic commitment of the speaker as a predicate to the statements allows the analyst to employ the same procedure for argument type identification as with the previous examples of first-order arguments. Example 3 now has the form 'q is T, because r is T' and can therefore be called a second-order subject argument. Given that the predicate 'is true' (T) is labelled within the framework of the PTA as a statement of value (V), the systematic type indicator is '2 sub VV' or 'γ VV'. Finally, the determination of the nature of the relationship between the subjects provides their traditional name (in this case, 'argument from disjuncts').

In revisiting Example 4, it suffices for the analyst to add 'is true' (T) to only the conclusion of the argument. For in so doing, it becomes clear that the argument has the subject *we only use 10% of our brain* (q) as its fulcrum and that its working is based on the relationship between *being said by Einstein* (Z) and *being true* (T).

> Example 4 – revisited
> *We only use 10% of our brain* (q) [*is true*] (T), *because that* (q) *was said by Einstein* (Z)

Given that the conclusion can be labelled as a statement of value and the premise as a statement of fact, this argument can now be identified as a second-order predicate argument with the form 'q is T, because q is Z' and the systematic name '2 pre VF' or 'δ VF'. Traditionally, such an argument is known as the 'argument from authority'.

In sum, the addition of the epistemic predicate 'is true' (T) allows the analyst to identify the type of argument on the basis of determining the common term in the statements expressing the conclusion and the premise of the argument. Following this strategy not only brings the classification of second-order arguments in line with that of first-order arguments, it also has the advantage of enabling the determination of their argumentative lever. In the case of second-order subject arguments, it reveals that their working is based on a relationship between complete propositions. This category thus covers all the arguments that are distinguished in propositional logic, such as the argument from disjuncts. In the case of second-order predicate arguments, the strategy reveals that their working is based on a relationship between something that is predicated of a complete proposition and the truth or acceptability of that proposition. This category thus covers all the arguments that depend in some way or another from the trustworthiness of their source, such as the argument from authority. It is in this sense that the theoretical framework of the *Periodic Table of Arguments* can be seen as a systematic and comprehensive framework that integrates the traditional dialectical accounts of argument schemes and fallacies and the rhetorical accounts of the means of persuasion (see Wagemans, 2016).

4. REPRESENTING SECOND-ORDER ARGUMENTS

Now that we have explained our method for representing first-order arguments by means of arg-adtrees and have indicated the differences and commonalities between first-order and second-order arguments, we turn to propose how to represent the addition of 'is true' (T) to the premise and/or the conclusion of second-order arguments in their corresponding arg-adtree.

Our proposal is based on the following reflections about the nature of the information that is covered in such an adtree. As we mentioned above, an arg-adtree first of all contains linguistic information about the two statements that make up the represented argument. This 'linguistic' information pertains to the morphosyntactic characteristics of these sentences. Second, an arg-adtree contains 'pragmatic' information, by which label we mean to indicate information pertaining to the use of language, in particular its argumentative use of trying to convince an addressee of the acceptability of the conclusion. As we explained by means of an example in Section 2, the pragmatic information covers various aspects of such argumentative language use: the argumentative function of the statements (conclusion or premise), the order of presentation, and the type of argument they substantiate (which includes information about the argument form, i.e., the specific constellation of subjects and predicates of the statements, as well as about the argument substance, i.e., the specific combination of types of statements).

In order to represent second-order arguments in an arg-adtree, it seems to be necessary to first determine whether the information about the epistemic commitment of the arguer to the truth or acceptability of the statements is of a linguistic or a pragmatic nature. If it is of a linguistic nature, as the addition of 'is true' (T) by the analyst suggests, it could be represented as an extra branch in the adtree. If it is of a pragmatic nature, as the notion of epistemic commitment suggests, it could be represented by introducing a symbol for this type of commitment that can be placed under the relevant hook or character in the adtree.

In our view, however, this is a false dilemma, for the simple reason that the analytical strategy of adding 'is true' (T) as a predicate to one or both of the statements that make up the argument can be seen as a *linguistic* expression of *pragmatic* information. In fact, one could add this predicate to the two statements that make up a first-order argument as well. From a pragmatic point of view, someone who puts forward a statement in order to support the acceptability of another statement is committed to the truth or acceptability of both statements as well as their connection (see van Eemeren & Grootendorst, 1992, p. 31). The only reason why this information is left out of the corresponding arg-adtree of a first-order argument such as the one pictured in Figure 1, is that the analyst does not have to add the epistemic commitments in order to identify the type of argument. For second-order arguments, as we explained in the previous section, such an addition is necessary.

Apart from this theoretical justification of why the expression 'is true' can be seen as a linguistic expression of pragmatic information, it is also actually used as such in argumentative discourse. Moreover, in classical rhetorical taxonomies of arguments (*topoi, loci*), one finds examples in which the epistemic commitment is expressed in exactly this way. Cicero, for instance, provides the following example of what he subsumes under the heading of the 'external loci' and can be identified as an argument from authority: 'This is true, for Q. Lutatius has said so'.[3]

The above considerations lead us to propose to represent the pragmatic information about the epistemic commitment of the arguer to the truth or acceptability of the statements in second-order arguments in the corresponding arg-adtrees by means of adding 'is true' as an extra branch in the adtree with the symbol 'T' right under it.

In Figure 2, we pictured the arg-adtree of *He must have gone to the pub (q) [is true] (T), because the interview is cancelled (r) [is true] (T)*, which has been identified as a second-order subject argument.

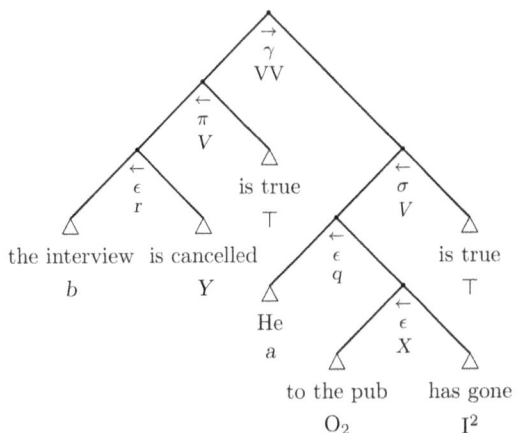

Figure 2 – The arg-adtree of Example 3 – revisited

---

[3] This example is also discussed in Wagemans (2019b, p. 63). For more information about classical rhetorical taxonomies of arguments, see van Eemeren et al. (2014, pp. 86-94).

In this case, the analyst adds the expression 'is true' as a predicate to both the conclusion and the premise of the argument. Also, in a similar way as this has been done for other pragmatic information such as that about the argumentative function of the statements as a conclusion or a premise, the symbol 'T' is placed under the expression.

In Figure 3, we pictured the arg-adtree of *We only use 10% of our brain (q) [is true] (T), because that (q) was said by Einstein (Z)*, which has been identified as a second-order predicate argument.

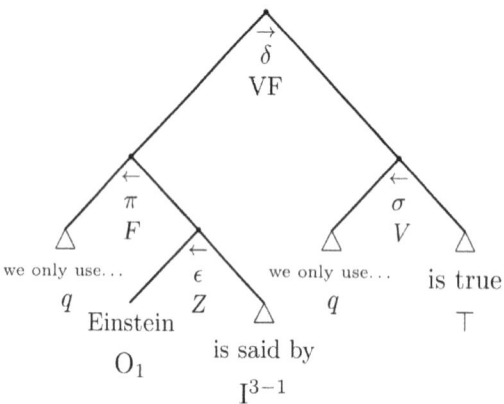

Figure 3 – The arg-adtree of Example 4 – revisited

In this case, the analyst adds the expression 'is true' as a predicate only to the conclusion of the argument. Like in the adtree of the previous example, the symbol 'T' is placed under the expression.

5. CONCLUSION

In this paper, we have proposed a method for representing so-called 'second-order' arguments within the framework of *Adpositional Argumentation* (AdArg). Our proposal is based on the starting points of our method for representing 'first-order' arguments in combination with an analysis of the difference between second-order and first-order arguments. The main conclusion of our research is that the additional pragmatic information about the epistemic commitment of the arguer regarding the truth or acceptability of the statements that make up the argument, which has to be added by the analyst in order to identify the type of argument, can be represented by means of its linguistic expression 'is true' (T). We have illustrated this proposal by providing the arg-adtrees of two examples of second-order arguments.

The considerations that underly our proposal give rise to several new challenges in our project of representing linguistic and pragmatic information about argumentative discourse in arg-adtrees. One question that needs to be addressed is how to add the epistemic commitment of the arguer to arg-adtrees representing first-order arguments. For the fact that adding the linguistic expression of this pragmatic information is not necessary for identifying the type of argument, does not imply that it should not be added at all. In a similar vein, it should be explored how to add other linguistic expressions of pragmatic information to arg-adtrees. An example is 'My conclusion is that [...]', which expression is sometimes used by arguers to indicate the argumentative function of the statement followed by it.

Another issue to be addressed in further research is whether it would be possible to add the negation of the epistemic commitment to arg-adtrees, for instance in the form of the predicate 'is not true' or 'is false' ($\bot$). If this can be done, our approach would cover not only those situations in which a conclusion is *supported* by a premise, but also those in which a conclusion is *refuted* on the basis of a premise. Apart from that, we think that such an extension would enable researchers to study the interrelations between *Adpositional Argumentation* (AdArg) and approaches working with formal argumentation frameworks – for example, those included in Modgil, Budzynska and Lawrence (2018). By conveniently hiding details of the information contained in arg-adtrees, they can be represented as nodes in a network, thereby resulting in something very similar to Dung graphs.

Finally, there is the question of how to represent different linguistic expressions of similar pragmatic information. Apart from inserting 'is true' after the statement, for instance, second-order arguments can also be reconstructed by inserting 'You should accept' before the statement. In the first case, what is added to the original text expresses an epistemic commitment of the arguer. In the second case, however, what is added expresses an epistemic directive towards the addressee. By studying the linguistic and pragmatic characteristics of these and other expressions in more detail, we hope to further develop our approach of *Adpositional Argumentation* (AdArg) as a high-precision method for representing argumentative discourse.

ACKNOWLEDGEMENTS: The authors thank Marco Benini for his thorough reading of the manuscript, and in particular for checking the formal aspects of the arg-adtrees of the two examples of second-order arguments.

REFERENCES

Eemeren, F.H. van, & Grootendorst, R. (1992). *Argumentation, communication, and fallacies*. Hillsdale, NJ: Lawrence Erlbaum.

Eemeren, F.H. van, Garssen, B.J., Krabbe, E.C.W., Snoeck Henkemans, A.F., Verheij, B., & Wagemans, J.H.M. (2014). *Handbook of argumentation theory*. Dordrecht: Springer.

Gobbo, F., & Benini, M. (2011). *Constructive Adpositional Grammars. Foundations of Constructive Linguistics*. Newcastle upon Tyne: Cambridge Scholars Publishing.

Gobbo, F., Benini, M., & Wagemans, J.H.M. (to appear). Adpositional Argumentation Annotation: Guidelines for a Gold Standard Corpus of argumentative discourse. *Intelligenza Artificiale*.

Gobbo, F., & Wagemans, J.H.M. (2019a). A method for reconstructing first-order arguments in natural language. In P. Dondio & L. Longo (Eds.), *Proceedings of the 2nd Workshop on Advances in Argumentation in Artificial Intelligence (AI^3 2018)* (pp. 27-41). Aachen: Sun SITE Central Europe. URL = http://ceur-ws.org/Vol-2296/.

Gobbo, F., & Wagemans, J.H.M. (2019b). Adpositional Argumentation (AdARg): A new method for representing linguistic and pragmatic information about argumentative discourse. In S. Doutre & T. de Lima (Eds.), *Actes 13èmes Journées d'Intelligence Artificielle Fondamentale (JIAF 2019)* (pp. 101-107). Association française pour l'Intelligence Artificielle.

Gobbo, F., & Wagemans, J.H.M. (2019c). Building argumentative adpositional trees: Towards a high precision method for reconstructing arguments in natural language. In B.J. Garssen, D. Godden, G.R. Mitchell & J.H.M. Wagemans (Eds.), *Proceedings of the Ninth Conference of the International Society for the Study of Argumentation* (pp. 408-420). Amsterdam: SIC SAT.

Modgil, S., Budzynska, K., & Lawrence, J. (2018). *Computational models of argument. Proceedings of COMMA 2018*. Amsterdam: IOS Press.

Wagemans, J.H.M. (2016). Constructing a *Periodic Table of Arguments*. In P. Bondy & L. Benacquista (Eds.), *Argumentation, Objectivity, and Bias: Proceedings of the 11th International Conference of the Ontario Society for the Study of Argumentation (OSSA), 18-21 May 2016* (pp. 1-12). Windsor, ON: OSSA.

Wagemans, J.H.M. (2019). Four basic argument forms. *Research in Language, 17*(1), 57-69. DOI: https://doi.org/10.2478/rela-2019-0005.

Argumentation, dissent, and luck

JOB DE GREFTE
University of Groningen
j.a.m.de.grefte@rug.nl

In this paper, I approach the practice of argumentation and the issue of dissent from the perspective of social and anti-luck epistemology. In particular, I show how dissent can exclude reflective luck, and argue that dissent is epistemically legitimate only if it does so.

KEYWORDS: argumentation, dissent, veritic luck, reflective luck, epistemically perverse dissent

1. INTRODUCTION

In this paper, I integrate findings from anti-luck epistemology, social epistemology and argumentation theory to provide a novel perspective on the epistemic value of argumentation. While the perspective developed in this paper is consistent with current veritistic approaches to argumentation (e.g. Goldman, 1994), it goes beyond existing work by incorporating insights from recent anti-luck epistemology. It is argued that argumentation helps us to exclude certain epistemically problematic forms of luck from our beliefs, and doings so provides epistemic value that exceeds the directly veritistic value of producing true belief.

The paper is structured as follows. In section 2, I provide an overview of the findings from anti-luck epistemology and social epistemology that I will be using in this paper. In section 3, I argue that argumentation is a belief-forming method. In section 4, I argue that argumentation eliminates error-possibilities, and doing so, helps us to eliminate certain kinds of luck. Section 5 concludes.

2. ANTI-LUCK EPISTEMOLOGY, SOCIAL EPISTEMOLOGY, ARGUMENTATION THEORY

Let us start by introducing the background theory relevant for the rest of the paper. In this section I focus on anti-luck epistemology and social epistemology, in the next on argumentation theory.

Anti-luck epistemology is a branch of epistemology that investigates the relation between important epistemological concepts and various forms of luck. Ever since Gettier (1963), epistemologists have recognized that certain kinds of luck are incompatible with knowledge. What prevents knowledge in Gettier cases is that one's belief-forming method produces a true belief but it is a matter of luck that it does. This kind of luck is commonly called *veritic luck* (Engel, 1992; Unger, 1968).[1]

Another potentially problematic form of luck is *reflective luck*. One is reflectively lucky in believing proposition *p* just in case it seems from one's reflective perspective that one's belief is veritically lucky. Reflective and veritic luck are distinct kinds of luck because it may seem that one's belief is veritically lucky without it actually being so, and vice versa. As Duncan Pritchard has argued, externalists about knowledge usually only require the absence of veritic luck. Internalists in addition require the absence of reflective luck (Pritchard, 2005). In this paper, unless otherwise specified, 'knowledge' shall refer to internalist knowledge.

What does it mean to say that one's belief-forming method produced a true belief by luck? On the modal analysis used in this paper, luck is given a modal gloss. One's belief-forming method luckily produces a true belief, if and only if it produces a true belief in the actual world, but there are 'close' possible worlds where it instead produces a false belief. Closeness is then defined in terms of world similarity, as it is in standard Lewesian sphere systems (Lewis, 1973).[2]

To determine whether our beliefs are subject to veritic luck, we need to determine whether they are formed in a way that could have easily produced a false belief. In section 4 I will argue that argumentation is one way to do so. Here I want to note that people have not always thought that eliminating nearby error-possibilities is sufficient for knowledge. Descartes, for example, thought that we could only know if every conceivable source of error has been eliminated (Descartes, 1996). Since that road quickly leads to skepticism, contemporary epistemologists usually demand less; instead of absolute certainty for our beliefs, they demand *safety,* where a belief is safe if and only if not produced by a method that could have easily produced a false

---

[1] Veritic luck has recently been subdivided into intervening veritic luck and environmental veritic luck (Carter & Pritchard, 2015). For our purposes this distinction is insubstantial.

[2] There are many interesting and puzzling open questions regarding the nature of world-similarity and in particular its measure. These issues are not directly relevant to the present paper, and so we leave them open. The only assumption I make is that some measure can be found.

belief. (Pritchard, 2008; Sosa, 1999; Williamson, 2000). By the definition of veritic luck, this entails that knowledge requires the elimination of veritic luck

A second field of relevance for this paper is social epistemology (e.g. Goldman, 1987, 1999). Argumentation is, after all a social process, and studying its epistemic properties is squarely within the purview of social epistemology, a field that studies the epistemic properties of social interactions.[3] It is thus not surprising that argumentation has been a focal point of social epistemology for quite some time (Goldman, 1994, 1997, 2003).

The approach in this paper is to expand on existing work on argumentation in social epistemology by incorporating insights from anti-luck epistemology. I will argue that there are clear benefits of doing so. First, it leads to a better understanding of the epistemic value of argumentation. Second, it enables a novel perspective of an open puzzle in argumentation theory: a puzzle concerning the persistent interlocutor.

So far, argumentation has been investigated from a veritistic point of view, where norms for good argumentation are derived from its ability to produce *true* beliefs (Goldman, 1994). Since knowledge requires truth, these veritistic norms have epistemic value. But knowledge requires more than just truth, among other things, as we saw above, the absence of certain forms of luck. We may thus extend and contribute to the understanding of the epistemic properties of argumentation by investigating the extent to which argumentation may not only produce true belief, but non-luckily true belief. That is the central aim of this paper.

3. ARGUMENTATION AS A SOCIAL BELIEF-FORMING PROCESS

In the previous section we stated that argumentation is a social belief-forming process. In this section, we will provide a more precise description of argumentation as a social belief-forming process, drawing on the pragma-dialectical theory of argumentation (Van Eemeren & Grootendorst, 2016; Van Eemeren, Grootendorst, & Eemeren, 2004).

A belief-forming process is a process that produces beliefs. If that process occurs within the skull of a single agent, we speak of individual belief-forming processes. If it crucially involves other people, we speak of a social belief-forming process. These processes can be mixed: if I form a belief about *p* on the basis of the social belief-forming process 'relying on testimony', then part of this social process will

---

[3] For recent overviews of field, see (Goldman & O'Connor, 2019; Goldman & Whitcomb, 2010; Haddock, Millar, & Pritchard, 2010)

consist of my individual belief-forming process 'relying on auditive stimuli'.

Argumentation is a social belief-forming process, since it crucially involves different parties arguing with each other. Sometimes, one of the parties' voices is internalized, and we 'argue with ourselves'. Still, the basic case is the social setting were two or more people are involved. That is at least the sense of argumentation with which we will be concerned in this paper.

Belief-forming processes can be described on different levels of generality. The same process may be described as 'relying on eyesight', 'relying on eyesight in good lighting conditions', 'relying on eyesight in good lighting conditions while awake and looking at medium-sized objects from medium distance', etc. Infamously, we lack principled reasons for saying one of these levels is epistemically the most relevant. (Conee & Feldman, 1998). Without such a principled distinction, there seems to be no answer to the question what *the* belief-forming process is in a given case. This problem, known as the generality problem, affects our paper because both veritic and reflective luck draw on the notion of a belief-forming process.

Besides the fact that this problem is not specific to anti-luck epistemology, and in fact plagues all major theories of epistemic justification (Bishop, 2010; Comesaña, 2006), I will sidestep this problem by stipulation. For the argumentation theory literature contains many fairly specific descriptions of the various roles and steps in the process of argumentation, and we may simply select one of those processes and ask whether *that* process will help us eliminate luck.[4]

In particular, the description of the argumentative process that we will draw on in this paper is known as the pragma-dialectical framework of argumentation developed by Frans van Eemeren and Rob Grootendorst (Van Eemeren & Grootendorst, 2016; Van Eemeren et al., 2004).[5] According to the pragma-dialectical theory, argumentation is "a complex speech act, the purpose of which is to contribute to the resolution of a difference of opinion, or dispute" (Van Eemeren & Grootendorst, 2016).[6]

---

[5] Not only is this a prominent recent approach, the roles of opponent and proponent, on which our argument primarily depend, are found in other approaches as well, indicating that our findings will generalize. Nevertheless, for reasons of concreteness and clarity, we focus on the pragma-dialectical theory.

[6] A complex speech act is a speech act that may consist of various sentences, contrary to simple speech acts, which can consist only of a single sentence.

According to van Eemeren and Grootendorst, argumentation is an attempt to resolve a difference of opinion. Differences of opinion can take multiple forms. Pragma-dialecticians distinguish two roles in argumentation: that of the proponent and that of the opponent. The role of proponent is to advance and defend a thesis, and the role of opponent is to cast doubt on the thesis. Van Eemeren and Grootendorst speak of an unmixed dispute if proponent only asserts a thesis and opponent merely tries to cast doubt on the thesis, and of a mixed dispute when opponent advances theses of her own (Van Eemeren & Grootendorst, 2016). Mixed disputes can always be broken down into separate unmixed disputes. If proponent asserts, for example, that it will rain tomorrow, and opponent advances the thesis that it will not rain tomorrow, this mixed dispute can be broken down into two separate unmixed disputes: one about proponents thesis that it will rain tomorrow (and opponent expressing doubt about that) and one about *another* unmixed dispute where opponent (now taking the role of proponent in this second unmixed dispute) advances the thesis that it will not rain tomorrow (and proponent taking the role of opponent and expressing doubt about this latter claim). In what follows we will focus on the unmixed dispute, since it is the basic case.

If argumentation is a process, what kind of process is it? Van Eemeren and Grootendorst characterize it as a process that normally proceeds according to a fairly specific set of stages, the first of which is the *confrontation* stage, in which a difference of opinion is recognized. One of the parties expresses an opinion and the other party at least doubts whether the position is tenable.[7]

Recognizing a difference of opinion is only the start of argumentation. After all, proponent and opponent may each go their own ways after recognizing their difference of opinion. For argumentation to start, the process has to move on to the next stage: the *opening stage*. In this stage, proponent and opponent form the intention to *resolve* their difference of opinion, and thus in effect, to engage in argumentation. In this stage, proponent and opponent lay out the rules of engagement: what premises may be assumed as background knowledge, and what inference patterns will count as valid?

Once the starting points and rules of the discussion have been agreed upon, proponent and opponent enter the third stage: the *argumentation stage*. In a unmixed dispute, proponent's sole role in this stage is to defend her standpoint, and opponents only role is to cast doubt on that standpoint, taking into account their common starting points and the agreed upon rules.

---

[7] Again, we look at unmixed disputes here as the basic case.

The result of the argumentation stage is either that proponent has successfully defended her standpoint, or that opponent has successfully been able to cast doubt on this standpoint. Once neither party is able to make further argumentative moves, the discussion moves towards the final stage: the *closing* stage, where the difference of opinion is resolved, either in favor of the standpoint of the proponent or in favor of the doubt concerning that standpoint expressed by the opponent.

Several remarks need to be made about this cursory overview of the stages recognized in pragma-dialectical theory of argumentation. The first is that these stages are rarely explicitly recognized in real argumentative settings, and even more rarely do parties to the debate run through these stages in order. Often, argumentation proceeds by leaving much background knowledge implicit, and parties may only find out that some thesis is not part of the background knowledge during the argumentation stage. This does not detract from the theoretical usefulness of the model, nor from the use we make of it in this paper. The pragma-dialectical model of argumentation is meant as an ideal, and in this paper, we are concerned with establishing how argumentation, *if it proceeds as it should*, can help us acquire knowledge. So, we may safely lay any worries about the idealizations in the model to one side.

The above should suffice to show that argumentation on the pragma-dialectical theory is a fairly regimented process or method. But is it a *belief-forming* method? In order to answer this question, we need to establish that the process of argumentation results in the formation of beliefs.

According to pragma-dialecticians, argumentation is a process with a particular aim; namely, to resolve a difference of opinion. Here, we assume that one's opinion about p can be modelled as a belief in either p or its negation. No difference of opinion can be resolved without one party changing their opinion. If opinions are beliefs, then this means the aim of argumentation is to change at least one party's beliefs. Again, this may not always occur in practice, but at least in the ideal case argumentation is concluded in the closing stage with either proponent maintaining their thesis, and thus opponent changing their beliefs about the dubitability of the standpoint, or opponent maintaining their doubt, meaning that proponent has changed her belief about the defensibility of the standpoint. On the pragma-dialectical theory, argumentation has the primary aim of forming new beliefs.

It must be recognized here that the above seems to conflict with one of the methodological principles professed by van Eemeren and Grootendorst, that is, the principle of *externalization*:

> Externalization is achieved by starting from what people have expressed, implicitly or explicitly, instead of speculating about what they think or believe. ... Insofar as implicit elements can be made explicit in an adequate reconstruction, they can also be used, so that everything that creates a commitment for the language users is taken into account. (Van Eemeren & Grootendorst, 2016)

According to this quote, the pragma-dialectical theory derives commitments for the parties involved in argumentation not from what is believed but from what is (implicitly) expressed. This is compatible with our claim above, that argumentation has the primary aim of changing people's beliefs. Even if the *legitimacy* of the moves made by proponent and opponent in argumentation depends only on their public commitments and not on their private beliefs, the intended aim of argumentation is still to change parties' private beliefs.

The public commitments are connected to the private beliefs in the following way. Some commitments incurred are commitments *to change one's private beliefs*. In particular, when the argumentation stage is over, and proponent has successfully defended their standpoint, then opponent incurs a commitment to change her belief about the dubitability of the standpoint. Conversely, if opponent has successfully cast doubt on the standpoint, then proponent incurs a commitment to change her belief (about the defensibility of the standpoint). No matter what happens, by the time of the closing stage, one of the parties has incurred a commitment to change their belief.

Two closing remarks are in order. First, actual argumentation may fall short of the ideal described by the pragma-dialectical theory. People are generally stubborn in their beliefs, and sometimes even the best arguments fail to produce an actual belief-change, even if the relevant party has incurred the commitment to make the change. Beliefs are in practice not always under such direct control.[8] But remember that I aim to show that argumentation, *if it lives up to the ideal specified by the pragma-dialectical theory*, will result in belief-change, and this requires the parties to the discussion to live up to the commitments they incurred during the process, and therefore, for at least one of them to change their beliefs.

Second, the closing stage is not the only stage in which the parties to the discussion may form new beliefs. In the argument stage, and even in the confrontation and opening stages, many new beliefs may be formed, for example, beliefs about the beliefs of the other party, about their background knowledge, about the considerations that speak in favor or against a given standpoint, and so on. But this need not

---

[8] Cf. (Alston, 1989, p. 91ff)

necessarily be the case. Perhaps opponent and proponent know each other and the considerations for and against the standpoint that they are discussion extremely well, and so they may gain no new beliefs in the process of arguing. But if argumentation is aimed at resolving a difference of opinion, this means that in successful argumentation, it has to be at least the case that in the closing stage one of the parties forms a new belief about the standpoint under consideration.

4. NEARBY ERROR

Let us briefly take stock. In the first section, I spelled out the aim of the paper: to combine anti-luck epistemology with social epistemology and argumentation theory to investigate the epistemic properties of argumentation, particularly in terms of epistemic luck. The project is to show that argumentation excludes certain kinds of luck and that, since knowledge is incompatible with this kind of luck, argumentation helps us to acquire knowledge. In the previous paragraph I argued that argumentation is a belief-forming method. In this section I will argue that it is a belief-forming method that helps us eliminate luck from our belief.

Let us start by looking at the different roles in argumentation. On the one hand, we have the proponent of the standpoint, whose task it is to defend her standpoint. On the other, we have the opponent, whose task it is to question the standpoint. How does opponent question a standpoint? There are two possibilities: opponent may pose a motivated or an unmotivated challenge. The difference between these is that a motivated challenge contains a reason for believing the standpoint is false, whereas an unmotivated challenge merely asks the proponent to provide additional support for the standpoint. Consider the following example:

> Jill: The earth is round
> Jack: Why would that be true?
> Jill: Because it appears that way from space.
> Jack: But aren't appearances sometimes deceiving?

Here, Jill is the proponent of the thesis that the earth is round, and Jack the opponent. Jack first poses an unmotivated challenge to the thesis, and then poses a motivated challenge. Both of these challenges are legitimate according to the pragma-dialectical theory, and if Jill wants to succeed in defending here standpoint she has to respond to both challenges.

Let us define error-possibilities as possible worlds in which one's belief is false. What I want to suggest is that both motivated and unmotivated challenges posit error possibilities, and a successful defense of the standpoint by the proponent requires both kinds of error-possibilities to be eliminated. What are the relevant error-possibilities? In the example above, it may not be immediately apparent that Jack's unmotivated challenge posits an error possibility, for Jack does not provide a concrete scenario in which what Jill said is false. But questioning the truth of a statement is tantamount to acknowledging the *possibility* that it is false. Error-possibilities are just that: possibilities of being wrong. If Jack questions the truth of Jill's standpoint, he is in fact raising the possibility that it is false. That Jack does not provide support for his doubt does not make his challenge illegitimate on the pragma-dialectical theory of argumentation: rather, puts a burden on the proponent of the standpoint to provide reasons for thinking it is true, to provide reasons, that is, for thinking that this error possibility does not obtain.

Motivated challenges also raise error possibilities. Consider again the example above. Jack's motivated challenge consists of raising the explicit possibility that the appearances used by Jill to support her main standpoint are deceiving, and her standpoint is wrong. To defend herself, Jill has to provide support for the claim that the appearances are not deceiving in this case, and therefore, Jill has to eliminate the error possibility raised by Jack's motivated challenge.

I submit that what goes for the case of Jill and Jack goes for argumentation generally. In the pragma-dialectical framework, opponents are allowed to raise both motivated and unmotivated challenges to the standpoint or standpoints put forward by the proponent. In both cases, such challenges involve positing error possibilities, and in both cases, the proponent incurs a commitment to eliminate these error possibilities. Argumentation (in the ideal case) is a belief-forming process in the course of which at least some error possibilities are excluded.

So far, the claim that argumentation excludes error possibilities may seem fairly obvious. After all, what is argumentation other than a game of giving and asking for reasons, and what are reasons other than considerations that show a standpoint is true, i.e. not false? What is the value of looking at argumentation through an anti-luck lens specifically?

There are two ways in which the present analysis improves upon existing work. First, I have argued that argumentation helps us to eliminate error-possibilities. Since argumentation is a social belief-forming process, this means we have found a belief-forming method that may not only produce true belief, but non-lucky true belief. Since knowledge requires non-lucky true belief, we may hope to acquire knowledge by argumentation. This provides a more detailed view of the epistemic benefits of argumentation than present in the literature.

Standard work on argumentation talks about the reliability of argumentation (e.g. Goldman, 1994, 2003), but we have sketched a picture where argumentation may not only be reliable, but *safe* as well. To the best of my knowledge, this possible epistemic benefit of argumentation has not been identified before.

The findings above allow us to sketch an even more detailed picture of the value of argumentation. For as we saw above, the difference between veritic luck and reflective luck is that for veritic luck, it matters whether one's belief-forming method could have *actually* produced a false belief in a nearby world, whereas for reflective luck it matters whether this appears to be so from one's reflective perspective. In defending her standpoint against raised error-possibilities, proponent can only draw on the information she has reflectively accessible. It is for this reason that the kind of luck that argumentation helps to eliminate is of the reflective kind. Since internalist, *reflective* knowledge requires the elimination of reflective luck, but externalist, animal knowledge does not (de Grefte, 2018; Pritchard, 2005; Sosa, 2007), this means that argumentation will be valuable particularly with respect to acquiring the former kind of knowledge.[9]

Aside from a better picture of the epistemic value of argumentation, our analysis also reveals something about dissent in argumentation. One of the main contributions that anti-luck epistemology has for argumentation theory is that it gives us a principles way of distinguishing between raised error-possibilities. For the anti-luck epistemologist, not all error-possibilities are created equal. As we saw above, a belief is subject to reflective luck just in case it is true and produced by a method that, from one's reflectively accessible perspective, could have *easily* produced a false belief. Crucially, this is not the same as requiring that the belief-forming method could not have produced a false belief at all. To determine whether a belief is reflectively lucky, we thus do not need to exclude *all* error-possibilities, but merely the close ones.

All challenges by opponent raise error-possibilities, but not all raise *nearby* error possibilities. As we saw in section 2, nearby error-possibilities are worlds that are substantially like our own where one's method produces a false belief. By contrast, a far-off error possibility involves a world substantially *unlike* our own where one's method

---

[9] This does not mean that it is impossible to acquire externalist knowledge by argumentation. As people like Goldman have stressed before, argumentation may simply be a reliable way of forming one's beliefs. Similarly, the reflectively accessible beliefs involved in argumentation may simply be true, and so the process may on occasion eliminate veritic luck as well. But we cannot say that it is an inherent feature of argumentation that veritic luck is diminished in argumentation.

produces a false belief. Opponent may raise either nearby or far-off error possibilities in the challenge to a standpoint. Consider the following exchange:

> Jeremy: Climate change is real
> Jaimy: But have you not heard of several studies concluding it is not?
> Jeremy: Those studies have weak methodologies.
> Jaimy: But what if there is no external world? In that case, there is no climate and consequently no climate change.

In the first case, Jaimy is raising a nearby error possibility. This is so because the error-possibility in question invokes a world where several scientific studies point towards the falsity of climate change and it is indeed false. A world where empirical research is a reliable guide to truth is a world that is (presumably) much like what the world according to Jaimy and Jeremy's reflective perspective is like. That this is so for many people is shown by the many adherents these scientific studies still have.

Now let us look at the second challenge of Jaimy. Here, he raises a far-off error possibility. This is so because it invokes the possibility that there is no external world at all, and such a possible world is presumably very much unlike our own.

We have seen an example of opponent raising a nearby error-possibility, and raising a far-off error-possibility. This suffices to show that it is possible for opponent to raise both kinds of error-possibility. Crucially, however, only the first case will aid in the exclusion of reflective luck, since for this only nearby error-possibilities are relevant. From the perspective of the pragma-dialectical theory, all challenges by opponent, as long as they respect the shared starting points and inference rules agreed upon in the opening state, are legitimate. So, raising far-off error possibilities is *argumentatively* legitimate. But challenges that involve far-off error-possibilities can be criticized from the epistemic point of view. The elimination of such error-possibilities is not necessary for reflective knowledge. This kind of knowledge is among the most valuable epistemic states, one that arguably includes other valuable epistemic states like animal knowledge and epistemic justification (Sosa, 2009, 2015). If this is so, then discussing far-off error-possibilities is not only irrelevant for reflective knowledge, but for animal knowledge and justification as well. it is hard to see what the *epistemic* value is of eliminating error-possibilities irrelevant for knowledge or justification.

This is not to deny that argumentation may serve other purposes than that of producing reflective knowledge. For such other purposes, raising far-off error-possibilities in argumentation may be beneficial. But from the epistemic point of view, there is something amiss with discussing such issues; the discussion takes up valuable cognitive resources and contributes nothing to the acquirement of our most coveted epistemic states. Accordingly, we may call argumentation that involves the discussion of far-off error-possibilities *epistemically perverse*, in contrast to epistemically legitimate argumentation that involves nearby error-possibilities.

5. CONCLUSION

Let us recap. In this paper, I have used findings from anti-luck epistemology to provide a novel perspective on the epistemic value of argumentation. In particular, I argued that argumentation is a social belief-forming process that helps to eliminate reflective error-possibilities. In doing so, argumentation excludes reflective luck. Since this kind of luck is incompatible with reflective knowledge, eliminating it will help us achieve this kind of knowledge. But not all dissent in argumentation will help fulfill this function. In contrast to epistemically legitimate challenges, which involve nearby error-possibilities, we may distinguish epistemically perverse dissent, dissent that involves far-off error-possibilities and that is irrelevant to the acquirement of reflective knowledge.

In this paper, we sought to clarify the connection between anti-luck epistemology, social epistemology and argumentation theory. We have seen there are close connections: if done properly, argumentation may help eliminate luck and provide knowledge. But we also saw that the relation between argumentative error and epistemic error is loose. Dissent may be epistemically perverse but argumentatively legitimate. Contrary to what people like Robert Brandom have argued, the game of rational belief and knowledge has different rules than the game of giving and asking for reasons.

ACKNOWLEDGEMENTS: I would like to thank the audience of the European Conference on Argumentation 2019 in Groningen for valuable feedback on an earlier version of this paper.

REFERENCES

Alston, W. P. (1989). *Epistemic Justification: Essays in the Theory of Knowledge*. Ithaca, NY: Cornell University Press.
Bishop, M. A. (2010). Why the generality problem is everybody's problem. *Philosophical Studies: An International Journal for Philosophy in the*

Analytic Tradition, *151*(2), 285–298.
Carter, J. A., & Pritchard, D. (2015). Knowledge How and Epistemic Luck. *Noûs*, *49*(3), 440–453.
Comesaña, J. (2006). A Well-Founded Solution to the Generality Problem. *Philosophical Studies*, *129*(1).
Conee, E., & Feldman, R. (1998). The Generality Problem for Reliabilism. *Philosophical Studies*, *89*(1), 1–29.
de Grefte, J. (2018). Epistemic justification and epistemic luck. *Synthese*, *195*(9), 3821–3836.
Descartes, R. (1996). *Meditations on First Philosophy: With Selections From the Objections and Replies*. Cambridge University Press.
Engel, M. (1992). Is Epistemic Luck Compatible with Knowledge? *The Southern Journal of Philosophy*, *30*(2), 59–75.
Gettier, E. L. (1963). Is Justified True Belief Knowledge? *Analysis*, *23*(6), 121–123.
Goldman, A. I. (1987). Foundations of social epistemics. *Synthese*, *73*(1).
Goldman, A. I. (1994). Argumentation and social epistemology. *The Journal of Philosophy*, *91*(1), 27–49.
Goldman, A. I. (1997). Argumentation and Interpersonal Justification. *Argumentation*, *11*(2), 155–164.
Goldman, A. I. (1999). *Knowledge in a Social World*. Oxford University Press.
Goldman, A. I. (2003). An epistemological approach to argumentation. *Informal Logic*, *23*(1).
Goldman, A. I., & O'Connor, C. (2019). Social Epistemology. In E. N. Zalta (Ed.), *Stanford Encyclopedia of Philosophy* (fall 2019). Metaphysics Research Lab, Stanford University.
Goldman, A. I., & Whitcomb, D. (2010). *Social Epistemology: Essential Readings*. Oxford University Press.
Haddock, A., Millar, A., & Pritchard, D. (Eds.). (2010). *Social epistemology*. Oxford: Oxford University Press.
Lewis, D. K. (1973). *Counterfactuals*. Cambridge, MA: Harvard University Press.
Pritchard, D. (2005). *Epistemic luck*. New York, NY: Oxford University Press.
Pritchard, D. (2008). Sensitivity, safety, and anti-luck epistemology. *The Oxford Handbook of Skepticism*, 437–455.
Sosa, E. (1999). How to Defeat Opposition to Moore. *Nous*, *33*(s13), 141–153.
Sosa, E. (2007). A Virtue Epistemology. In *A Virtue Epistemology*. Oxford: Oxford University Press.
Sosa, E. (2009). *Reflective Knowledge: Apt Belief and Reflective Knowledge, Volume II*. Oxford: Oxford University Press.
Sosa, E. (2015). *Judgment and Agency*. Oxford: Oxford University Press.
Unger, P. (1968). An analyis of Factual Knowledge. *The Journal of Philosophy*, *65*(6), 157–170.
Van Eemeren, F. H., & Grootendorst, R. (2016). *Argumentation, communication, and fallacies: A pragma-dialectical perspective*. Routledge.
Van Eemeren, F. H., Grootendorst, R., & Eemeren, F. H. (2004). *A systematic theory of argumentation: The pragma-dialectical approach*. Cambridge University Press.
Williamson, T. (2000). *Knowledge and its Limits*. Oxford, NY: Oxford University Press.

ARGUMENT SCHEME THEORY[1]

HANS V. HANSEN
CRRAR, *University of Windsor, Canada*
hhansen@uwindsor.ca

I propose to study what we might call the theory or the meta-theory of argument schemes. Such a study will highlight not only the theoretical problems but also the practical problems for matters of understanding and argument making. The subject is studied under the headings of comparison with formal logic, functionality, comprehensiveness, completeness, genesis, normativity and effectiveness. Previous publications by Walton Garrsen, Hitchcock, Blair, Pinto and Prakken, serve as points of departure.

KEYWORDS: argument kinds, argument schemes, descriptive schemes, logical constants, schematic constants.

In this essay I want to pursue two questions belonging to Argument Scheme Theory. Part 1 considers the differences and similarities between logical forms and argument schemes; Part 2 speculates about the relationship between argument schemes and argument kinds.

1. SCHEMES AND FORMS

Walton, Reed and Macagno, in the most comprehensive study of the subject, define argument schemes as follows:

Df W1 ARGUMENTATION SCHEMES are [i] forms of argument (structures or inferences) [ii] that represent structures of common types of

---

[1] I dedicate this essay to the memory of my friend and teacher, Dr Robert C. Pinto (1934 - 2019). Earlier versions were presented in 2019 at CRRAR in the University of Windsor and the European Conference on Argumentation at the University of Groningen. For helpful discussion of the issues in this essay I thank J.A. Blair, Jose Gasćon, David Hitchcock, Catherine Hundleby, Douglas Walton, Christopher Tindale, and Waleed Mebane.,

arguments used in everyday discourse as well as [iii] in special contexts like those of legal argumentation and scientific argumentation. [iv] They include the deductive and inductive forms of argument that we are already so familiar in logic. However [v] they also represent forms of argument that are neither deductive nor inductive, but that fall into a third category, sometimes called defeasible, presumptive, or abductive. (Walton et al. 2008, p. 1)

Just two years earlier Walton gave a slightly different definition of argument schemes that neglected some of these conditions but introduced some others. Argumentation schemes, he wrote,

Df W2 [i] are distinct forms of argument ... [ii] [that are] ... inherently presumptive and defeasible, and thus [iii] they are different in nature from deductive and inductive arguments. Each of the forms of argument ... [iv] is used as a presumptive argument in a dialogue that carries a weight of plausibility. [v] If the respondent accepts the premises, then that gives him a good reason also to accept the conclusion ... [vi] They are used to shift a burden of proof to one side or the other in a dialogue ... . (Walton 2006, p. 84)

Combining the ideas contained in these two definitions, we can make an overview of the several proposed aspects of argument schemes as follows.

ARGUMENT SCHEMES AND THEIR INSTANTIATIONS	
PROPERTIES	ASPECT
1. schemes are **patterns** of arguments	- meta-logical aspect
2. schemes are ordered **sequences of sentence** forms	- syntactic aspect
3. instantiations of schemes are **defeasible arguments**	- logical aspect
4. instantiations of schemes are **common types** of arguments	- social (empirical) aspect

5. instantiations of schemes **have a weight of plausibility**	- normative aspect
6. conclusions of instantiations of schemes are **presumptions**	- linguistic / semantic aspect
7. instantiations of schemes are used to shift a **burden of proof**.	- interactional / dialogical aspect

It is especially properties 3 to 7 that are claimed to be characteristic of the kinds of arguments that have drawn the attention of scheme theorists studying argumentation. Properties 1 to 3 hold for inductive argument schemes as well as for presumptive argument schemes.

There are, however, noticeable differences between Df-W1 and Df-W2. The first definition informs us, in neutral terms, that argument schemes are of common types of arguments and also of some of the types of arguments in specialized fields. This is not emphasized in the second definition. More striking it is that whereas Df-W1 explicitly includes deductive and inductive logical forms as argument scheme structures, Df-W2 is equally explicit in excluding them. As well, it is noticeable that whereas the second definition tells that argument schemes have a weight of plausibility, i.e., that arguments that fit the schemes give some support to their conclusions, the first definition makes no such claim. These two definitions were given at different times by different sets of authors in a rapidly evolving field of research. Nitpicking would not be appropriate. Nevertheless, the points of difference in the two definitions does bring two questions to the fore that are of great importance to informal logicians. Question 1 concerns the difference between logical forms and argument schemes, if any there is. Question 2 is about whether or not arguments instancing the schemes are normative, i.e., have a weight of plausibility in virtue of being instances of the schemes. It is then the second and third dimensions of schemes mentioned above that I want to explore: the syntactic and normative aspects.

1.2

Why are we interested in the difference between logical forms and argument schemes? It is because they are similar kinds of patterns of arguments that are used for the analysis and evaluation of natural language arguments, but it is unclear what it is that makes them different from each other.

In Robert Audi's *Dictionary,* John Corcoran (1999, p. 818) writes that a scheme is a "a metalinguistic frame or template used to specify an infinite set of sentences, its instances, by finite means, often taken with a side condition on how its blanks or placeholders are to be filled." Corcoran's definition does not mention arguments; he speaks only broadly of metalinguistic frameworks that can be used to generate sentences. David Hitchcock (2010, p. 157) adopts schemes to arguments like this:

Df H   An ARGUMENT SCHEME[2] is (i) a pattern of argument, (ii) a sequence of sentential forms with variables, with (iii) the last sentential form introduced by a conclusion indicator like 'so' or 'therefore'.

We must be clear that schemes are not themselves arguments, but patterns of arguments. They are patterns built not from sentences but from sentence forms. A sentence form is not a sentence itself (and thus neither a premise nor a conclusion) but a pattern or structure made of fixed words (constants) and variables such that when the variables are appropriately replaced by words, a well-formed sentence results. One of the sentence forms in the scheme is designated as the one that will become an argument's conclusion once it is turned into a sentence; the other sentence forms in the scheme when they become sentences will turn into premises. When all the variables in all the sentence forms in a scheme are replaced by appropriate words (there are restrictions on this), the sentences collectively become an argument as is made explicit by the presence of a conclusion indicator word.

However, what Hitchcock has said about argument schemes applies as well to the logical forms of symbolic logic. How, then, shall we distinguish forms and schemes?   One suggestion is that argument schemes are "not so abstract that they become purely formal schemes like the valid scheme for modus ponens arguments" (Hitchcock 2010, p. 157).   The difference between forms and schemes is to be found in the level of generality of the argument patterns, the forms being too general to be schemes, and the schemes insufficiently general to be forms. Nevertheless, this distinction is vague. Let us consider what might underlie it.

1.3 About logical forms

Modus ponens and tollens, hypothetical syllogisms, disjunctive syllogisms and constructive and destructive dilemmas are well-known valid forms of sentential logic. They are forms (patterns, structures) of arguments or

---

[2] I have changed 'argumentation scheme' to 'argument scheme'.

inferences consisting of logical constants and sentential (propositional) variables. (Going forward I will try to stick with the language of sentences.) In sentential logic, the most basic kind of modern formal logic, the only constants are sentence connectives and the only variables are sentential variables. The sentence connectives are constants because they are specified as always have the same (constant) meaning; the sentential variables are variables because they can be replaced by any indicative sentences, long or short, true or false. When the constants are combined with sentences substituted for the variables, compound (or molecular) sentences result; when all the variables in a logical form have been replaced with sentences, an argument results. The constants chosen for sentential logic are truth-functional constants, most often these: '&', 'v', '~', and '⊃'. They are selected because each one of them resembles a natural language sentence connective: 'and', 'or', 'not', and 'if-then', respectively.   When the logical constants are combined with propositional variables, $p, q, r, ...$, logical forms result, e.g.,

If p then q	p **or** q	p **and** q
p	**Not**-p	so, q
so, q	so, q	

When we move up to the next level of formal logic, predicate logic, (or quantification theory, or first-order logic), the vocabulary is still divided between constants and variables. Here predicate letters, F, G, H, etc., are introduces and assigned a constant meaning in a given context of argument. They can stand for properties (one-place predicates, e.g., 'Fx') and relations (multiple-place predicates, 'Gxy', 'Hxyz', etc.) Also needed are variables for individual things (u, v, w, ...). Since predicate logic incorporates propositional logic it already has a stock of constants and variables with which to begin. It adds more: the new logical constants are the quantifiers '∃', and '∀' – they always have the same meaning, 'some' and 'all'[3]. Also added is identity ('='). When the ambit of logic reaches even further into modalities – alethic, epistemic, temporal, deontic, etc. – it is just a matter of adding more constants and variables. In alethic modal logic, for example, we add the constants 'necessary' (□) and 'possible' (◊), and a new range of variables for possible worlds (w_1, w_2, w_3, etc.). I apologize for compressing so much technical detail into a very few sentences. The details are not important, it is the general point

---

[3] These logical symbols are not at all necessary for the deployment of forms in logic, they are just very convenient, and because their meanings are stipulated they avert discussion of what they mean.)

that matters: logical forms, of all manner of complexity, are made of constants and variables (and punctuation devices).

1.4 About argument schemes.

Let us then consider schemes also in their syntactical aspects. Like argument patterns in the various formal logics, argument schemes also consist of constants and variables. However, the constants of *scheme logic* (as we might call it) are different than the constants of formal logic. We can call them 'schematic constants'. Examples of argument schemes are:

X's **cause** Y's	X **indicates that** p	*It is false that*
X **obtains**	So, p.	**there is evidence that** p
So, Y will **obtain**.		So, *not-p*.

In the left-most scheme are 'cause' and 'obtains' (is /will be present) which have the roles of constants; X and Y are variables ranging over individual events or states. In the centre-scheme, 'indicates that' is a schematic constant, and X is a variable for a person and p is a variable for a proposition. In the example on the far right, 'there is evidence that' is a schematic constant combining with the truth-functional constants in italics, 'it is false that' and 'not', and p is a variable for propositions. In the first example it is clear that 'cause' and 'obtains' are not truth-functional constants because the variables in their range are not propositions; hence there is no truth-value to compute. In the second and third example, although the inputs are propositions, at least one of the constants in each instance is not truth functional: even with substitutions made we would not be able to calculate the truth value of p, or come to a justified belief about p, from 'A indicates that p', or from 'it is false there is no evidence for p'. Thus, each of the schemes above contains at least one constant that does not belong to logic (as reviewed above) but some of them do contain a logical constant. Let us call these non-logical constants, schematic constants. Like the logical constants, the schematic constants are to be considered as always having he same meaning[4] in whatever argument results from when substitutions are made on the variables.

Some schemes are not distinguished by the nature of their constants, but by the nature of their quantifiers. Consider this scheme for argument from sign, modelled on Walton et al., 2008 (p. 329).

---

[4] Doug Walton, in conversation, has cautioned me against taking this constancy too strictly.

Generally when A (is true) then B (is true);
A is true;
So, (presumably) B is true.

'Generally' is not the universal quantifier of formal logic (∀), and it is not a statistical quantifier like 'most' or '*n* %' either. Some quantifiers seems especially well suited for presumptions and plausible propositions, and 'Generally' is one of them. It allows for some exceptions to the conditional in its scope, as do alternative quantifiers 'normally' and 'usually'. Since these quantifiers are found in connection with presumptive reasoning, I will call them 'presumptive quantifiers'.

1.5 Comparison of logical forms and argument schemes.

	Formal Logic	**Schematic Logic**
Constants	and (&); or (v); not (~); if-then (⊃) predicates (F, G, H, ...)	says that; causes; is an expert; is similar to; means that; correlates with; is between; is committed to; is a part of; is widely accepted; is plausible; is a part of; can be classified as; is a means to; is a sign of ... etc. properties / predicates
	QUANTIFIERS: some (∃) ; all (∀) ; identity (=) ; necessarily (□) ; possible (◊)	QUANTIFIERS: generally ; normally ; usually
Variables	propositions (p, q, r, ...) individuals (u, v, w, ...) world variables (w1, w2, w3, ...)	propositions persons; objects; events (states); actions; cases (situations)

There are no important differences in the variables of formal logic and schematic logic: they both use propositional variables and the variety of variables occurring in schemes (for causes, events, cases, persons, etc.) can all be subsumed under the variables for individuals in predicate logic. Hence, the difference between formal logic and schematic logic is found in the difference of their respective constants and quantifiers.

If we want to define 'argument scheme' in such a way that it does not include logical forms we will need to introduce a distinction. It is between using and mentioning a schematic constant. The sentence form "X said that p" uses the schematic constant '... said that ...' . But the sentence form "X's speech was a boring analysis of ... *said that* ..." only mentions it. Thus, argument schemes are those argument patterns that use at least one schematic constant or one schematic quantifier; whereas logical forms are argument patterns that use no schematic constants or quantifiers (only the specified logical constants and quantifiers). Let us then define 'argument scheme' as follows.

Df Sc. An ARGUMENT SCHEME is (i) a pattern of argument, (ii) made of a sequence of sentential forms with variables, of which (iii) at least one of the sentential forms contains a use of a schematic constant or a use of a schematic quantifier, and (iv) the last sentential form is introduced by a conclusion indicator like 'so' or 'therefore'.

Marking the distinction between schemes and forms with reference to their quantifiers and constants is better than doing it with reference to levels of generality. One reason is that it is more precise. Once the constants and quantifiers of each kind of logic are distinctly delineated, the distinction can be clearly drawn. Another reason is that the syntactic distinction between forms and schemes can explain the vague distinction based on generality. To understand this we must first appreciate that it is a mis-analogy to compare argument schemes to the logical forms of sentential logic, like modus ponens. Schemes are much more like the logical forms of predicate logic because they too involve quantifiers and individual variables. Since sentential logic parses discourse by the unit of the atomic sentence it is more general than predicate logic, which delves into the internal structure of atomic sentences. So, since schemes are more like the forms of predicate logic than they are like the forms of sentential logic, schemes are less general than the forms of sentential logic.

2 SCHEMES AND KINDS.

In this section, I argue for three claims: (1) that there are neutral argument schemes; (2) that neutral schemes give the identity conditions for argument kinds; and, (3) that argument kinds are illatively neutral.

2.1 The argument for the existence of neutral (descriptive) schemes.

A distinction often mentioned but seldom elaborated is the one between descriptive and normative argument schemes. Martin Kienpointner is reported to hold that whereas normative schemes have a normative conclusion and at least one normative premise, descriptive schemes have neither normative premises nor normative conclusions (see Blair 2001, pp. 374-75). This is a distinction between those schemes that come to a normative conclusion and those that don't; but both kinds are normative in that they give *prima facie* support for their conclusions. Another attempt to characterize descriptive schemes is Perelman and Olbrechts-Tyteca's approach, which Hitchcock says, is descriptive, meaning that it classifies kinds of arguments without including any direction for their evaluation.

> They are describing how people actually argue outside demonstrative contexts, and how rhetorical handbooks recommend that they ague. It is of no concern to them whether a form of argument actually establishes the truth of a factual statement, the wisdom of a recommendation or the merits of an evaluation. (Hitchcock, 160; see *The New Rhetoric*, p. 188)

This we may call the anthropological approach. It is reportive/descriptive about those kinds of argument schemes in use which are thought to promote adherence to theses. However, Perelman and Olbrechts-Tyteca do not concern with themselves with evaluation at this point. It is J.A Blair who draws a distinction between schemes that describe an actual instance of reasoning, whether it be good or bad, with normative schemes "that portray a supposedly valid or cogent pattern of inference or argument" (Blair 2001, p. 375). However, on Blair's view, some schemes can belong to both categories because some of the actual instances of reasoning found in the descriptive group may portray a cogent pattern of reasoning and therefore fit the normative category as well.

Let us explore another possibility. It is based on distinguishing argument kinds, to be identified by descriptive argument schemes, from positive or negative tokens of argument kinds, to be identified by normative argument schemes. How far-fetched is this suggestion that there are descriptive (normatively neutral) argument schemes? I think the seeds of the idea were sown some time ago. One of the important insights in the history of informal logic is that fallacies are related to patterns of argument; witness this passage from the *Fundamentals of Argumentation Theory*:

> ... it is now generally conceded that patterns of argument once considered uniformly fallacious are, in fact, fallacious only in some cases. ... The defining characteristics of a pattern or a type of argument are therefore to be distinguished from the defining conditions of the fallacious occurrences of that pattern. (van Eemeren et al., 1996, pp. 181-82.)

Here there is a distinction between "a pattern or a type of argument" – an argument scheme – and the added conditions needed for an argument to be a fallacious token of the type. Walton takes the story a step further.

> Many of the most common forms of argument associated with major fallacies, such as argument from expert opinion, *ad hominem* argument, argument from analogy and argument from correlation to cause, have now been analyzed using the devise of argumentation schemes. ... We need to recall that although the traditional logic textbooks mainly treated these forms of argumentation under the heading of informal fallacies, in many instances they are reasonable but defeasible arguments. (Walton, 2013, p. 220)[5]

If we can distinguish fallacious instances of an argument kind from the kind itself, so should we be able to distinguish positive instances of the kind from the kind itself. The scheme for an argument kind, which leans neither toward the positive, or the negative is a descriptive or neutral scheme. In short, the matter of neutral, descriptive, argument schemes is what good and bad instances of an argument kind have in common and no more.

As an illustration, compare these two schemes, one for a fallacy, the other for a presumptively good argument. Johnson and Blair (1994, p. 129) give a scheme for the Improper Appeal to Practice.

> 1. Someone defends an action against criticism by arguing that the conduct is widely practiced, is a customary practice, or is a rational practice.
> 2. Either there is, in fact, no such practice, or, in these circumstances, the existence of the practice is not relevant or not sufficient to justify or excuse the conduct being criticized.

---

[5] See also Hitchcock, 2010, pp. 160-61.

Compare that with Walton's scheme (2006, p. 93) for Argument from Popular Practice by which he seems to have the same kind of argument in mind of which Johnson and Blair were speaking.

> 1. A is a popular practice among those who are familiar with what is acceptable or not with regard to A.
> 2. If A is a popular practice among those who are familiar with what is acceptable or not with regard to A, that gives a reason to think that A is acceptable.
> C. Therefore. A is acceptable in this case.

The contrast here is between the appeal to popular practice arguments that are fallacies and those that are presumptively acceptable. What do they have in common? A plausible answer is given in the first line of Johnson and Blair's characterization: it is the kind of argument in which *that X is widely practised, or customary, or rational, is offered as a reason for the acceptability of X*. Notice that this characterization of the argument kind is neutral.

Here is another example. Johnson and Blair (p. 101) characterize the *ad hominem* (abusive) fallacy as follows.

> 1. The critic responds to the position of an arguer by launching a personal attack on the arguer (ignoring the arguer's position).
> 2. The personal attack on the arguer is irrelevant to any assessment of the argument.

Compare that description of a fallacy with Walton's scheme (2006, p. 123) for abusive *ad hominem* argument. (He calls it 'direct' to distinguish it from the circumstantial *ad hominem*, but it is often called 'abusive' *ad hominem*.)

> *a* is a person of bad character.
> So, *a*'s argument should not be accepted.

Here again the descriptive content for what the two schemes have in common is part of what we find in the first line of Johnson and Blair's formula: the argument kind abusive *ad hominem* is *an argument in which a critic ... launch[es] a personal attack on the arguer which may or may not be relevant*. Again, this is a neutral characterization of an argument kind. If the attack is not relevant it is a fallacy; if having a bad character is relevant it may be a good presumptive argument as Walton's scheme intimates. But these judgments will be based on considerations beyond what the scheme provides. For each argument kind, whatever all the good and bad tokens of the kind have in common, that is what the descriptive (or neutral) schemes for the kind will be. Therefore, descriptive schemes are not abstracted from anyone's actual reasoning or arguing, nor are they anthropological data, nor are they schemes that bar the use sentence forms for normative sentences. Descriptive schemes are rather discovered by distilling what possible good and bad instances of an argument kind have in common. What they are is something we infer.

     Of course, it is possible to talk of only good arguments of a certain kind (*ad verecundiams*, e.g.) and then say that these good arguments constitute a kind of argument. We can also talk of bad *ad verecundiam* arguments and consider them to be another argument kind. So, good *ad vercundiams* can be one argument kind and bad *ad verecundiams* another argument kind. But what makes them both *ad verecundiam* arguments? The question presupposes that there is something these two kinds of arguments – the good and the bad kinds – have in common, something which is general and contributes neither to the goodness nor to badness of arguments, but does serve to individuate argument kinds.

2.2 Neutral schemes and argument kinds.

I propose that these neutral descriptive schemes give the necessary and sufficient conditions for argument kinds. Yet, how shall we individuate argument kinds?   It seems best to do it on the basis of the kinds of reasons (premises) for their conclusions. There are two different ways to understand 'reason' in this context. One possibility is a relational approach which sorts reasons in arguments by the kind of link they have to their conclusions, irrespective of the content of those reasons. This view, the relational-view, is preferred by David Hitchcock[6] who illustrates it with these two arguments: (1) "This is red, so this is coloured", and (3) "This is square, so this is shaped", holding that they are arguments of the same kind because both are arguments from a determinate to a determinable, that is, the premise-conclusion link is of the same kind. Another possibility, the content-view or subject-matter view, is to sort premises into kinds by the content of their reasons; for example, all *ad misericordiam* arguments will have reasons of sympathy among their premises making them arguments of the same kind; and since reasons of fear (*ad baculum*), are different from reasons of sympathy, they will be a different kind of argument. The content view distinguishes kinds of evidence rather than kinds of support relations. It is

---

[6] Personal communication, June 2019.

not clear whether, all things considered, the one way of individuating argument kinds is better than the other, but for the purpose of this study, I will explore the content view of distinguishing argument kinds. This seems consistent with contemporary work on argument schemes and has historical antecedents in the works of the Port Royal logicians, and John Locke and Isaac Watts, who divided arguments on the basis of their subject matter.

There are some consequences to the view that neutral schemes give the necessary and sufficient conditions for argument kinds. If any of the necessary conditions for belonging to the argument kind were left out, the definition would fail to capture the character of the kind of argument in question. Hence, the instantiations of any argument scheme (defining an argument kind) will have to include all the necessary conditions of belonging to the argument kind – no defining condition can be left out. Hence, the premises of any instantiation of such a scheme will have linked premises. (Premises are linked just in case if anyone of them is removed, or is false, support for the conclusion is lost or at least significantly diminished; arguments whose premises are not linked but are convergent with respect to a conclusion suffer less drastically when one of their premises is false or is removed (See Walton 2006, pp. 149-51)). Consequently, an argument scheme is one in which no proper subset of its sentential forms constitute an argument scheme.

2.3 Argument kinds are illatively neutral.

Arguments are often classified as either deductive or inductive. This view parallels the idea that there are two kinds of logic, deductive and inductive, one for each class of arguments. This view presupposes that we can always identify arguments as deductive or inductive before we set about making our logical evaluations. How is that to be done?   One suggestion is that we must appeal to the intentions of argument makers. Do they indicate illative modalities by using illative adverbs like 'necessarily' or 'probably'? Sometimes yes, but when they don't we are then left in the position of not knowing by what standard to evaluate arguments – for example, is this a bad deduction or a good induction? Responding to this difficulty, Brian Skyrms (2000, p. 22) holds the view that "deductive and inductive logic are not distinguished by the different types of arguments with which they deal, but by the different standards against which we evaluate arguments". In other words, 'deductive' and 'inductive' do not name two kinds of arguments but rather different standards by which to evaluate arguments. This is a point about argument kinds not about arguments – there are, of course, particular arguments which necessitate their conclusions, and others that make their conclusions probable. Skyrms' point is about "types of arguments" and it should hold as well for the supposed class of "presumptive arguments" with which many closely identify argument schemes. From the Skyrmsian perspective some arguments may be used to establish presumptions but there is no argument kind, Presumptive Arguments, anymore than there is a kind deductive argument. But perhaps there is a third standard, by which we can evaluate arguments, the presumptive standard.

Now, if 'deductive,' 'inductive' and 'presumptive' do not name argument kinds, then argument kinds do not have illative modalities such as 'necessarily,' 'probably' or 'presumably', although individual arguments may well include such indicators. Where these adverbs are included they will advise arguers by which logical standard an argument is proposed to be evaluated. Argument receivers/evaluators not given any guidance by an argument maker must choose what they deem to be the most appropriate illative modality in judging arguments. However, since neutral (descriptive) schemes represent argument kinds, those schemes will not include illative adverbs. Descriptive schemes are, we may say, illatively neutral. On this view, an inductive appeal to authority and a presumptive appeal to authority will be different arguments because different standards of evaluation are in play but, on the content view of argument identity, they will be arguments of the same kind. Interestingly, it turns out that on this view of argument kinds, it is misleading to say that some argument kinds are defeasible. To be defeasible means that initial judgments are subject to possible revision in light of new information. However, this cannot be said of argument kinds since they are illatively neutral and make no normative claim: they do not, in themselves, depend on standards. Hence, it is standards that may be defeasible, not argument kinds (neutrally defined). When defeasible standards are combined with instances of argument kinds, defeasibility is a factor in argument evaluation. Nevertheless, when we are focussed on neutral schemes, the term 'defeasible scheme' is a category mistake.

REFERENCES

Blair, J.A. (2001). Walton's argumentation schemes for presumptive reasoning: A critique and development. *Argumentation* 15, 365-79.
Corcoran, J. (1999). Scheme. *The Cambridge Dictionary of Philosophy*, 2nd ed. R. Audi, (ed.). Cambridge: Cambridge University Press.
Eemeren, F.H. van, R. Grootendorst and F. Snoeck Henkemans. (1996). *Fundamentals of Argumentation Theory*. Mahwah, NJ: Lawrence Erlbaum.
Hitchcock, D. (2010). The generation of argument schemes. In C. Reed and C.W. Tindale (Eds.), *Dialectics, Dialogue, and Argumentation*. (pp. 157-166.) Milton Keynes: College Publications.

Johnson, R.H., and J.A. Blair. (1994). *Logical Self-Defence*. New York: McGraw-Hill.
Pinto, R.C. (2001). *Argument, Inference and Dialectic*. Dordrecht: Kluwer.
Skyrms, B. (2000). *Choice and Chance*, 4th ed. Belmont, CA: Wadsworth.
Walton, D.N. (2006). *Fundamentals of Critical Argumentation*. New York: Cambridge University Press.
Walton, D.N. (2013). *Methods of Argumentation*. Cambridge: Cambridge University Press.
Walton, D.N., C. Reed and F. Macagno (2008) *Argumentation Schemes*. Cambridge: Cambridge University Press.

Resolution of deep disagreement: not simply consensus

LEAH HENDERSON
Department of Theoretical Philosophy,
University of Groningen
l.henderson@rug.nl

> Robert Fogelin has argued that in deep disagreements resolution cannot be achieved by rational argumentation. In response Richard Feldman has claimed that deep disagreements can always be resolved by suspension of judgment. I argue that Feldman's claim is based on a relatively superficial notion of "resolution" of a disagreement and that the real concerns behind Fogelin's argument are more substantive.
>
> KEYWORDS: deep disagreement, Fogelin, Feldman, suspension of judgment, imprecise probability.

1. INTRODUCTION

What should the role of rational argumentation be in addressing the deepest disagreements that arise in our society? Robert Fogelin has argued for the rather pessimistic conclusion that deep disagreements cannot be resolved by rational arguments (Fogelin, 1985). If he is right, this would have significant implications for how deep disagreements should be approached. In particular, it might be thought that an emphasis on exchanging reasons and arguments may sometimes be misplaced, or even, as some have suggested, "dangerous" (Campolo, 2005).

On the other hand, there has also been considerable resistance to Fogelin's argument, and a number of good points have been raised in response (Aikin, 2018; Lugg, 1986; Memedi, 2007; Phillips, 2008; Ranalli, 2018; Siegel, 2019). In this paper, I will focus on the reply given by Richard Feldman (Feldman, 2005). Feldman argues that we may always achieve a rational resolution, even in a deep disagreement, if both parties suspend judgment on the issue. I will argue that this reply by Feldman really misses the point of Fogelin's argument, and fails to touch the interesting issues that it raises.

The plan for the paper is the following. In section 2, I briefly summarise Fogelin's argument and Feldman's reply. In section 3, I introduce some distinctions which will be helpful in analysing Fogelin's argument. I also discuss how these work in a context where agents have degrees of belief in propositions, not just full beliefs or full commitments. In the light of this analysis, I critique Feldman's reply in section 4. In section 5, I try to indicate where the interesting issues raised by Fogelin's argument really lie.

2. DEEP DISAGREEMENT: FOGELIN'S ARGUMENT AND FELDMAN'S REPLY

2.1 Fogelin's argument

Robert Fogelin has put forward a skeptical position about the power of informal logic and critical thinking to resolve disagreements (Fogelin, 1985). In particular, he suggests, there are some disagreements which are "deep", and fail to be resolvable by rational argumentation, or any rational means. He puts the point as follows:

> if deep disagreements can arise, what rational procedures can be used for their resolution? The drift of this discussion leads to the answer NONE' (Fogelin, 1985, p. 9)

What makes a disagreement "deep"? According to Fogelin, in a deep disagreement, the parties involved disagree at a profound level over "framework propositions" in the Wittgensteinian sense. These framework propositions, he claims, are deeply enmeshed in

> a whole system of mutually supporting propositions (and paradigms, models, styles of acting and thinking) that constitute, if I may use the phrase, a form of life" (Fogelin, 1985, p. 9).

As an example, he offers the case of abortion, where disagreement centres around the moral status of the foetus. The idea that the foetus has a certain relevant kind of personhood, he suggests, is often grounded in a much broader tradition of religious belief which involves many other commitments. This broader network of beliefs and commitments may not be shared with those who deny the foetus has such a status.

Fogelin's thesis is that "deep disagreements cannot be resolved through the use of argument, for they undercut the conditions essential to arguing" (Fogelin, 1985, p. 8). The reason that he gives is that argumentative exchange is "normal" when "it takes place within a context of *broadly* shared beliefs and preferences' (Fogelin, 1985, p. 6). When the context becomes less normal, argument becomes impossible, because the 'conditions for argument do not exist" (Fogelin, 1985, p. 7).

> The language of argument may persist, but it becomes pointless since it makes an appeal to something that does not exist: a shared background of beliefs and preferences. (Fogelin, 1985, p. 7).

Those who have a deep disagreement then lack the shared background required to make argument work.

2.2 Feldman's reply

In the face of this pessimism, Richard Feldman has argued that there always is a way to resolve deep disagreements. He says that a disagreement has a "rational resolution available" when there are "some arguments and evidence which could be put forward to which the rational response is agreement" (Feldman, 2005, p. 16). In some cases, he says, there will be a resolution of a disagreement "if two people begin by disagreeing about something and then one person comes round to the other's point of view". If this happens on the basis of the presentation of arguments and evidence, then this counts as a "rational resolution" of the disagreement.

Feldman's main point is that this is not the only way a rational resolution may be achieved. Another possibility is that both parties suspend judgment about the issue in question. Feldman admits that this does not amount to a "resolution of the issue", but he thinks it does count as a "resolution of their disagreement" (Feldman, 2005, p. 17). Feldman contends that in normal disagreements, there is always such a resolution of disagreement available. Either the evidence and arguments make it rational to agree, or the parties should suspend judgment. Feldman then extends the argument to cases of deep disagreement. He argues that there is no reason why this kind of resolution could not be applied to the framework propositions that are implicated in deep disagreements. Even if there are complex evidential connections to systems of propositions, one can still evaluate whether one's evidence supports the proposition, goes against it, or is neutral. Thus suspension of judgment is always there also an option.

3. CONSENSUS AND COMMON GROUND

In this section, I will draw a distinction between consensus and common ground, which will prove helpful in analysing the significance of Fogelin's argument.

3.1 Consensus

It is first useful to draw on Isaac Levi's distinction between different ways of using the notion of "consensus". When two agents find themselves in a disagreement, they may initiate an investigation or discussion to try to resolve it. As Levi says

> an early step in such a joint effort is to identify those shared agreements which might serve as the noncontroversial basis of subsequent inquiry' (Levi, 1985, p. 145).

This gives us one notion of consensus: the "consensus of the participants at the beginning of inquiry which constitutes the background of shared agreements on which the investigation is initially grounded" (Levi, 1985, p. 145). I will call this "$consensus_1$". This should be distinguished from the consensus that participants may sometimes achieve as the outcome of inquiry, which we will refer to as "$consensus_2$".

Levi discusses how these different types of consensus may be represented both in the setting which concerns knowledge, and in the setting where agents have states of partial belief, or "credal states" (Levi, 1974). In the first case, each agent involved in a disagreement has some corpus of propositions which they take to be certain, and which they might say they "know". For example, agent A might take proposition h to be certain. In doing so, she does not regard $\sim h$ as a serious possibility. Agent B, on the other hand, might take $\sim h$ to be certain, and not regard h as a serious possibility. After discovering that they disagree, both agents may revise their commitments by "contraction": that is, by removing propositions from the set to which they are fully committed. Thus A may contract by removing h and B may contract by removing $\sim h$. They are both then in the state of shared agreement where their corpus contains neither h nor $\sim h$. This may be seen as a state of "suspension of judgment' regarding the truth of h. Such a suspended state may be taken as a $consensus_1$.

A and B may now continue to investigate. They might gather more evidence, swap evidence, or attempt to convince one another with arguments. This process may lead to them both "expanding" their commitments again by adding h (or $\sim h$). If they converge in this way, this would be a consensus that they achieve as the outcome of inquiry ($consensus_2$). It is also possible that nothing further is gained in the process of inquiry and the $consensus_2$ achieved at the end of the inquiry does not go beyond the original $consensus_1$.

The difference between types of consensus can also be specified in the setting of credal states. Now a disagreement may not be between what agents take themselves to know, but rather the agents may have different personal probabilities for a proposition. For example, agent A thinks h quite unlikely, and might assign $p_A(h)=0.2$, whereas agent B thinks it quite likely, and assigns $p_B(h)=0.8$. What do consensus$_1$ and consensus$_2$ amount to? In particular, what is the analogue of the suspended judgment which characterised consensus$_1$ in the knowledge case? Levi suggests that we should here make use of imprecise probability[1]. A state of suspended judgment can be represented not by one probability distribution alone, but by a set of probability distributions. A set of probabilities can be specified by giving the "lower" and "upper" probabilities, which are defined as the lowest and highest probabilities in the set respectively. How such a set can represent shared agreement can be seen by considering the behavioural interpretation which can be given of the lower and upper probabilities (Elkin, 2018; Walley, 1991). The idea here is that an agent may be offered gambles – for instance, a gamble which will pay 1 unit if h turns out to be true, and 0 units otherwise. The behavioural interpretation of a lower probability for h is that it is the highest price that the agent would be prepared to pay for such a gamble on h. The agent may not, however, be prepared to sell the gamble at that price. The upper probability is interpreted as the lowest price for which the agent would be inclined to sell the gamble. When an agent's lower and upper probabilities coincide, they have a "precise" probability for h, where there is one price they regard as fair for both buying and selling the gamble. In general, though, the lower and upper probabilities may come apart, giving a set of probabilities whose degree of imprecision about a proposition h can be measured by the difference between the upper and lower probability.

In the simple example above, the two parties initially have precise probabilities $p_A(h)=0.2$ and $p_B(h)=0.8$. Then the first agent is disposed to buy the gamble on h that pays 1 if h is true and 0 otherwise for prices less than $p_A(h)=0.2$, and is disposed to sell the gamble for prices above $p_A(h)=0.2$. The second agent has similar dispositions with respect to $p_B(h)=0.8$. Their shared dispositions can be represented by an imprecise probability with lower probability $\underline{p}(h) = 0.2$ and upper probability $\overline{p}(h) = 0.8$, because both are disposed to buy at prices below 0.2 and to sell at prices above 0.8. One may also think of the set [0.2, 0.8] as the set of probability measures which are regarded as permissible for the purposes of evaluating different options with respect to expected value (Levi, 1974).

Just as in the knowledge case, after the parties identify their shared agreements in a consensus$_1$, inquiry and dialogue may take place, leading to a possible consensus$_2$ after the inquiry. Further information

---

[1] Levi refers to this as "indeterminate probability", but "imprecise probability" is the more common term now.

will generally remove imprecision (Walley, 1991).[2] The disagreeing parties might reach a consensus where they both adopt the credence that one of the parties held initially. Alternatively the final result might be some kind of combination of their opinions. For example, the opinions could be combined in a "linear pool", where the final opinion is a linear combination of the initial credences with some weights. For example, it might be $p(h) = w_1 p_A(h) + w_2 p_B(h)$, where w_1 and w_2 are weights that sum to one. Again, it is also possible that the inquiry or argument does not succeed in moving the parties at all, and they remain in their initial state of consensus$_1$.

3.1 Common Ground

It is important to distinguish the notion of consensus from the notion of the "common ground" that the agents share. The common ground of the agents is the *content* of the shared state that represents their consensus. If two parties have achieved consensus by contracting to a state of suspended judgment over h, then their state of consensus has no content. They agree only on the proposition $h \vee \sim h$, so their state of agreement is completely non-informative. We will say then that they do not have any common ground regarding h. This would contrast with a case where the two agents disagree over h, but are both committed to another proposition g. Then the consensus of shared agreement that they come to has some content, namely g.

We can also see how in the probabilistic case it is possible to form a consensus state which contains some content in the case of a disagreement. If A and B have a consensus state represented by the set of probabilities between 0.2 and 0.8, then they do have some common ground. They agree not to assign probabilities between 0 and 0.2 or between 0.8 and 1. By contrast, if the disagreeing parties initially have precise probabilities 0 and 1 respectively, then the state of shared agreement of these commitments is the completely vacuous set of all probabilities in the interval [0,1]. In this case, although there is a consensus on this state, there is no common ground. This is the probabilistic analogue of the case discussed above where A is initially committed to h and B to $\sim h$. In general, an advantage of the probabilistic representation is that it allows us to represent more nuanced states of consensus, with different degrees of common ground. The amount of common ground is reflected in the precision of the set of probabilities. If

---

[2] Though not always. There is a phenomenon known as "dilation" in which further information leads to an increase in the imprecision of the state (Seidenfeld, 1993; Walley, 1991).

the two parties agree on one precise probability, they then have the maximal amount of common ground, and in many cases this would count as a resolution of the disagreement.

4. THE LIMITS OF FELDMAN'S REPLY

Let us now return to Feldman's response to Fogelin's argument. Fogelin's thesis is that resolution of disagreement cannot be achieved by rational argumentation in cases of deep disagreement. Feldman understands "resolution of disagreement" purely in terms of achieving agreement or consensus, and his point is that this is relatively easy to achieve by suspending judgment. However, the concern behind Fogelin's argument is arguably more substantive. The core issue here concerns whether people can make progress towards substantive agreement on important matters like abortion or affirmative action by means of rational argument. Another way to put it is, can rational argument be "productive" on these deep questions (Phillips, 2008). In order to understand the notion of "progress" or "productive" here, we need more than simply the notion of consensus. We will say that an argument is "productive" if it increases the amount of common ground between the parties involved, in relevant ways. Thus, an argument will be productive if the consensus$_2$ represents more common ground than the initial consensus$_1$. A full "resolution" will be achieved if the parties achieve common ground – or come to agree – on all the propositions which are important to the issue at hand.

Feldman's "resolution" achieves agreement, and this may correspond to the consensus$_1$ that parties may achieve by suspending judgment. However, in some cases the common ground that such a consensus achieves is empty. This is in fact the case in the examples Feldman himself discusses. He considers cases where both parties suspend judgment about h and thus wind up with a completely non-informative state of opinion. This misses the point that Fogelin is getting at. Fogelin is asking whether productive argumentation can proceed from such a starting point. Certainly such a completely non-informative state cannot serve as the resolution of the disagreement – as Feldman himself acknowledges when he says that it is not a "resolution of the issue", but only a way of achieving agreement.

5. THE REAL ISSUES RAISED BY FOGELIN'S ARGUMENT

The real concerns behind Fogelin's argument are two-fold. First, in cases of deep disagreement, it may be hard to have much common ground in consensus$_1$. As we have seen, Fogelin explicitly focuses on the lack of "a shared background of beliefs and preferences". Second, a lack of common ground in consensus₁ hinders the pursuit of rational argumentation. The reason is simply that common ground is used as a resource in argumentation. Convincing another person using an argument usually requires that you find at least some premises for your argument that you can get them to agree to. The concern is that if the common ground is sufficiently empty at the beginning of inquiry, then arguments cannot tap into shared commitments in order to make progress.

Thus we may reconstruct Fogelin's argument as follows:

1. There are situations where people are committed to such different frameworks, hinge propositions, etc. that they lack substantial common ground. Call these "deep disagreements'.
2. If parties lack substantial common ground, this undercuts (substantially) the conditions for coming to agreement by means of rational argumentation.
C. In cases of deep disagreement, reaching agreement, or even making progress, by rational argumentation is not possible.

So understood, Fogelin's argument does raise important and interesting issues, several of which have already been discussed in the literature.

One set of issues concerns the second premise above. What are the minimal requirements on common ground for rational argumentation? Can, contrary to what Fogelin claims, rational argumentation proceed without it, or with a very minimal and readily achieved common ground? Some authors have argued that there are ways that rational argumentation can proceed that do not rely (so heavily) on the possession of common ground. For example, Andrew Lugg suggests that

> the strategy of reverting to neutral ground is only one strategy among many. Individuals can also bring about a shift in one another's allegiances by demonstrating hidden strengths of their own views and by eliciting hidden weaknesses of alternative views. Furthermore, they may find themselves having to shift ground as a result of their discovering things *wrong* with the views they accept and things *right* with the ones that they reject (Lugg, 1986, p. 48).

In a similar vein, Phillips (2008) suggests that arguments can be profitably pursued without any common ground in terms of shared beliefs and preferences as long as there is a certain shared commitment to procedural norms of argumentation.

Another set of issues concerns the first premise. Are there ever really situations where sufficiently substantial common ground is missing? Is there not some level — perhaps a more general level — at which people can find common ground even on matters of value? Although this may be true, the question that is relevant to Fogelin's argument is whether the disagreeing parties can access their common ground in a way which makes it available for use in argumentation. Situations of deep disagreement do seem to be ones where the identification of common ground can be particularly difficult (Phillips, 2008). Not all our commitments may be transparent to us, particularly when they concern very fundamental beliefs and values, so it can be hard to extract these for the purposes of lining them up with the commitments of another. And this is made even harder by the entanglement of these commitments in a whole system of propositions, as will typically be the case in a deep disagreement. Even if common ground does exist, then, it may be difficult to identify it in these sorts of situations. Some authors have suggested that simply pursuing the usual procedures of arguing will result in the common ground being suitably brought to the surface, eg. (Lugg, 1986; Siegel, 2013). However, pursuing argumentation with the aim of persuasion may not be the only or the most effective way of finding common ground. More in-depth analysis of the processes that lead to the identification of common ground and the conditions that make them effective seems to be in order.

It is worth noting that Fogelin's argument would be rather uncontroversial had the conclusion simply been that rational argumentation is *difficult*, or even *more difficult*, for deep disagreements. It is the claim that rational argumentation is *impossible* in cases of deep disagreement which is striking (Turner, 2005). Harvey Siegel has suggested that, in Popper's words, Fogelin "exaggerates a difficulty into an impossibility" (Siegel, 2013, p. 16). In order to defend the stronger conclusion that rational argumentation is impossible, it is necessary to establish either that rational argumentation is completely impossible without common ground, or that common ground of the required type is completely impossible to identify in cases of deep disagreement, or both. This is a considerably more demanding task than simply showing that arguing without common ground is more difficult or that common ground is harder to find in cases of deep disagreement, both claims which seem rather plausible, perhaps even obvious.

6. CONCLUSION

Fogelin's argument raises the important question of what the role of critical thinking and rational argument should be in dealing with difficult disagreements. His argument primarily concerns the role of common ground in argumentative practice, and the main point is that since argumentation normally makes use of common ground as a kind of resource, it may be crippled when that resource is lacking.

Feldman has attempted to argue, contra Fogelin, that deep disagreements can always be resolved by suspending judgment. But this reply understands "resolution of disagreement" only in terms of achieving consensus. It is important in discussion of deep disagreement to distinguish between a consensus and common ground. Suspending judgment can result in consensus, but it does not mean that the parties have any substantial common ground. Fogelin is pessimistic about how productive arguments can be in situations where the parties lack common ground at the outset. This point is left completely untouched by Feldman's reply. The more substantive issue raised by Fogelin's argument is whether the difficulties in identifying and exploiting common ground in cases of deep disagreement render progress or resolution actually impossible, as opposed to merely more difficult.

ACKNOWLEDGEMENTS: Thanks to Jan Albert van Laar for helpful comments on a draft of this paper.

REFERENCES

Aikin, S. (2018). Deep disagreement, the dark enlightenment, and the rhetoric of the red pill. *Journal of Applied Philosophy, 36*(3), 420-435.
Campolo, C. (2005). Treacherous ascents: on seeking common ground for conflict resolution. *Informal Logic, 25*(1), 37-50.
Elkin, L. a. W., Gregory. (2018). Resolving peer disagreements through imprecise probabilities. *Noûs, 52*(2), 260-278.
Feldman, R. (2005). Deep disagreement, rational resolutions and critical thinking. *Informal Logic, 25*(1), 13-23.
Fogelin, R. (1985). The logic of deep disagreements. *Informal Logic, 7*, 1-8.
Levi, I. (1974). On indeterminate probabilities. *Journal of Philosophy, 71*(13), 391-418.
Levi, I. (1985). Consensus as shared agreement and outcome of inquiry. *Synthese, 62*(1), 3-11.
Lugg, A. (1986). Deep disagreement and informal logic: no cause for alarm. *Informal Logic, 8*, 47-51.
Memedi, V. (2007). Resolving deep disagreement. In H. V. Hansen (Ed.), *Dissensus and the search for common ground* (pp. 1-10). Windsor, ON: OSSA.
Phillips, D. (2008). Investigating the shared background required for argument: a critique of Fogelin's thesis on deep disagreement. *Informal Logic, 28*(2), 86-101.
Ranalli, C. (2018). Deep disagreement and hinge epistemology. *Synthese*, https://doi.org/10.1007/s11229-018-01956-2.
Seidenfeld, T. a. W., Larry. (1993). Dilation for sets of probabilities. *Annals of Statistics, 21*, 1139-1154.
Siegel, H. (2013). Argumentation and the epistemology of disagreement. *OSSA Conference Archive, 157*, 1-22.

Siegel, H. (2019). Hinges, disagreements, and arguments: (rationally) believing hinge propositions and arguing across deep disagreements. *Topoi*, doi:10.1007/s11245-018-9625-6.

Turner, D. a. W., Larry. (2005). Revisiting deep disagreement. *Informal Logic, 25*(1), 25-35.

Walley, P. (1991). *Statistical Reasoning with Imprecise Probabilities*: Peter Walley.

Appeals to Popularity: Roles and Functions of 'Everyone knows X'

THIERRY HERMAN
Universities of Lausanne and Neuchâtel
Thierry.Herman@unine.ch

What is the rhetorical effectiveness of arguing with an expression like "Everyone knows X", which is a literally wrong hyperbole? Is "everyone knows X" a good example of a premise used for an appeal to popular opinion? I will argue here that studying more closely this expression in its pragmatic context reveals that "appeal to popularity" may not be the main scheme: "everybody knows X" can ridicule those who don't know X, or just recall a shared fact or opinion. A typology of different cases is drawn to cover the rhetorical effects of this expression.

KEYWORDS: appeal to popularity or to popular opinion, appeal to common knowledge, everybody knows X, rhetoric and pragmatics, rhetorical effects, facts and opinions.

1. INTRODUCTION

The "*ad populum* appeal" "has not yet received a great amount of attention in the literature" (Jansen, 2018, p. 425). This scheme seems to be quite obvious, and easy to teach to students, for example. There is some debate on the fallaciousness of this scheme (Godden, 2008; Jansen, 2018), but I will not tackle this issue here. Yet, I will argue as Jansen (2018) and van Leeuwen (Jansen & van Leeuwen 2019), grounding their analyses in actual examples of language in context, that *ad populum* appeals are far more complex than expected[1]. It seems that there is a wide agreement around Walton's rendition of the scheme's structure: "everybody (in a particular reference group G) accepts A. Therefore, A is true (or you should accept A)" (see Jansen & van Leeuwen, 2019 for a synthesis; Walton, 1999, p. 200). Three observations need to be made, though: 1. Walton uses "everybody" and

---

[1] This paper is an abridged and modified version of Herman & Oswald's paper (to appear in 2020). I thank Steve Oswald for letting me rewrite some parts of this article for these proceedings.

not "most people" for example, while the latter expression seems to be a more exemplary mean to express appeals to popularity; 2. "accept A" is not really the same as knowing A; 3. Walton considers "A is true" as synonymous with "you should accept A". These points raise some questions which are crucially relevant for this investigation.

It should first be noted that "everybody" bears, more often than not, an intensifying feature. In an open context, like in public discourses, it is literally wrong that everybody knows X; even an obvious truth might not be known by some persons in an audience. Now, if indeed this expression is used in a closed context, which is the case of example (1) where the exclusive audience is set by "here", "Everyone knows X" could be considered, *prima facie*, as a not relevant expression in argumentation. In this case, its role seems to be to remind the audience of a state of affairs:

> (1) OK so as everyone knows here, I'm waiting to hear back from my welding test to see if I got certified.[2]

If "everybody knows *P*" is reduced to the reactivation of previously shared knowledge, one could then ask why a speaker would take the trouble of phrasing it with the universal quantifier. "Everybody knows X" must find its own relevance compared to "as many of you know". It is astonishing that Walton uses "Everybody knows X" as a paragon of an "appeal to popularity". Indeed, it may exemplify the fallacy of "begging the question":

> (2) (...) We need Border Security, and as EVERYONE knows, you can't have Border Security without a Wall. (...) (Donald Trump, on Twitter, 23 December 2018)

In this example, the premise "Everyone knows that you can't have a border security without a wall" implies the truth of the conclusion "you can't have a border security without a wall". Since everyone knows X, it is theoretically not necessary and not relevant to convince anyone about the truth of X. I will elaborate on this point in the first section below.

But Walton's quote raises another topic of interest. Different types of propositional contents are likely to fill the variable *X*, in the expression "everybody knows *X*". In principle, the verb 'to know' introduces issues related to knowledge and should, *ipso facto*, only scope over facts/states of affairs. Therefore, in "everybody knows (fact) *P*", we should infer that the corresponding state of affairs is already known and that the expression is there only to remind us of it, at the

---

[2] https://scitexas.edu/welding/woman-welder-sci-spotlight/

time of utterance. Then, it can be concluded by "X is true". Nevertheless, the expression is often used to introduce a personal opinion, which, by definition, cannot be a piece of widespread knowledge, as in the following example, where a prediction is said as already known:

(3) This isn't about the Wall, *everybody knows that a Wall will work perfectly* (In Israel the Wall works 99.9%). This is only about the Dems not letting Donald Trump & the Republicans have a win. (Donald Trump, on Twitter, 28 December 2018, 14:10; italics are mine)

Moreover, it cannot be concluded anything about the truth of a future event: it is more a matter of agreement/disagreement. The use of the verb "to know" rather than "to agree" could be considered as a strategy for concealing an opinion, which is by nature open to discussion and counter-argument, in a proclaimed knowledge that says its own, and indisputable, obviousness. Of course, "everybody knows X" is not an indicator of general agreement. For example, one of the replies of the above-mentioned Trump's Tweet says: "So many wrong things. 1. Israel is the size if Rhode Island; 2. *A wall will not work perfectly* (...)" (Even Lieberman, on Twitter, Dec 28 2018, italics are mine).

I will try to observe in the second section whether appeals to popularity have differences of forms and functions if X is a fact or an evaluative opinion. For example, Jansen discusses a similar idea when she distinguishes descriptive from prescriptive standpoints to conclude that "*ad populum* arguments supporting a descriptive standpoint are always fallacious", while adding that "this judgement does not hold for a prescriptive standpoint" (Jansen, 2018, p. 435).

As it may appear by the preceding words, I'm a linguist and a discourse analyst who is interested by the philosophical realm of argumentation. As such, I am driven by the duty to recall how important and relevant micro-observations of linguistic details can be in an argumentative dialogue or in a rhetorical situation. I consider my researches in a rhetoric-pragmatic framework: I want to explain how the use of linguistic devices may trigger persuasive effects in argumentative contexts, taking into account stylistic and rhetoric traits of a linguistic form and the cognitive processes at play in communication (Sperber & Wilson, 1995). This framework is not a normative framework in the argumentative sense, since it is not meant to help us judge *a priori* the fallaciousness of a scheme; it is neither a merely descriptive framework, as it does not assist us in drawing a typology of different linguistic forms of the *ad populum* scheme. It is predominantly an explanatory framework in order to account for the effectiveness or ineffectiveness of discursive moves, stylistic designs or argumentative schemes.

2. WHO IS "EVERYBODY", AFTER ALL?

As shown in Herman & Oswald (2020) and briefly recalled here, "everybody" only rarely refers to the whole audience. It appears, therefore, that the expression "everybody knows *P*" is a form of rhetorical amplification or intensification. Even in a closed context, like example 1, where "everybody knows *P*" is reduced to the reactivation of previously shared knowledge, one could then ask why a speaker would take the trouble of phrasing it with the universal quantifier instead of "as you know", "as many of you know", etc. In (1), the possible existence of a minority who was unaware of the depicted fact seems to be excluded from the denotation, which is not completely the case with "as you know". In other words, "everybody knows *X*" makes the fact that *X* could not be ignored salient. An ignorant member in the audience should quickly accommodate this information. With this expression, the speaker may thus signal that she has delivered a piece of *prominent* information, and this is where the rhetorical potential linked to the denotation of the expression may very well be taken to lie. I contend that it could very well compete with the role played by the inherent intensification of the expression, while used in public contexts where "everybody knows" is false: a rhetorical strategy through which a speaker only pretends to recall a unanimously shared piece of knowledge when in fact she is distorting reality and going for an *ad populum* argument, like in examples (2) and (3).

     As a discourse analyst who is interested in the socio-historical context of an utterance and its background knowledge, I would like to measure the gap, to assess the degree of amplification, between the probable denoted referent and the totality expressed by the universal quantifier. For me, there are three possibilities: (1) the gap is not large at all and "everybody knows X" is an amplification of "the large majority of people knows X"; (2) The gap cannot be assessed: the addressee cannot be sure whether X is true of false; (3) The gap is obviously large: everybody refers to a minority of people. In this case, the rhetorical gap is far too big to bridge to hope to be persuasive.

     In the following example (4), even if "deadly" is a hyperbole, the gap seems relatively small, especially because "everybody knows" is used to recall a fact that seems completely obvious and already known in order to highlight a lesser-known corollary fact:

> (4) Another strategy is to speed up — or slow down — the pace of your remarks. Everyone knows that speaking in a monotone voice is deadly. But a corollary mistake is that, even if your voice has plenty of range, speakers often use the same rate of speech all the time. (Dorie Clark, Harvard Business Review, 19

September 2019, https://hbr.org/2019/09/what-to-do-when-youre-losing-your-audience-during-a-presentation)

Measuring the width of the gap between what is said as known and what is actually known can pose a number of problems for the analyst, of course. This is why, in the preceding example, I have first used linguistic clues that instruct the addressee to interpret an utterance as obvious. Then, background knowledge may also be used to help measure the width of the gap.

In example (5), the gap cannot be assessed, in particular since Orrinh Hatch did not attend to the hearing. Moreover, in the background knowledge, the Russian interference in the US election is still under investigation.

(5) Sen. Orrin Hatch, R-Utah, didn't attend the hearing, but said in an interview that everyone knows that Russia interfered in the election and it has admitted as much. (https://tinyurl.com/y35royub)

This case seems to be the only one which truly conveys a *standard ad populum* argument. The hyperbolic nature of "everybody knows *P*" is still active, but that it will additionally fulfil an *ad populum* role in case our lack of investment in the personal investigation of the issue pushed us to trust what many people – people who are likely to be well informed – are said to know.

Finally, example (6), a Donald Trump's tweet, I'm note sure that the *ad populum* effect is at the core of the rhetorical strategy. Trump plays on the already obvious character of the state of affairs he mentions, which is by nature completely unknown (a prediction) and in which "everybody knows" should be pragmatically enriched to "everybody with good sense". "Everybody knows" amplifies the gap between the ingroup (Trump's allies) and the outgroup.

(6) The Fake News Media is doing everything they can to crash the economy because they think that will be bad for me and my re-election. The problem they have is that the economy is way too strong and we will soon be winning big on Trade, and everyone knows that, including China! https://www.politico.com/story/2019/08/15/trump-economy-recession-stock-market-media-1464577

3. WHAT DO WE KNOW?

With « everyone knows x », only factual claims are expected in principle. And these claims are either true or false. But *ad populum* schemes are not always dealing with facts, and even « everybody knows X » can perfectly be used with opinions, which are more assessed by agreement or disagreement rather than truth or untruth. For example, Jansen shows a strong difference between descriptive and prescriptive standpoints in *ad populum* schemes (2018). This is typically not an observation that has been made in the literature on arguments appealing to common knowledge. Godden, for instance, in relaying a widespread distinction between appeals to shared knowledge and appeals to popular opinion, does not make any such distinction since he takes the conclusion of both arguments to be identical (Godden, 2008, pp. 106–107):

> (7) Basic form of appeal to popular opinion (bandwagon)
> It is widely held among S that P
> Therefore, P is true
>
> *Basic form of appeal to common knowledge*
> It is widely known among S that P
> Therefore, P is true

At the risk of weakening this account, I argue that while in cases in which P is presented as a state of affairs, the conclusion drawn by Godden above is admissible, when P is an opinion (the bandwagon case in (7)), further qualification seems to be in order.

I will study here a broad difference between facts and opinions (which combine evaluative and prescriptive standpoints). Broadly speaking, four (in fact six) types of examples can be specified. The criteria are the difference between facts and opinions (which I won't describe in detail here) and the previously depicted levels of amplification between what is presumably assumed as shared or known and what is acknowledged by the speaker as not universally known or shared. I let aside here facts that are not so known and opinions which are not universally shared, because those cases are classical examples of ad populum schemes with "everyone knows X". I will here focus on more extreme cases.

> *Case I: widely known or shared facts*
>
> (8) "Obviously natural gas, as everybody knows, is very, very flammable. There's a quarter-mile radius evacuation zone as

> safety, in case there was any fire or explosion that happened, we want to make sure everyone is safe and away from that," said Mike Ponticello, Broome County Emergency Services Coordinator. (https://spectrumlocalnews.com/nys/central-ny/news/2019/09/23/tractor-trailer-crash-on-i-88-prompts-evacuation-of-nearby-homes)

When the expression "everybody knows *P*" targets the representation of a state of affairs, one could imagine, *a priori*, that the function of the utterance is merely to recall said state of affairs; this would in turn annihilate the appeal to shared knowledge, taken as an argumentative move of support. Indeed, in such a case we would not get the previously mentioned premise "it is widely known", but instead "it is known by everyone". That changes everything: as such, this premise, presumably supporting the conclusion "it is true", could be described as a *petitio principii* whose conclusion fails to carry any informational import. In case I, one could wonder about the relevance of mentioning that "everybody knows" it. But an interesting observation emerges from the deletion of the expression: if we delete it, and assuming that "natural gas is very flammable" is indeed widely known, the sentence becomes completely trivial; moreover, "obviously" in example (8) points to the same triviality. The expression "everybody knows" may here serve to weaken the triviality of the sentence and to acquit, in an anticipatory move, the speaker of recalling such trivialities.

Case II: Doubtful or controversial facts

> (9) As everybody knows, but the haters & losers refuse to acknowledge, I do not wear a "wig". My hair may not be perfect but it's mine. (https://twitter.com/realDonaldTrump/status/327077073380331525)

If wearing a wig is subject to debate, that means that the referring expression "everybody knows" manifestly fails to denote the totality of people, and thus that it fails to appropriately describe reality. It would indeed be completely irrelevant to utter (9) or to simply talk about the wig issue if, in the first place, it was obvious to everyone that Trump does not wear a wig.

Interestingly, the effect of the appeal to popularity seems twofold in this example. Not only does it attempt to present as true something that many people could legitimately doubt, but it also reinforces the obviousness of a state of affairs that is presumably known, thereby *ridiculing* the hating minority who would refuse to admit an obvious fact. It seems that in cases in which "everybody knows

X" blatantly fails to include "everybody", in addition to appealing to popular opinion in order to increase belief in the truth of X, the speaker also attacks those who doubt X – as an *ad personam* attack of sorts.

>   *Case III: widely known or shared opinions*
> (10)   Gastronomy in Spain is, as everybody knows, outstanding. We all love, paella and jamón, but there are other excellent gastronomic choices unknown to the general public that are equally exceptional and Madrid offers the possibility to discover them all. (http://www.my-little-madrid.com/2016/07/find-murcia-region-in-madrid-restaurant.html)

It appears that "everyone knows" is rather used to support premises, here a general one about gastronomy in Spain. However, the idea that *x* is reputedly shared is said as obviously shared. In this case, it seems that "everyone knows X" mostly serves to recall and perhaps stabilize some *doxa*. Of course, a speaker will also exert pressure on anyone in disagreement with *x* to adhere to X. The goal is presumably not to make the audience accept *x*, but to rely on X's obviousness. It is therefore not obvious that "everybody knows opinion X" functions as a canonical appeal to popularity.

>   *Case IV: doubtful or controversial opinions*
> (11)   "The fact is we need the wall. The Democrats know it. Everybody knows it. It's only a game when they say you don't need the wall" (Donald Trump, tweet, Dec 21st, 2018,

The obviousness effect introduced by Donald Trump with " everybody knows it" seems to be meant to mock any attempt to refute his opinion – which is certainly an indirect way of increasing the adherence to the main claim, namely, While the opinion introduced by "everybody knows" is manifestly controversial and does not trigger shared agreement, I argue that its rhetorical purpose resides in taking advantage of universal quantification to target potential adversaries who disagree with X by *ridiculing* them. It therefore seems that "everyone knows X" should be treated more like an *ad personam* attack targeting those who doubt *x* than like an appeal to popularity meant to reinforce the likelihood of X.

4. CONCLUSION

While appeal to popular opinion seemed to be quite a straightforward argument scheme, which can be described by the use of the expression "everybody knows X" in its premises, it turns out that the rhetoric-pragmatic study (Oswald & Herman 2016) of this expression makes things more complex. In particular, the use of this expression when it is clear that it cannot cover all people, or even a majority, requires finding pragmatic effects other than those described by the appeal to popularity.

Then, the epistemological and evidential nature of the expression "everybody knows X" allowed us to ground a distinction between instances where X denotes a fact and instances where X denotes an opinion. When dealing with propositions denoting already shared or known states of affairs (or facts), "everybody knows X" seems to function as a device meant to recall widespread information. When it is used to denote opinions, it draws on the evidential and epistemic properties of widespread information to immunise X from being called into question. And when facts and opinions are overtly not shared, it seems that the *ad populum* appeal may combine with an *ad personam* attack or a threat against those who are not included in "everybody knows X", ridiculing the people who are not sharing the "obvious fact or opinion".

ACKNOWLEDGEMENTS: Thanks to Steve Oswald for letting me rewrite a new version of our common paper (to appear in 2020). While some parts are identical, many examples, the three degrees of amplification in chapter 2 and the typology in chapter 3 are new.

REFERENCES

Carston, R. (2002). *Thoughts and utterances: the pragmatics of explicit communication*. Oxford, U.K.; Malden, Mass.: Blackwell Pub.
Carston, R. (2010). Explicit Communication and 'Free' Pragmatic Enrichment. In B. Soria & E. Romero (Eds.), *Explicit Communication: Robyn Carston's Pragmatics* (pp. 217–285). https://doi.org/10.1057/9780230292352_14
Godden, D. M. (2008). On common knowledge and ad populum: Acceptance as grounds for acceptability. *Philosophy & Rhetoric*, *41*(2), 101–129.
Herman, T., & Oswald, S., (to appear in 2020)"Everybody knows that there is something odd about ad populum arguments"
Jansen, H. (2018). Ad Populum Arguments in a Political Context. In S. Oswald & D. Maillat (Eds.), *Argumentation and Inference: Proceedings of the 2nd*

European Conference on Argumentation, Fribourg 2017 (Vol. 2, pp. 425–437). London: College Publications.

Jansen, H., & van Leeuwen, M. (2019). Presentational choice in ad populum argumentation. In B. Garssen, D. Godden, G. Mitchell, & J. Wagemans (Eds.), *Proceedings of the 9th Conference of the International Society for the Study of Argumentation* (pp. 573–582). Amsterdam: Sic Sat.

Oswald, S., & Herman, T. (2016). Argumentation, conspiracy and the moon: a rhetorical-pragmatic analysis. In M. Danesi & S. Greco (Eds.), *Case studies in discourse analysis* (pp. 295–330). Münich: Lincom Europa.

Sperber, D., & Wilson, D. (1995). *Relevance: Communication and Cognition*. Oxford: Blackwell.

Walton, D. N. (1999). *Appeal to popular opinion*. University Park, PA: Penn State Press.

Towards a Theory of Informal Argument Semantics

MARTIN HINTON
University of Łódź, Poland
Martin.hinton@uni.lodz.pl

In this paper I set out the framework for a theory of informal argument semantics which is designed to make the assessment of the language of arguments easier and more systematic than is currently the case. The framework, which attempts to identify arguments suffering from linguistic confusion, is intended to complement existing approaches to argument appraisal and is envisaged as a third stage of assessment after procedural and inferential analyses have been conducted.

KEYWORDS: argument evaluation, argument semantics, definition of argument, fallacy theory, language fallacies.

1. INTRODUCTION

Argumentation theory has been greatly informed by insights from linguistics and the study of discourse, most obviously in the pragma-dialectical approach championed by Frans van Eemeren and Rob Grootendorst (2004). However, these insights have been rather from the field of pragmatics than from semantics. Problems of meaning in argumentation theory are largely considered only in the study of certain 'fallacies of language', generally taken to be cases of some kind of ambiguity, although there has been interest recently in considering other traditional fallacies as language rather than reasoning based errors (see Visser et al., 2018).

    Ambiguity, particularly in the form of equivocation, is a serious matter in argument; but it is also an inherent part of language, and it is far from the only concern caused by the meanings of words. Language contains within it argumentative content, whether the "argumentativity" of Anscombre & Ducrot (1989) or the implicatures of Grice (1975). Words also have relationships: some of them cannot be used together, what I refer to as semantic incompatibility; and often sentences are imprecise to the point at which no real propositional value can be found in them.

In this paper, I set out the framework for a theory of informal argument semantics which is designed to make the assessment of the language of arguments easier and more systematic than is currently the case. The framework attempts to identify arguments which are based on linguistic confusion; arguments which feature linguistic confusion, be it ambiguity, imprecision or meaninglessness; and arguments that lead to linguistic confusion in their conclusions. This is a challenging task with a broad scope and this paper represents the early steps towards a settled theory. It will, however, feature a number of examples of argument from philosophy and politics, and illustrate how they can be better understood through a thorough semantic analysis. This analysis, it should be stressed, is not designed to replace, but rather to complement existing approaches to argument appraisal and is envisaged as a third stage after procedural and inferential analyses have been conducted.

2. A THEORY OF ARGUMENTATION

In order to establish a system of informal argument semantics, certain background theoretical assumptions are necessary. Firstly, I should make it clear that the semantic assessment scheme which is the end product of this work is an informal scheme, and that it is designed to be applied to informal arguments: the informal argument semantics I refer to, therefore, is both an informal semantics and a semantics of informal argument. The word 'semantics' is not used in exactly the same sense as in more formal work where a semantics constitutes a list of what is acceptable within a particular system; something which would hardly be possible when dealing with natural language. However, the 'semantics' of this paper does mean a scheme for determining whether the textual input provided by an apparently argumentative utterance is acceptable and meaningful, based on the semantic qualities of the words used, rather than their pragmatic force within the discourse situation of which they are part.

As I mentioned in the introduction, there is a sense in which all language is, or at least contains, argument. What we say is what we have inferred to be the truth, or what we have inferred to be the appropriate or advantageous thing to say at this moment. We expect others to be able to follow our line of thought, without making each step explicit, and to make further inferences on the basis of our utterances. Language use, then, can be said to inherently contain argumentative content. That does not mean, however, that every piece of language can be construed as being an argument. When someone makes an assertion without providing any support for it, he is simply asserting, not arguing. That his words hint at previous reasoning and lead to implied inferences does not make them, in themselves, an argument. When an utterance is used as part of an argument, however, that implied reasoning and those further implicatures do become part of the overall argument structure and their meanings become part of its semantic content.

The mere assertion of some standpoint, then, is not an argument; but what is? Definitions of both argument and argumentation abound: in their textbook, Copi, Cohen & McMahon, (2014, p.5), claim that:

> argument refers strictly to any group of propositions of which one is claimed to follow from the others, which are regarded as providing support for the truth of that one. For every possible inference there is a corresponding argument.

Which is a rather awkward way of putting it. Tindale makes it clearer by describing arguments as structures "where one or more statements (premises) are given in support of a conclusion" (2007, p.1). Apart from its clarity, Tindale's version has an advantage in that it refers to statements which are "given". This introduces both the idea of the form of expression of the argument, how it is given, and the necessity of their being some context in which the giving takes place.

Argument can, of course, mean one such structure or a type of discourse occurrence in which any number of such structures are expressed and exchanged. This ambiguity can be eased by employing the word 'argumentation' for the latter, or it can be compounded by using 'argumentation' for both: "An argumentation consists of one or more expressions in which a constellation of propositions is expressed" (van Eemeren & Grootendorst, 2004, p. 2), but also, one page earlier:

> Argumentation is a verbal, social, and rational activity aimed at convincing a reasonable critic of the acceptability of a standpoint by putting forward a constellation of propositions justifying or refuting the proposition expressed in the standpoint.

Still, taking these suggestions together, we can safely conclude that arguments are "expressed" and that argumentation is an "activity", what Ralph Johnson succinctly calls "an exhibition of rationality" (2000, p.13). Building upon these, and with my own purposes in mind, I propose to define an argument as an expression of reasoning, and argumentation as the expression of reasoning within a process. By what that process may be constituted, I am more flexible than the pragma-dialecticians, and am happy to include activities which do not much resemble the model of the critical discussion, but what other modes of argumentation are possible is not a subject to take further in this essay.

The proposed definition of argumentation, which I do not claim to be better from a theoretical viewpoint than some other suggestions, has an attractive practical consequence in the consideration of the poor practice of the activity, known as fallacy theory. The three stages of the definition translate easily into three varieties of fallacy, rendering unnecessary the contortions sometimes performed to include all the well-known fallacies within categories, or, indeed, within any conception of what fallacies actually are. On my model, an argument can go wrong, and therefore deserve rejection, at any one of three stages (sometimes more than one at once): it can contain unsound reasoning, it can be erroneously expressed, or it can be unsuitable to the process. In this way, we arrive at three varieties of fallacy: reasoning fallacies, linguistic fallacies, and process fallacies. While some of the frequently discussed examples may be capable of rejection on the basis of more than one of these areas for analysis, none is left outside the typology and no general category is needed as a collection point for all those which do not fit comfortably elsewhere. Any argument which is rejected must be rejected at one point of the analysis, and that rejection will automatically assign it to the fallacy group associated with that assessment stage.

3. A SCHEME FOR THE ASSESSMENT OF ARGUMENTS

The scheme for the analysis of arguments which I set out in the following section is, therefore, divided into four sections: an initial analysis, which establishes whether, in fact, the chosen text does contain an argument; a process analysis to determine the suitability of the argument to the argumentation discourse in which it has been offered, the details of which, obviously, are governed by the standards of that process and the goals of the participants; a reasoning analysis which includes the investigation of formal logical fallacies, considers the truth or otherwise of the premises, implicit and explicit, and assesses the strength of the inference, retaining a reference to the earlier process stage and the requirements of the given situation as regards argument strength; and, finally, a linguistic analysis, which is the main focus of this article, and is described in detail below. An argument which passes through all these stages without being rejected must be accepted, at least presumptively, by any reasonable arguer. At each stage, rejection may actually mean reformation, and where it is possible to avoid the fallacy which has been committed by changing the argument in some way, linguistically or structurally, it can be so-altered and re-submitted to the initial stage.

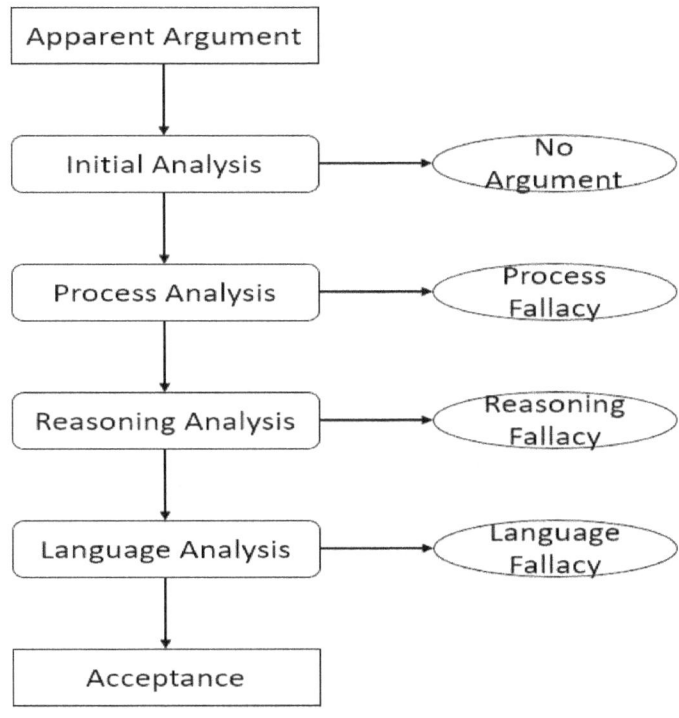

Figure 1 – A general scheme for argument assessment.

There are two points to be made about this ordering of the stages of analysis. Firstly, the order is implied by the practicalities of the process of assessment and a desire to save time. The initial stage discovers whether there is an argument to assess, this is, I assume, an uncontroversial beginning, especially as a great many texts, such as newspaper opinion columns, political speeches, and, sadly, academic essays, which are expected to contain arguments, do not, in fact, do so. This is followed by consideration of whether the argument is admissible and relevant to the process. It makes sense to look at this first since there is no point in carrying out detailed analysis on arguments which are of no use even if they are strong and clear. If an argument can be admitted to the process, then its reasoning is checked, and only then once that has been accepted, is the most detailed and complex level examined, the informal argument semantics.

An analyst whose main interest is in forms of inference might be tempted to reverse these last two on the principle that it is not worth doing detailed inferential analysis on arguments which are flawed linguistically. This is a point of little consequence to the overall framework since the scheme can be thought of as having a nature more circular than linear, with the proviso that once a full lap has been completed, there is no just cause to refuse acceptance of the argument. It should be obvious that the initial stage is, largely, a stage of linguistic analysis. It is an analysis at the level of normal reading comprehension which would be expected to weed out examples of arguments expressed very badly indeed: the informal semantics stage is intended to be applied to arguments which look good and may only be suffering from some deeply hidden flaw, not obvious earlier.

This is a point related to the second justification for this ordering of the stages which is its reflection of the levels of sophistication of argument as set out by Harald Wohlrapp. Wohlrapp identifies Natural, Scientific, and Philosophical argument. The first is what people engage in most of the time, it is largely unrestrained by any rules and he describes it as "a confusing mess" (2014, p. 384). The second level is more organised, it is "for making claims and for validating them with justifications or invalidating them with objections" (2014, p. 385). Arguments at this level may descend to the Natural if good principles of reasoning are ignored, or may ascend to the Philosophical where the very grounds of validity of justification at the Scientific level are called into question.

There are, I believe, parallels to be drawn with the three elements of my definition of argumentation and the three levels of analysis at which fallacies may be detected. Natural argumentation is characterised by a lack of agreed process which means that whatever standards of process one introduces in the assessment of such arguments (most of the rules of pragma-dialectics can be applied at this stage), one is likely to find that standards are not adhered to and fallacies are committed constantly: Natural arguments will often be irrelevant, sometimes threatening, frequently unsupported. The Scientific level demonstrates the use of reason and evidence, so at this level one may expect to find fewer process violations, scientists (hopefully) don't insult one another or refuse to defend their work, but there will certainly be examples of poor logic and questionable conclusions; correlations taken as causes, weak inferences turned into strong claims, statistics misused and misunderstood. This level of argument, then, corresponds to the reasoning level of analysis and would be expected to produce examples of the fallacies found therein. The highest, most abstract level, the Philosophical, is most vulnerable to errors of language. A host of philosophers have criticised their peers and rivals for mistakes in their systems which result from misconceptions of language. These misconceptions are generally of either of two types: attempts to shape language use by redefining words; or attempts to find the truth about the world through reasoning upon the accidents of language. In both cases, it is a misunderstanding of the very nature of language which leads the philosopher into error.

The reason for the emphasis in this paper on the analysis of language is thus revealed: not only is there an obvious lack of an informal argument semantics, which needs to be filled, but it is the assessment of the subtlest errors in the most fundamental arguments for which it is needed. Describing and evaluating the everyday argumentative discourse of society and opinion makers in order to help people understand the process of which they are a part and raise the level of public discussion is a noble task, but how much nobler to expose the errors and misapprehensions which riddle the beliefs and assumptions which underpin that society's culture and hold it in ignorance and superstition!

4. INFOMAL ARGUMENT SEMANTICS

The third main stage of analysis is carried out according to the informal argument semantics depicted in figure 2 below.

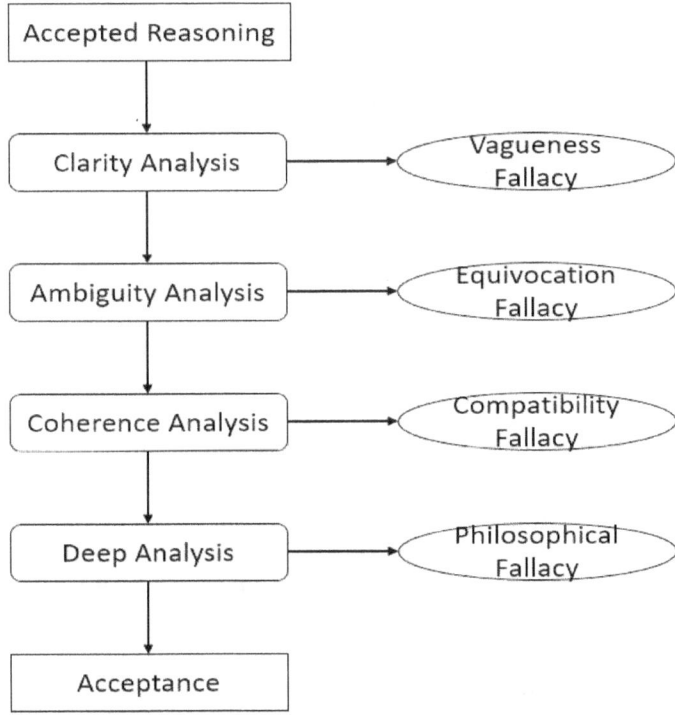

Figure 2 – A scheme for the linguistic assessment of arguments

Traditionally, those argument forms considered linguistic fallacies are either condemned as displaying species of vagueness or some kind of ambiguity. In both cases the problem is an ability to properly discern the meaning of the argument in question. It should also be noted that in both cases the question is whether the meaning is precise enough for the task at hand: statements can always be further clarified, and ambiguity is an essential feature of language, without which we should require as many words as we have objects of sense.

Vagueness comes in three main forms: language may be obscure, obtuse, twisted and contorted, full of jargon, or odd usage; it may be clear as far as it goes but insufficiently precise for its purpose, its key terms under-determined; or, separate statements which are acceptable in themselves may not add up to a clear and coherent argument for a comprehensible conclusion. Such language is not to be confused with nonsense, of course. Nonsense is not really language at all, it just looks like it, and would not be accepted past the initial stage of analysis. The difficulty in gaining the real sense of the argument with which we are concerned at this point is far subtler.

Ambiguity, in which I shall include amphiboly for present purposes, can disrupt an argument in two ways: either by making it hard to know which of two meanings was actually meant, or by masking an error in logic. In the latter case, what appears to be a common term in different parts of the argument structure is, in fact, not one at all, and an equivocation has occurred.

Less commonly recognised, and, therefore, more interesting, are what I refer to as compatibility fallacies. These are cases where words have been used with a clear sense and according to the rules of syntax, but have been combined in ways which make the final sentence a semantic impossibility. I can offer two simple examples of this. One comes from the UK Labour Party 2017 general election manifesto. The policy of the party was to accept the result of the referendum which decided in favour of the UK's leaving the European Union, but also to "reject 'no deal'" (Labour Party, 2017, p. 24). The problem with this standpoint is that the lack of a deal is not the kind of thing which one can reject. Rejecting leaving without a deal is only possible if one is prepared to reject leaving full stop, which was not the policy at the time, or accept absolutely any deal on offer, which the party declined to do. The same incoherent position has been advanced by many British politicians in the succeeding years.

While politicians may not always be expected to make a lot of sense, a second example constitutes one of the key flaws in one of the most important ethical systems in the history of philosophy. According to Jeremy Bentham, in deciding how to act we should: "Sum up all the values of all the pleasures on the one side, and those of all the pains on the other" (Bentham, [1789]1962, p. 66). Whether we make the object of our arithmetic pleasure, pain, utility or happiness, is of no importance: the fact remains, that only numbers can be summed. There is a fine line, and it may be no line at all, between such examples and those I consider a result of the fetishisation of language and, thus, concept fallacies. The linguistic fact that one pleasure can be said to be greater than another, that one pain is small and another enormous, has seduced many otherwise intelligent philosophers into thinking that they might treat them as though they could sensibly be given numerical values.

This is leading us into the fourth level of linguistic evaluation, the deep analysis which exposes philosophical fallacies, and is depicted in more detail in figure 3.

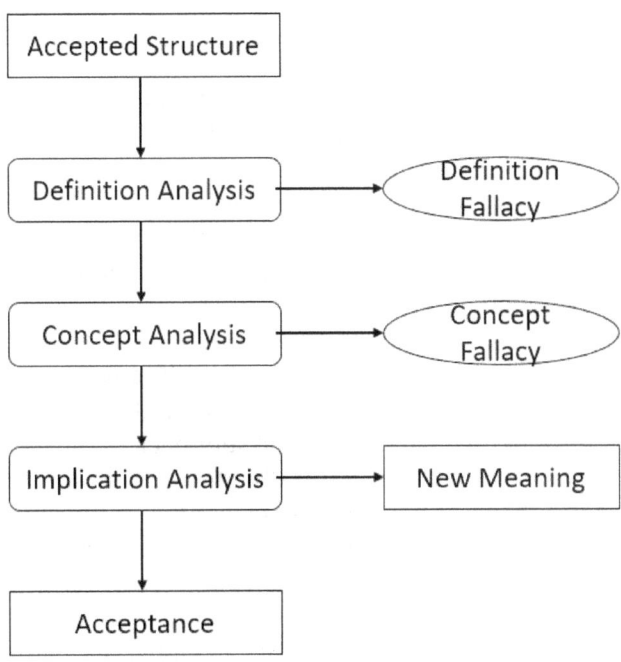

Figure 3 – A scheme for the 'deep' analysis of the language of arguments.

Definition fallacies are most commonly varieties of "persuasive definitions" a term introduced by Charles Stevenson (1944) and best described by Andrew Aberdein's phrase "gerrymandering a term" (2006), since the fallacy involves shifting the borders of what a particular word can be used for in order to suit one's own purposes. The 'no true Scotsman' fallacy is an example of this nefarious practice and, thus, also an example of a fallacy which is not usually grouped with fallacies of language, but ought to be. Other varieties of suspect definition are G.E. Moore's Naturalistic Fallacy (1903) and Leonard Nelson's Philosophical Fallacy (2016).

Something has already been said of concept fallacies, brought on by an undue faith in the ability of language to represent the world as it is. To avoid such pitfalls, one would do well to remember Arne Naess's warning "that the existence of some concept term in no ways guarantees that something falls under the concept" (1966, p. 67). Wittgenstein also draws attention to this phenomenon, noting in the Blue Book (1958: 1) that:

> The questions "What is length?", "What is meaning?", "What is the number one?" etc. produce in us a mental cramp. We feel that we can't point to anything in reply to them and yet ought to point to something.

Later, he states: "This kind of mistake recurs again and again in philosophy" (1958, p. 6), and I suspect outside of it as well.

The final layer of analysis is different from that all have come before in that it does not identify any fallacious moves in an argument, rather it serves to fully explicate what that argument is. In the implication analysis, all of the inherent argumentativity and the Gricean implicatures of the statements making up the argument structure are considered, and it is also here that ethical assumptions lying behind any evaluative or emotional language are exposed. If anything of relevance to the argumentation process is discovered, then it is made explicit and added to the argument which can then be sent back to the initial stage to begin its evaluation once more. Naturally, it is a question of judgement whether or not a given implicature is relevant, and this applies all the way through any analysis employing this scheme. The entire scheme, including the informal argument semantics section, is a guide for a human analyst, not an automated process. It cannot take the place of the analyst because it is concerned with meaning, its subtleties and contradictions, which are of infinite variety and known only to a language user. The purpose of the scheme is to show which questions we should ask when assessing arguments. The experienced analyst will be able to jump straight from the initial stage to the relevant questions which are likely to expose the frailties of the text before him: there is no requirement to follow the scheme step-by-step when one can already see which stage will prove decisive.

5. AN EXAMPLE

Space will allow for only a partial analysis of one example, in which I shall concentrate on linguistic considerations. Below is a statement made by Ruth Halperin-Kaddari, who is vice-chair of the UN committee on the elimination of discrimination against women, and which, I suggest, does appear to have merit on an initial view. Her comments concerning the law governing access to abortion in Northern Ireland were quoted in the Guardian newspaper (Gentleman, 2018):

> Denial of abortion and criminalisation of abortion amounts to discrimination against women because it is a denial of a service that only women need.

The use of the word "because" makes it clear that this is an argument in which the first part of the sentence is the conclusion, or standpoint, and the second part is the supporting premise. The phrase "amounts to" gives an early indication that the argument may be making some kind of definition or categorisation.

First, we ask if the argument is sufficiently clear. There are two apparent problems here: first the switch from "denial [...] and criminalisation" to "is a denial" suggests that criminalisation is not important, or more likely, not relevant to this particular argument. The statement can easily be rephrased without the word. Secondly, abortion is referred to as "service", a usage which might be designed to hide the reality of what it involves. Even if such euphemistic language is readily understood, it can create a further problem. There does not appear to be any equivocation in the argument, but there is uncertainty over the ambiguous word "need". Since abortion in cases of danger to the mother's life is not illegal in Northern Ireland, the "need" must refer to something other than physical survival. Once emotional and self-realisational needs are brought into consideration, however, it is no longer clear that they are relevant only to women, and the use of the word "service" also implies something other than the medical procedure which only women can undergo. This is an example of how a piece of reasoning considered sound at the earlier stage of analysis can begin to look doubtful when the language of the premises is studied more closely.

All of these concerns can be overcome with some rephrasing of the argument, although it may lose a little force. We are left with:

> Denial of abortion amounts to discrimination against women because it is denial of a procedure only women can undergo.

There does not appear to be any semantic incompatibility here, so we can move on to the deep linguistic analysis stage. First, we ask: is there an attempt at redefinition? Clearly there is, but that does not automatically disqualify the argument. A definition may ask us to look at the use of a word anew, but we may agree that certain cases are covered by that word even if we had not realised it previously. 'Discrimination' we normally understand to mean different treatment, usually worse, of some individuals on the basis of some feature they possess. We understand it to mean that differential treatment is unfair. In this case we are being asked to include in our understanding of the word cases not where two groups are treated differently, but where one is denied something which the other cannot have. Whatever one's view on abortion or discrimination, that is certainly stretching the accepted meaning of the word.

One way in which this definition, or rather categorisation, could be maintained is if we take it to imply that men, generally speaking, are not denied procedures which only they can undergo. This would require further supporting arguments and would be difficult to show given the lack of anything which might be considered a male equivalent of the right to abortion. We have now moved into the final stage, the implication analysis. One other interesting inference which the argument invites us to draw is that if we do accept the principle that denial of some good which is only available to women constitutes discrimination against women, then presumably, we are committed to maintaining that such denial against any group amounts to discrimination against that group. That is a general principle which might lead to some awkward places, perhaps awkward enough to cause a supporter of our original argument to reconsider.

There is more to say about this argument, but the intention here was to show how the scheme works in practice. The clarity and ambiguity assessments led to some rephrasing of the original text; the deeper analysis highlighted a possible definitional fallacy which itself leads to some important implications. The full argument, rephrased and carrying what it implies explicitly, can then be re-submitted for full analysis from the beginning.

6. CONCLUSION

The definition I have offered of argumentation, as the expression of reasoning within a process, allows the tidy division of fallacious argument moves into three categories, and the arrangement of the analysis of arguments into three corresponding stages.

The assessment of suitability to process and soundness of reasoning I have mentioned, but not expanded upon, and I invite others to fill out those sections of the analysis scheme. My own focus has been on the discovery of fallacies of expression through the development of an informal argument semantics of which I have given only an outline in this paper. The previous section in which I analysed a short example of an argumentative utterance illustrated the power of the scheme based on that semantics to draw out weaknesses and hidden implications in arguments which appear to have acceptable form.

REFERENCES

Aberdein, A. (2006). Raising the tone: definition, bullshit, and the definition of bullshit. In G. Reisch & G. Hardcastle (Eds.), Bullshit and Philosophy (pp. 151–169). Chicago: Open Court.

Anscombre, J-C., & Ducrot, O. (1989) Argumentativity and informativity. In M. Meyer (Ed.), From metaphysics to Rhetoric (pp. 71-97). Dordrecht: Kluwer Academic Publishers.

Bentham, J. ([1789] 1962). The works of Jeremy Bentham. New York: Russell & Russell.

Copi, I., Cohen, C., & McMahon, K. (2014). Introduction to logic. Harlow: Pearson.

Eemeren, F.H. van, & Grootendorst, R. (2004). A systematic theory of argumentation. Cambridge: Cambridge University Press.

Gentleman, A. (2018). Northern Irish abortion law violates women's rights, say UN officials. The Guardian 23/02/2018. https://www.theguardian.com/uk-news/2018/feb/23/northern-ireland-abortion-law-violates-womens-rights-says-un-committee

Grice, H. P. (1975). Logic and conversation. In P. Cole & J. Morgan (Eds.), Syntax and Semantics, 3: Speech Acts (pp. 41–58). New York: Academic Press.

Johnson, R.H. (2000). Manifest rationality, A pragmatic theory of argument. Mahwah, NJ: Lawrence Erlbaum Associates.

Labour Party. (2017). For the many, not the few. The Labour Party manifesto 2017. http://labour.org.uk/wp-content/uploads/2017/10/labour-manifesto-2017.pdf

Moore, G.E. (1903). Principia ethica. London: Cambridge University Press.

Naess, A. (1966). Communication and argument. London: Allen and Unwin.

Nelson, L. (2016). A theory of philosophical fallacies. Translated by F. Leal & D. Carus. Cham: Springer.

Stevenson, C. (1944). Ethics and language. New Haven, CT: Yale University Press.

Tindale, C. (2007). Fallacies and argument appraisal. Cambridge: Cambridge University Press.

Visser, J., Koszowy, M., Konat, B., Budzynska, K., & Reed, C. (2018) Straw man as misuse of rephrase. In S. Oswald, & D. Maillat (Eds.), Argumentation and inference: Proceedings of the 2nd European Conference on Argumentation, Fribourg 2017. Vol. II, (pp. 941-962). London: College Publications.

Wittgenstein, L. (1958) The blue and brown books, preliminary studies for the Philosophical Investigations. London: Blackwell.
Wohlrapp, H. (2014). The concept of argument. Dordrecht: Springer.

Stereotyping

DAVID HITCHCOCK
Department of Philosophy, McMaster University, Canada
hitchckd@mcmaster.ca

Logic textbooks ignore stereotyping, even though 'stereotyping' is the most commonly used fallacy label. This paper defines the term, and discusses:
- under what conditions stereotyping is mistaken,
- whether it can be morally objectionable even if epistemically warranted,
- its relationship to other forms of reasoning,
- how to recognize and respond to stereotyping, and
- how to avoid committing the fallacy.

KEY WORDS: argument schemes, biases, debiasing, fallacies, Implicit Association Test, mistaken stereotyping, profiles of dialogue, stereotypes, stereotyping

1. INTRODUCTION

A search in July 2016 turned up 12.6 million Web pages with the term 'stereotyping', far more than for any of 61 other fallacy terms (Hitchcock, 2017, pp. 428-429). Despite this common use, theoretical and textbook treatments of fallacies ignore stereotyping. It does not occur in what Woods (1992; 2004) calls "the gang of eighteen" traditional fallacies surveyed by Hamblin (1970, pp. 9-49) or among the 26 traditional fallacies listed by van Eemeren and Grootendorst (1992, pp. 212-215). It is not treated in *Fallacies: Classical and contemporary readings* (Hansen & Pinto, 1995) or in introductory textbooks that discuss fallacies, such as (Johnson & Blair, 1993), (Copi & Cohen, 2002), (Hurley, 2006), (Bailin & Battersby, 2010), and (Groarke & Tindale, 2012).

I propose therefore to provide an account of stereotyping, one that could be used as the basis of a treatment of the phenomenon in an introductory textbook. I start by defining the term 'stereotyping' as the label for a certain type of reasoning. I then discuss under what conditions this form of reasoning is mistaken, and consider when non-mistaken reasoning of this form is nevertheless morally objectionable. I discuss the relationship of this form of reasoning to other forms of reasoning. Finally, I discuss how to recognize and respond to mistaken stereotyping and how to avoid mistaken stereotyping in one's own thinking and communicating.

2. A DEFINITION OF 'STEREOTYPING'

2.1 Stereotypes

According to the *Oxford English Dictionary* (Murray et al., 1971/1933), the word 'stereotype' came into the English language at the end of the 18th century as the name for a new process of printing from a solid plate, the name combining two Greek words, '*stereos*' (solid) and '*typos*' (type, outline). The word became used for the plate itself, and then figuratively for something constantly repeated without change, such as a stereotypical phrase. The dictionary does not record a use of 'stereotype' as the name of a fixed belief about a human group, even though this use occurs as early as Walter Lippmann's *Public Opinion* (Lippmann, 1922), which devotes one-fifth of its pages to the role of stereotypes in forming public opinion. In keeping with the word's etymology, its various uses have in common a reference to something fixed that produces the same result whenever it is used.

Nowadays, people use the word 'stereotype' quite broadly. In an online list of sentences using the word 'stereotype',[1] people are said to have stereotypes about individual people (Jesus), institutions (the church), places (modern London), doctrines (moderate liberal feminism), and activities (using the Internet, cruises). But the most frequent objects of stereotypes in these examples are human groups, distinguished in various ways: by their recreational activities (gaming, sports, abusing drugs), their clothing (sweaters, vests), their ethnic group (Moor, white man, Asians, the Irish, Jersey Italian families), their views (bigots), their occupation (cowboys), their sexual orientation (male heterosexuals, lesbians), or their medical diagnosis (children with autism spectrum disorder). In these examples, calling a characterization a stereotype generally goes with a challenge to its accuracy. In the quoted sentences, the holders of the stereotypes are not identified; rather, the mentioned stereotypes are treated as generally accepted. The attributed features are predominantly mental traits (treacherous, noble, overly anxious and insecure) or ongoing behavioural patterns (neither quiet nor thoughtful, a casanova, ready to move in with a sexual partner after a couple of dates, unaffectionate, fond of drink, having mob links). There is one attribution of a physical feature (being short) and

---

[1] At https://sentence.yourdictionary.com/stereotype; accessed 2019-09-12.

one of skin colour and sex (white male). The last-mentioned example attributes membership in a certain group to people with an ongoing mental trait, rather than vice versa, as in the other examples. We can thus extract from these examples a central or prototypical meaning of 'stereotype' in one of its current senses as *a fixed association in a person's mind of one or more ongoing behavioural or mental traits with membership in a human group identified by such factors as sex, ethnic status, stage of life, occupation, sexual orientation, clothing or recreational activities.*[2] Stereotypes are expressed linguistically as "bare plural generics"—i.e. quantified sentences with no explicit quantifier, such as 'the Irish are fond of drink' or 'bigots are white men' (two examples in the cited list of sentences). In the mind of its holder, a stereotype is a belief, possibly implicit, which philosophers and psychologists characterize variously as a concept (conceived as a prototype), a stage on the way to forming a concept, or an informational structure associated with a concept (Beeghly, 2015). People use stereotypes to categorize individual human beings and form expectations of them based on those categorizations.

The proposed definition of the term 'stereotype' deliberately uses the phrase 'human group'. Hacking (1991, p. 118) introduces in a discussion of natural kinds a useful distinction between a sub-kind and a kind-derived sub-group, a distinction that we can apply to people, who are a natural kind. A sub-kind, he writes, has a large and plausibly inexhaustible set of properties that its complement lacks, whereas a kind-derived sub-group does not. Thus, for example, Manx cats are a sub-kind of cats, because they share a large set of properties that non-Manx cats lack. But white cats are merely a sub-group derived from the kind cats, because they do not share a large set of properties that non-white cats lack. Applying this distinction to people, one might suppose that men share a large set of properties that women lack but that "white" (i.e. fair-skinned) people do not share a large set of properties that non-white people lack. If so, men would be a sub-kind of human beings, whereas white people would be a kind-derived sub-group. Since people hold stereotypes both about men and about white people, the

---

[2] Schneider (2005, p. 24) defines a stereotype as "qualities perceived to be associated with particular groups or categories of people". Like the above definition, his definition omits any theoretical explanation or evaluation of stereotypes and includes associations in either direction between group membership and qualities. It differs from the above definition in leaving unspecified how long a stereotype is held, what sorts and duration of qualities it attributes, and how the stereotyped human groups are characterized. It also differs in requiring more than one associated quality and in calling the association "perceived" rather than merely "in a person's mind". The differences seem inconsequential.

definition proposed in this paper uses the phrase 'human group' rather than, for example, the phrase 'social kind', which some definitions of 'stereotype' use.

The definition says nothing about whether people acquire stereotypes mostly by personal experience or mostly by cultural influences, how inflexibly people apply them to individual members of human groups about which they have stereotypes, or how easily and in response to what sorts of influences people change their stereotypes—issues that have been studied extensively by social psychologists, as reported for example by Schneider (2005).

2.2 Stereotyping

Stereotyping comes in stages (Beeghly, 2015). It begins with activation of a stereotype—e.g. thinking of Americans as optimistic or of Canadians as apologetic. The activated stereotype may then function as a reason for, or an unconscious causal influence on, some conclusion or decision. At this stage, it may be the primary influence, in the sense that the conclusion or decision would not have been reached without the activation of the stereotype. At the final stage, the stereotype may receive public expression, for example in a cartoon or an offhand remark. The activation and mental use of a stereotype may go on quite unconsciously and more or less simultaneously. A particularly problematic kind of stereotyping is the application of a stereotype to an individual without any other basis for judging that person. We can define this kind of stereotyping as follows:

> To stereotype a person is to attribute to them a pattern of behaviour or mental trait on the sole basis of a stereotype of a group to which the person is assumed to belong or to attribute to them membership in a human group on the sole basis of an exhibited pattern of behaviour or mental trait.

3. EVALUATION

Stereotyping a person is a kind of inference, in which the stereotype is used to license a move from the assumption that the person belongs to a certain human group to the conclusion that the person fits the stereotype, or vice versa. The first kind of inference has the following form:

> <Person x> belongs to <human group G>.
> Therefore, <person x> has <behavioural or mental trait F>.

The second kind of inference has the converse form:

<Person x> has <behavioural or mental trait F>.
Therefore, <person x> belongs to <human group G>.

To evaluate an inference of either type, one needs to determine whether the inference has a covering generalization that holds in all or most or all normal cases—including counter-factual cases (Hitchcock, 2011). Stereotypes however are not expressed as universal or for-the-most-part or *ceteris paribus* generalizations. As mentioned earlier, they are expressed as bare plural generics, i.e. as unqualified generalizations. Hence one needs a theory about the truth-conditions for bare plural generics to determine (a) whether an assumed inference-licensing stereotype is true and (b) whether it is equivalent to a counter-factual-supporting universal or for-the-most-part or *ceteris paribus* generalization.

There is broad consensus about which bare plural generics concerning natural kinds are true and which are false. For example, everyone with minimal knowledge about ducks and pigeons would agree that it is true that ducks quack and false that pigeons give birth to live young. But there is no consensus on the underlying explanatory logical form and semantics of bare plural generics. Some accounts treat bare plural generics as predications about the kind, either simple kind predications (Liebesman, 2007) or sophisticated ones (Teichman, 2015). Most treat them as predications about members of the kind, either saying what they are normally like (e.g. Nickel, 2016) or what most of them are like (e.g. Tessler & Goodman, 2019). Although these accounts treat them as *ceteris paribus* or for-the-most-part generalizations, they are hedged with qualifications, in order to fit test cases of bare plural generics with a known truth-value. True bare plural generics tolerate exceptions, sometimes a large percentage of exceptions. For example, it is true that ducks lay eggs, but the only individual ducks that lay eggs are adult female ducks, who are a minority of ducks. It also true that sea turtles live to 80 or 90 years, but only about one in 1,000 sea turtle hatchlings survives to adulthood. Another striking fact about true bare plural generics about natural kinds is that their truth has a theoretical explanation, whether biological, chemical or physical. Their truth is not just an accidental fact.

It is a reasonable assumption that bare plural generics about human groups have the same logical form, semantics and truth conditions as bare plural generics about natural kinds. Why would the truth conditions be different in the case of "black" (i.e. dark-skinned) people than in the case of black cats? Without taking sides in the ongoing scholarly debate about the underlying logical form and semantics of bare plural generics, we can take advantage of Hacking's distinction between kind-derived groups that are sub-kinds and kind-derived groups that are not. On his account, a human group is a sub-kind if and only if it has a large and plausibly inexhaustible set of properties that its complement lacks. Black people, for example, are a sub-kind if and only if they have a large and plausibly inexhaustible set of properties that non-black people lack. If black people are a sub-kind in this sense, then there are true bare plural generics about black people—those that attribute to them the properties that black people typically share but non-black people typically lack. Conversely, if black people are not a sub-kind in this sense, then probably the only true bare plural generic about them is the uninformative tautology that black people are black. According to a current consensus in physical anthropology (Sinha, 2011), although human populations that have evolved separately differ in various genetically based respects from one another, quite distinct populations include people with dark skin. Hence, in all probability the only bare plural generic true of black people is that they are black. Hence no stereotype about black people is true.

On the other hand, one's sex or racial/ethnic sub-group entails (or at least makes probable) distinctive characteristic properties, which may include behavioural or mental traits (Freedman & Freedman, 1969). One's occupation, religious affiliation, nationality or educational specialization may also entail (or make probable) distinctive behavioural or mental traits; if so, groups defined by these variables would be human sub-kinds on Hacking's criterion, but distinguished culturally rather than biologically.

Thus many stereotypes used to license inferences are true. But do they license those inferences? To serve as inference-licenses, they need to support counter-factual instances. Their truth cannot be merely accidental. Suppose for example that Canadians really are apologetic. To license an inference, this stereotype would have to hold counter-factually, in the sense that (for example) Germans would be apologetic if they were Canadians. Further, the truth of an inference-licensing stereotype would have to amount to the truth of a *ceteris paribus* or for-the-most-part generalization, given that stereotypes are taken to have exceptions and so not to amount to universal generalizations. It should not tolerate such a large percentage of exceptions as the true generic statements that ducks lay eggs and sea turtles live to 80 or 90 years.

We can thus conclude that stereotyping a person is a valid form of reasoning if a stereotype that would license the inference is true of most people in the stereotyped group or is normally true of people in the stereotyped group, provided that the truth of this stereotype has a theoretical basis and is not merely accidental. If the premiss of such reasoning is correct, then the conclusion is either probably or presumably correct, relative to this premiss. Additional true information compatible with the truth of the premiss and of the inference-licensing stereotype can rebut this conclusion, showing that it is false. In particular, since all stereotypes have exceptions, the person about whom the inference is drawn may not fit the stereotype. These conditions for justified stereotyping can be represented in the standard form of an argument scheme with critical questions (Walton, Reed & Macagno, 2008). One critical question concerns the truth of the premiss, three concern the adequacy of the assumed stereotype as an inference-license, and one concerns whether the stereotyped person is an exception to a generally or provisionally accurate stereotype.

The criteria for valid stereotyping that are proposed in the preceding paragraph hold independently of the mental state of a person doing the stereotyping or of a person evaluating their reasoning. But people do not have direct access to them. They must rely on beliefs that the criteria are met, which should be justified if the inference is to be epistemically legitimate.

Stereotyping a person is on the present account not a fallacy. The fallacy is not stereotyping, but mistaken stereotyping. Stereotyping is mistaken when one of the conditions for valid stereotyping is not met or a defeater of valid stereotyping is ignored. These failures consist in (1) falsehood of the premiss or (2) falsehood of the stereotype used to license the inference or (3) non-equivalence of the stereotype to a *ceteris paribus* or to a for-the-most-part generalization or (4) merely accidental truth of this stereotype or (5) the fact that the stereotyped person does not fit the stereotype. Epistemically, a person who applies a stereotype to a person is mistaken in doing so if they either (a) lack adequate justification for holding either (1) that the stereotyped person belongs to the stereotyped group or (2) that members of the stereotyped group have the inferred feature or (3) that the stereotype is equivalent to either a *ceteris paribus* or a for-the-most-part generalization or (4) that the stereotype holds counter-factually rather than merely accidentally, or (b) fail to take into account practically available evidence that the stereotyped person does not fit the stereotype, or (c) draw the conclusion as holding definitely rather than in a qualified way as holding with probability or provisionally. These six conditions constitute the fallacy of mistaken stereotyping.

Stereotyping can unjustly harm the stereotyped person (Banaji and Greenwald, 2016/2013). Even if such stereotyping is epistemically justified, it is morally unjustified. In general, epistemically justified stereotyping is morally unjustified when morality requires definite (rather than probable or presumptive) attribution of the inferred property and an inference-licensing universal generalization is not epistemically justified.

Stereotyping is a kind of argument from sign, where membership in a specified human group is taken to be a sign of an ongoing behavioural or mental specified trait, or vice versa. Since even true stereotypes have exceptions, stereotyping is never a conclusive argument from sign. It can make it probable that the stereotyped person has the inferred feature, or establish a presumption to that effect. The conditions for such successful reasoning or argument are similar to those for a successful non-conclusive argument from sign. Formation of a false stereotype may be due to hasty generalization on the basis of limited experience of members of the stereotyped group or to unjustified reliance on other people's say-so (i.e. the fallacy of *argumentum ad verecundiam*). Definitive attribution of a behavioural or mental trait on the basis of group membership (or vice versa) is due to confusion of a for-the-most-part or provisional generalization with a universal generalization, which could be treated as a kind of *secundum quid* fallacy (dropping the qualification).

4. RECOGNIZING, RESPONDING TO, AND AVOIDING STEREOTYPING

Stereotyping is hard to recognize, because it is often implicit, for example in a person's emotional response to, or behaviour towards, another person. Banaji and Greenwald (2016/2013) report, for example, that on average participants in a study of characteristics preferred in a quiz show teammate traded nine IQ points for a partner who was thinner; these participants are unlikely to have verbalized their implicit stereotype of fat people as dumber than thin people. If someone makes a judgment about another person that appears to be based on their membership in some human group, a tree of possible responses and counter-responses opens up. A diagram of this tree in a "profile of dialogue" (Krabbe, 1999) would be too complicated for display here. The following sequence should give some idea of the possibilities:

> *Proponent*: Person x has behavioural or mental trait F.
> *Opponent*: You are stereotyping x.
> *P*: What makes you think that?
> *O*: You are basing your judgment solely on the fact that x is a member of group G.
> *P*: I am. So what? Gs have trait F.
> *O*: All of them?
> *P*: No.
> *O*: Most of them?

P: Well, I'm not sure.
O: And is it anything more than an accident that some Gs have trait F?
P: I'm not sure.
O: Well, you need more evidence that x has trait F, don't you?
P: I guess I do.

It is hard to avoid mistaken stereotyping in one's own thinking and one's responses to others, because all people stereotype and they often do so quite unconsciously (Schneider, 2005; Banaji & Greenwald, 2016/2013, p. 89). Social psychologists generally interpret the universal human tendency to slot people they meet into groups distinguished by sex, age, skin colour, and other characteristics as a side-effect of a tendency to categorize living organisms and to associate properties with each kind—a tendency that is likely to have been selected for in human evolutionary history. The tendency cannot be turned off. One can take various Implicit Association Tests online[3] to discover one's implicit automatic associations, starting with discovering whether one has an implicit automatic association of flowers with pleasant things, and if so how strong it is.[4] The experience of taking the flower-insect test gives one a sense of the way in which automatic associations make sorting tasks easier or harder, and thus provides evidence of the validity of Implicit Association Tests. Having had this experience, one can then test one's implicit associations with groups distinguished by sex, age, skin tone, race, sexual orientation, and other characteristics. Such implicit associations are not necessarily stereotypes in themselves, but they are likely to reflect stereotypes.

What can one do to counteract such "hidden biases"? Beaulac and Kenyon (2014) usefully distinguish four levels of "debiasing", ranging from the most internal and least effective to the least internal and most effective. Level 1 debiasing involves upbringing or education that prevents formation of the bias or eliminates it. According to the "contact hypothesis", bringing people of different groups together to get to know each other will change false stereotypes and reduce prejudice. Schneider (2004) reports that contact has been shown to change false stereotypes if people have mutually positive experiences from their interaction, have roughly equal status, and contact each other in a context that has institutional support for change. It helps, he says, if contact is intimate enough to lead to personalization. There is a

---

[3] At https://implicit.harvard.edu/implicit/; accessed 2019 09 12.
[4] This test is available at https://implicit.harvard.edu/implicit/user/agg/blindspot/indexflowerinsect.htm; accessed 2019 09 12.

typicality paradox: if a person is seen as typical of the stereotyped group, they will not produce enough disconfirming evidence to change the stereotype; but a person seen as atypical may be sub-typed as an exception and provide no incentive for change. Even if effective, Schneider claims, contact may only work at the surface to change the association between categories and features. He thinks that fundamental change is likely to occur when people's theories about why the category and its features hang together change—a change best promoted, he thinks, by educational efforts.

Level 2 debiasing involves training to recognize and correct for mistaken stereotyping. According to Schneider (2004) and Beaulac and Kenyon (2014), however, trying to suppress mistaken stereotypes can be counter-productive. Further, despite the training, one may fail to notice that one is stereotyping. Level 3 debiasing addresses this problem by supplementing training with situational nudges that prompt recognition that one is stereotyping; for example, members of a hiring committee may be reminded of the employer's policy of non-discrimination on such grounds as sex, race, ethnic origin, or religion. Level 4 debiasing makes biases (whether or not they are due to mistaken stereotyping) inoperative by concealing the group membership of individuals who are being evaluated; examples of such debiasing include blind peer review, blind grading of students' work, and having applicants for an orchestra position play behind a screen. Such concealment is not always feasible.

5. SUMMARY

A stereotype, in the sense discussed in this paper, is a fixed association in a person's mind of one or more ongoing behavioural or mental traits with membership in a human group identified by such factors as sex, ethnic status, stage of life, occupation, sexual orientation, clothing or recreational activities. An example is the common stereotype of athletes as dumb (i.e. not very smart). To stereotype a person is to attribute to them a pattern of behaviour or mental trait on the sole basis of a stereotype of a group to which the person is assumed to belong or conversely to attribute to them membership in a human group on the sole basis of an exhibited pattern of behaviour or mental trait. An example is assuming that the driver of a car that is being driven in a recklessly aggressive manner is a young man. Stereotyping in this sense is a kind of argument from sign, in which membership in the human group is taken to be a sign of the behavioural or mental trait, or vice versa. The inference is correct if and only if a stereotype that would license the inference is true of most people in the stereotyped group or is normally true of people in the stereotyped group, provided that the truth of this stereotype has a theoretical basis and is not merely accidental. It is epistemically legitimate to stereotype a person if one has good reason for thinking that this inferential condition is met, one has good reason for thinking that the premiss is true of the person, one has no good reason to think that the person is an exception to the stereotype, and one draws the conclusion in a qualified way as holding probably or provisionally. The fallacy of mistaken stereotyping consists of stereotyping a person when one either (a) lacks adequate justification for holding either (1) that the stereotyped person belongs to the stereotyped group or (2) that members of the stereotyped group have the inferred feature or (3) that the stereotype is equivalent to either a *ceteris paribus* or a for-the-most-part generalization or (4) that the stereotype holds counter-factually rather than merely accidentally, or (b) fails to take into account practically available evidence that the stereotyped person does not fit the stereotype, or (c) draws the conclusion as holding definitely rather than in a qualified way as holding with probability or provisionally. Formation of a false stereotype may be due to hasty generalization from experience or to cultural influences (whose acceptance involves an *ad verecundiam* fallacy). Epistemically justified stereotyping is not morally justified when morality requires definite attribution of the inferred property (not just probable or presumptive attribution) and an inference-licensing universal generalization is not epistemically justified.

Because stereotyping is often implicit and unconscious, it is hard to recognize when other people are mistakenly stereotyping someone and even harder to avoid doing so oneself. A profile of dialogue in response to a charge of (mistaken) stereotyping would be quite complex, given the variety of ways in which one could substantiate such a charge and the variety of possible responses to each possible substantiation. Contact has been shown to change false stereotypes if the people contacting each other have mutually positive experiences from their interaction, have roughly equal status, and contact each other in a context that has institutional support for change. Attempts to avoid mistakenly stereotyping another person are most successful if they rely either on such external influences as situational nudges to prompt recognition that one is stereotyping or on blindness to what group the other person belongs to.

REFERENCES

Bailin, S., & Battersby, M. (2010). *Reason in the balance: An inquiry approach to critical thinking.* Toronto: McGraw-Hill Ryerson.

Banaji, M. R., & Greenwald, A. G. (2016/2013). *Blindspot: Hidden biases of good people*. New York: Bantam. Hardback edition first published in 2013.
Beaulac, G., & Kenyon, T. (2014). Critical thinking education and debiasing. *Informal Logic, 34*(4), 341-363.
Beeghly, E. (2015). What is a stereotype? What is stereotyping? *Hypatia, 30*(4), 675-691.
Copi, I. M., & Cohen, C. (2002). *Introduction to logic*, 11th edition. Upper Saddle River, NJ: Prentice-Hall.
Eemeren, F. H. van, & Grootendorst, R. (1992). *Argumentation, communication, and fallacies: A pragma-dialectical perspective*. Hillsdale, NJ: Lawrence Erlbaum Associates.
Freedman, D. G., & Freedman, N. C. (1969). Behavioural differences between Chinese-American and European-American newborns. *Nature, 224*(5225), 1227.
Groarke, L. A., & Tindale, C. W. (2012). *Good reasoning matters: A constructive approach to critical thinking*, 5th edition. Don Mills, ON: Oxford University Press.
Hacking, I. (1991). A tradition of natural kinds. *Philosophical Studies, 61*(1), 109-126.
Hamblin, C. L. (1970). *Fallacies*. London: Methuen.
Hansen, H. V., & Pinto, R. C. (1995). *Fallacies: Classical and contemporary readings*. University Park, PA: Pennsylvania State University Press.
Hitchcock, D. (2011). Inference claims. *Informal Logic, 31*(3), 191-228.
Hitchcock, D. (2017). *On reasoning and argument: Essays in informal logic and on critical thinking*. Dordrecht: Springer.
Hurley, P. (2006). *A concise introduction to logic*, 9th edition. Belmont, CA: Wadsworth/Thomson.
Johnson, R. H., & Blair, J. A. (1993). *Logical self-defense*, 3rd edition. Toronto: McGraw-Hill Ryerson.
Krabbe, E. C. (1999). Profiles of dialogue. In J. D. Gerbrandy, M. J. Marx, M. de Rijke & Y. Venema (Eds.), *JFAK: Essays dedicated to Johan van Benthem on the occasion of his 50th birthday* (Vol. 3, pp. 25-36). Amsterdam: Amsterdam University Press.
Liebesman, D. (2011). Simple generics. *Noûs, 45*(3), 409-442.
Lippmann, W. (1922). *Public opinion*. New York: Harcourt, Brace and Company.
Murray, J., et al. (1971/1933). *Oxford English dictionary*, compact edition, 2 vols. Oxford: Oxford University Press. Micrographic reproduction of the 12-volume edition first published in 1933.
Nickel, B. (2016). *Between logic and the world: An integrated theory of generics*. New York: Oxford University Press.
Schneider, D. J. (2005). *The psychology of stereotyping*. New York: Guilford Press.
Sinha, R. (2011). Unit 2: Distribution and characteristics. In S. M. S. Chahal, R. Sinha, K. D. Sharma, & A. N. Sharma, *Biological diversity* (pp. 17-31). New Delhi: Indira Gandhi Open University. Available at http://14.139.40.199/bitstream/123456789/41421/1/MANI-002B4E.pdf; accessed 2019-09-12.

Teichman, M. (2015) *Characterizing kinds: A semantics for generic sentences.* (Doctoral dissertation). Retrieved from http://home.uchicago.edu/~teichman/dissertation.pdf.

Tessler, M. H., & Goodman, N. H. (2019). The language of generalization. *Psychological Review 126*(3), 395-436.

Walton, D., Reed, C., & Macagno, F. (2008). *Argumentation schemes.* New York: Cambridge University Press.

Woods, J. (1992). Who cares about the fallacies? In F. H. van Eemeren, R. Grootendorst, J. A. Blair, & C. A. Willard (Eds.), *Argumentation illuminated* (pp. 23-48). Amsterdam: SICSAT.

Woods, J. (2004). *The death of argument: Fallacies in agent-based reasoning.* Dordrecht: Kluwer.

Changing Norms of Argumentation

BETH INNOCENTI
University of Kansas
bimanole@ku.edu

What strategies do social actors use to try to change norms of argumentation, and why do they expect those strategies to work? I submit that the normative structure of strategies can at least partly account for why using the strategies can reasonably be expected to change norms of argumentation. To illustrate, I use normative pragmatic theory to explain how Audre Lorde's "The Uses of Anger" attempts to influence how academic colleagues respond to her anger.

KEYWORDS: anger, Audre Lorde, Black feminism, normative pragmatics, teaching

1. INTRODUCTION

How can social actors change norms of argumentation? On one hand, a wholly conceptual approach has not yielded a satisfactory answer. Asking what rules ought to regulate discussion about rules defers the question indefinitely. On the other hand, a difficulty with a wholly empirical approach, where scholars recommend norms based on observation of actual practices, is that "getting from what people typically do to what they ought to do requires a leap" (Tracy, 2011, p. 172). Alternatively, we could define the argumentation scholar's task as "just to describe a certain system of discussion rules and [. . .] not include the description of rules that govern the decision to select the very system he describes" (Krabbe, 2007, p. 240).

Given that "[m]any decisions on how to interact are themselves taken interactively" (Mercier & Sperber, 2017, p. 185), another approach is to analyze strategies social actors use to try to change norms of argumentation. The starting points of this approach include the following. First, as theorizing is itself a communication practice, so is communication practice theorizing (Craig, 1996; Jacobs & Jackson, 2006). Second, communication design is apparent at all levels, even in informal conversations (Jackson & Jacobs, 1980). Third, design principles inherent in acts of communication are theories of communication (Aakhus, 2007; Jackson & Aakhus, 2014). I submit that the normative structure of strategies can at least partly account for why using the strategies can reasonably be expected to change norms of argumentation.

To support that claim, I first explain how to describe the normative structure of a strategy. I then analyze the primary strategy used in a well-known, exemplary attempt to change a norm of argumentation: teaching. I argue that undertaking and discharging responsibilities incurred just by openly, deliberately intending to teach a norm of argumentation, creates practical reasons for addressees to make efforts to learn the norm.

2. NORMATIVE PRAGMATICS

A well-established method for describing the normative structure of strategies is normative pragmatics (e.g., Goodwin, 2001; Goodwin & Innocenti, 2019; Innocenti & Miller, 2016; Jacobs, 2000; Kauffeld, 1998; Kauffeld & Innocenti, 2018).[1] The normative structure of strategies refers to responsibilities undertaken in the open, deliberate use of strategies. The normative structure generates pragmatic force or, put differently, creates practical reasons for addressees to respond as the speaker openly, deliberately intends.

To illustrate, consider practical reasons created by holding a "Slow" sign in a road construction zone. What is the normative structure of that strategy? By holding the sign, the worker openly, deliberately displays her intent to influence drivers to drive slowly through the construction zone. The bigger, brighter, and better-positioned the sign, the more well-designed the context for holding all accountable for not driving responsibly through the road construction zone. Other things being equal, the sign-holder cannot plausibly disclaim her intent to influence drivers to drive slowly so can be held accountable if, say, she allows red cars to speed through with impunity; and drivers cannot plausibly deny seeing or understanding the sign so can be held accountable if they speed. So holding a big, bright, conspicuous sign creates two practical reasons for drivers to pass through the construction zone slowly. Drivers can now reason: (1) the sign-holder would not risk getting somebody killed, getting herself fired or imprisoned, or getting a reputation as reckless or worse, unless she planned to meet the responsibilities she undertook by holding the "Slow" sign; and (2) to avoid killing somebody or getting fined or imprisoned, or to display an identity as a prudent, courteous driver,

---

[1] See Kauffeld (2009) for a discussion of the Gricean speech act theory underlying normative pragmatics.

they can drive slowly through the construction zone.[2] Notice the sign-holding strategy here is not reason-giving; the words on the sign are not "Drive slowly because you do not want to kill a construction worker." Instead, the normative structure of sign-holding creates practical reasons.

The same basic story accounts for why using other kinds of communication strategies—speech acts like commands, and non-discursive features like size and color—can reasonably be expected to influence addressees as the speaker openly, deliberately intends. The more conspicuous the strategy, the greater the possibility of holding speaker and addressees accountable for failing to live up to responsibilities incurred by using the strategy, so the stronger the practical reasons created.

Pragmatic force is not compulsion. Social actors routinely act ingeniously to avoid, overcome, dismantle, replace, structures guiding or impeding action, including material structures such as a border wall and the normative structure of speech acts such as promising or warning. Normative pragmatic theory explains moral, ethical affordances and constraints created by communicatively designed contexts.

3. METHOD

To address the question of how social actors can change norms of argumentation, I analyze an exemplary attempt to influence addressees to begin having a discussion at all. In "The Uses of Anger," Audre Lorde tries to influence white women to respond to Black women's anger about racism not by disengaging from dialogue due to fear or guilt but by "recognizing the needs and the living contexts of other women" (2007, p. 126); Lorde notes, "Any discussion among women about racism must include the recognition and the use of anger" (2007, p. 128). This is just one norm she attempts to change; I discuss just this one because of its significance and for the sake of time. Lorde is trying to change the rules of the game, so to speak—to constrain white women from dismissing with impunity Black women's anger about racism as killing the mood, creating guilt, disrupting discussion, and more, and

---

[2] This practical reasoning accounts for why a speaker can reasonably expect using a strategy to influence addressees as intended. A sign-holder's internal cognitions may differ. If asked, she may say she is holding the sign because her boss told her to or because she has an injury preventing her from performing other road construction tasks. But the normative structure of the strategy nonetheless creates a context where she can be held accountable for allowing red cars to speed with impunity.

instead get them to engage Black women's arguments (2007, pp. 127, 131, 132; see also Cooper, 2018; Griffin, 2012; Olson, 2011).

Lorde's immediate audience comprised primarily academics and white women, Black women, and women of color attending her keynote address at the 1981 National Women's Studies Association conference, the theme of which was "Women Respond to Racism." Presumably they would not want to display, perpetuate, or exacerbate racism. That same year Lorde published the essay in *Women's Studies Quarterly*, and then in 1984 published it in a collection of her speeches and essays entitled *Sister Outsider*. The ongoing resonance of Lorde's essay with feminist advocates and thinkers (e.g., Cooper, 2018; Howes & Hundleby, 2018) indicates widespread recognition that it is a fitting response to dissent about how white women ought to respond to Black women's anger about racism.

Because normative pragmatic theory assumes social actors self-regulate their communication practices and that rationales for persuasion are inherent in their messages (Jacobs, 2000), researchers analyze messages for both strategy and metadiscourse about how strategies are designed to work. Strategies can be identified from the macro- to micro-level: uses of argument may be subordinate to some master speech act (Jacobs, 1989; Kauffeld, 1998), and stylistic devices from word choice to sentence structure to broader units of composition, or images such as a border around an advertisement, contribute to the overall persuasive design of messages (Fahnestock, 2011; Jacobs, 2000). Lorde's essay is a vivid, conspicuous sign directing action for avoiding the perpetuation and exacerbation of racism.

4. INTENT AND RESPONSIBILITIES

Lorde's many references to teaching and learning in the essay display her intent to teach white women how to respond to Black women's anger about racism; in this case teaching is a master speech act. But Lorde does not intend the essay "to be merely another case of the academy discussing life within the closed circuits of the academy" (2007, p. 127). She openly intends to teach civic action.

Teaching involves responsibilities not incurred by using other kinds of strategies. For example, Lorde does not incentivize by, say, offering a cookie. Lorde even disclaims responsibility for persuading or, as she puts it, "for altering the psyche of her oppressor, even when that psyche is embodied in another woman" (2007, p. 133). What responsibilities does Lorde undertake by teaching?

First, Lorde undertakes responsibility for the primary intent of getting addressees to try to learn. It would be incoherent to say, "I intend to teach you how to respond to Black women's anger about racism, but I am indifferent about whether you learn how to respond." If Lorde could plausibly disclaim that intent, then addressees could avoid learning with impunity, perhaps by just admiring Lorde's literary prowess.

But simply making declarative statements about what addressees should learn in order to get them to learn would be comparable to simply making declarative statements about what addressees should believe in order to get them to believe. Social actors routinely use additional strategies to get addressees to believe, learn, and so on. They can use any number of strategies to teach: arguing, illustrating, explaining, demonstrating, and more. A central strategy Lorde uses is "speak[ing] about anger, my anger, and what I have learned from my travels through its dominions" (2007, p. 127). Lorde says she speaks about her experiences in part because she does "not want this to become a theoretical discussion" (2007, p. 124).

5. DISCHARGING TEACHING RESPONSIBILITIES

By speaking about her experiences, Lorde vividly, conspicuously discharges four responsibilities undertaken in teaching. For the sake of time, I mainly focus on the opening of Lorde's essay where she lists eight experiences involving interactions with white women. The experiences are designed to show that "Women responding to racism means women responding to anger" (2007, p. 124).

5.1 Responsibility to know what you are talking about

First, Lorde undertakes responsibility to know what she is talking about. It would be incoherent to say, "I intend to teach you how to respond to Black women's anger about racism, but I don't know anything about that topic."

Lorde discharges that responsibility in part by listing her experiences. She begins with this one: "I speak out of direct and particular anger at an academic conference, and a white woman says, 'Tell me how you feel but don't say it too harshly or I cannot hear you'" (2007, p. 125). The penultimate example Lorde lists is this: "A white academic welcomes the appearance of a collection by non-Black women of Color. 'It allows me to deal with racism without dealing with the harshness of Black women,' she says to me" (2007, p. 126). These are just two of the experiences Lorde lists, analogous to holding up a "racism" sign, to create a context where all can be held accountable for failing to know what they are talking about.

First, because addressees can now hold Lorde accountable if she does not know what she is talking about, they can reason that Lorde would not risk her credibility unless she had made efforts to understand racism. Second, addressees can now be held accountable for not knowing what they are talking about. If white women were to say, "Those statements aren't racist. You are misunderstanding us"—not a far-fetched possibility given that Lorde also speaks about an experience asking a white woman what a week-long forum on Black and white women has given to her and the woman says, "'I think I've gotten a lot. I feel Black women really understand me a lot better now; they have a better idea of where I'm coming from,'" about which Lorde comments, "As if understanding her lay at the core of the racist problem" (2007, p. 125)—they would risk displaying just the sort of "defensiveness" that Lorde describes as one reason why Black women are angry about racism. Lorde describes defensiveness as one of the "bricks in a wall against which we all flounder" (2007, p. 124) and as "destructive of communication" (2007, p. 130). A defensive response is a fallible sign of "not dealing with" and "preserving racial blindness, the power of unaddressed privilege, unbreached, intact" (2007, pp. 131, 132). To avoid that criticism, addressees can try to learn.

5.2 Responsibility to understand what addressees do not understand

A second responsibility Lorde undertakes by teaching is to have made efforts to understand what addressees do not see, know, or understand. It would be incoherent to say, "I intend to teach you how to respond to Black women's anger about racism, and I believe you know how to respond."

Listing the eight experiences vividly displays that Lorde understands what white women do not see or understand. For example, she speaks about hearing "on campus after campus, 'How can we address the issues of racism? No women of Color attended.' Or, the other side of that statement, 'We have no one in our department equipped to teach their work,'" and comments, "In other words, racism is a Black women's problem, a problem of women of Color, and only we can discuss it" (2007, p. 125). In addition, Lorde speaks of a time when at a supermarket a little white girl exclaims about Lorde's two-year-old daughter, "'Oh look, Mommy, a baby maid!' And your mother shushes you, but she does not correct you. And so fifteen years later, at a conference on racism, you can still find that story humorous. But I hear your laughter is full of terror and dis-ease" (2007, p. 126). After listing experiences, Lorde directly addresses "the white women present who recognize these attitudes as familiar" (2007, p. 127). By speaking of "familiar" experiences and their pervasiveness, Lorde displays that she understands what will sound familiar and can even anticipate how some will respond. Now all can be held accountable for failing to accurately gauge addressees' understanding.

First, addressees can now hold Lorde accountable if she insults their moral, ethical intelligence, so can reason that Lorde would not risk their resentment unless she had made efforts to understand what they do not see, know, or understand. Second, addressees can now be held accountable for not recognizing or acknowledging their own ignorance, or for avoiding self-scrutiny and consideration of how they support racist structures not of their own making. Lorde's list of illustrations makes the risk serious as she displays the systemic pervasiveness of racism in popular media and everyday interactions by saying, for example, "You avoid the childhood assumptions formed by the raucous laughter at Rastus and Alfalfa, the acute message of your mommy's handkerchief spread upon the park bench because I had just been sitting there, the indelible and dehumanizing portraits of Amos 'n Andy and your daddy's humorous bedtime stories" (2007, p. 126). If white women dismiss Lorde's message as something they already know in order to attend to their own oppression, they risk the kind of criticism Lorde displays when she asks, "What woman here is so enamoured of her own oppression that she cannot see her heelprint upon another woman's face? What woman's terms of oppression have become precious and necessary to her as a ticket into the fold of the righteous, away from the cold winds of self-scrutiny?" (2007, p. 132). To avoid that criticism, addressees can own their ignorance and try to learn.

5.3 Responsibility to understand addressees' constraints

A third responsibility Lorde undertakes by teaching is to have made efforts to understand and appreciate constraints inhibiting learning. It would be incoherent to say, "I intend to teach you how to respond to Black women's anger, and I do not know or care how that may be difficult for you."

By listing experiences, Lorde conspicuously shows she recognizes constraints white women face in learning how to respond to Black women's anger about racism. For example, she writes: "I have seen situations where white women hear a racist remark, resent what has been said, and become filled with fury, and remain silent because they are afraid" (2007, p. 127). In addition, she writes about a time when she experienced the anger of a woman of color:

> The woman of Color who is not Black and who charges me with rendering her invisible by assuming that her struggles with racism are identical with my own has something to tell me that I had better learn from, lest we both waste ourselves fighting the truths between us. [. . .] And yes, it is very difficult

> to stand still and to listen to another woman's voice delineate an agony I do not share, or one to which I myself have contributed (2007, pp. 127-128).

By displaying vivid signs of understanding constraints white women face in learning to listen to Black women's anger about racism, Lorde designs a context where all can be held accountable for misunderstanding constraints to learning.

First, addressees can hold Lorde accountable if she fails to appreciate the difficulties they face and alienates them, so can reason that Lorde would not risk their resentment and give them reasons for turning away and not listening to her unless she had made efforts to understand their constraints. Second, addressees can now be held accountable for not making efforts to overcome their difficulties in listening to Black women's anger about racism. Lorde displays anger and resentment that her efforts and the efforts of other people of color are not reciprocated when she writes, "Oppressed peoples are always being asked to stretch a little more, to bridge the gap between blindness and humanity. Black women are expected to use our anger only in the service of other people's salvation or learning. But that time is over" (2007, p. 132). Lorde displays that failure to learn how to respond to anger about racism begets further anger about not living up to a reciprocal responsibility of meeting her efforts to understand their constraints—for perpetuating and exacerbating racism and injustice. To avoid criticism for moral apathy and perpetuating the problem, addressees can try to learn.

5.4 Responsibility to understand addressees' interests

A fourth responsibility Lorde undertakes by teaching is to have made efforts to understand addressees' interests in learning. It would be incoherent to say, "I intend to teach you how to respond to Black women's anger about racism, and I cannot say why it is in your interest to learn that."

Lorde openly takes responsibility for understanding addressees' interests when she writes at the beginning of the essay that guilt and defensiveness in response to Black women's anger about racism "serve none of our futures" (2007, p. 124). She discharges that responsibility by speaking about her experiences: "We have had to learn to move through them [furies] and use them for strength and force and insight within our daily lives. Those of us who did not learn this difficult lesson did not survive" (2007, p. 129). She also writes about experiences that all of them share as she displays that "It is not the anger of Black women which corrodes into blind, dehumanizing power, bent upon the annihilation of us all" (2007, p. 133). Lorde describes the "context of opposition and threat" (2007, p. 128) they all work in. She mentions "the size and complexity of the forces mounting against us and all that is most human within our environment" (2007, p. 128). She describes "the pressing need to make clear choices" and "the approaching storm that can feed the earth as well as bend the trees" (2007, p. 130). She singles out "the teeth of a system for which racism and sexism are primary, established, and necessary props of profit" (2007, p. 128). Lorde displays vivid signs that create a context where all can be held accountable for acting in their own interests.

First, addressees can hold Lorde accountable if she fails to understand their interests, so can reason that Lorde would not risk their resentment for alienating them or wasting their time unless she had made efforts to understand what is in their interests. Second, addressees cannot say it is not in their interest to learn how to respond to anger about racism without risking criticism for imprudence. Lorde raises the stakes from acting imprudently to a serious moral failure for endangering Black lives and the planet. To avoid criticism for participating in oppression, wasting energy, and endangering Black lives and the planet, addressees can make efforts to learn.

6. CONCLUSIONS

In sum, Lorde uses the strategy of teaching to change a norm of argumentation. Specifically she uses teaching to change how white women respond to Black women's anger about racism, from fear and guilt to recognizing the needs and living contexts of other women. She conspicuously puts herself in a position where she can be held accountable if she falls short in meeting responsibilities undertaken just by teaching, and at the same time creates a context where addressees can also be held accountable if they fall short in meeting reciprocal responsibilities to try to learn. The more she puts herself out there—the more conspicuously she displays she knows what she is talking about, understands what it is that addressees do not understand, appreciates constraints inhibiting addressees' learning, and understands their interests in learning—the better she creates a context where addressees can be held accountable for not meeting reciprocal responsibilities—such as knowing what they are talking about and owning their ignorance—so the more practical reasons addressees now have to try to learn.

Of course even the best teaching cannot compel anybody to learn. Addressees may choose to accept the risks of not learning, or call out what they see as the speaker's ignorance or blind spots, or explain why learning something is not in fact in their interest, and more. But the more a speaker displays that she has met responsibilities undertaken just by teaching, the more addressees become accountable for failing to make reciprocal efforts to learn, so the more practical reasons addressees now have to try to learn. This normative structure explains why Lorde's teaching could reasonably be expected to change a norm of argumentation. The strong normative structure also explains why the essay is a touchstone for Black feminism. These findings show that normative pragmatic theory offers a promising approach to opening discussion about how to change norms of argumentation.

ACKNOWLEDGEMENTS: Thanks to incisive questions and comments by David Godden, Michael Hoffmann, Catherine Hundleby, Scott Jacobs, and others at ECA, I have revised the essay's research question and title for the proceedings. My initial question and title—how to make norms of argumentation normative—collapsed distinctions between creating, changing, and abandoning norms.

REFERENCES

Aakhus, M. (2007). Communication as design. *Communication Monographs,* 74(1), 112-117.
Cooper, B. (2018). *Eloquent Rage: A Black Feminist Discovers Her Superpower.* New York: St. Martin's Press.
Craig, R. T. (1996). Practical-theoretical argumentation. *Argumentation,* 10(4), 461-474.
Fahnestock, J. (2011). *Rhetorical Style: The Uses of Language in Persuasion.* New York: Oxford University Press.
Goodwin, J. (2001). One question, two answers. *OSSA Conference Archive.* 40. https://scholar.uwindsor.ca/ossaarchive/OSSA4/papersandcommentaries/40/
Goodwin, J., & Innocenti, B. (2019). The pragmatic force of making an argument. *Topoi,* https://doi.org/10.1007/s11245-019-09643-8.
Griffin, R. A. (2012). I AM an angry black woman: Black feminist autoethnography, voice, and resistance. *Women's Studies in Communication,* 35(2), 138-157.
Howes, M., & Hundleby, C. (2018). The epistemology of anger in argumentation. *Symposion,* 5(2), 229-254.
Innocenti, B., & Miller, E. (2016). The persuasive force of political humor. *Journal of Communication,* 66(3), 366–85.
Jackson, S., & Aakhus, M. (2014). Becoming more reflective about the role of design in communication. *Journal of Applied Communication Research,* 42(2), 125-134.

Jackson, S., & Jacobs, S. (1980). Structure of conversational argument: Pragmatic bases for the enthymeme. *Quarterly Journal of Speech*, 66(3), 251-265.

Jacobs, S. (1989). Speech acts and arguments. *Argumentation,* 3(4), 345-365.

Jacobs, S. (2000). Rhetoric and dialectic from the standpoint of normative pragmatics. *Argumentation,* 14(3), 261-286.

Jacobs, S., & Jackson, S. (2006). Derailments of argumentation: It takes two to tango. In P. Houtlosser & A. van Rees (Eds.), *Considering Pragma-Dialectics: A Festschrift for Frans H. van Eemeren on the Occasion of His 60th Birthday* (pp. 121-133). Mahwah: Erlbaum.

Kauffeld, F. J. (1998). Presumptions and the distribution of argumentative burdens in acts of proposing and accusing. *Argumentation,* 12(2), 245-266.

Kauffeld, F. J. (2009). Grice's analysis of utterance-meaning and Cicero's Catilinarian apostrophe. *Argumentation,* 23(2), 239-257.

Kauffeld, F. J., & Innocenti, B. (2018). A normative pragmatic theory of exhorting. *Argumentation,* 32(4), 463-483.

Krabbe, E. C. W. (2007). On how to get beyond the opening stage. *Argumentation,* 21(3), 233-242.

Lorde, A. (2007). *Sister Outsider.* New York: Crossing Press.

Mercier, H., & Sperber, D. (2017). *The Enigma of Reason.* Cambridge: Harvard University Press.

Olson, L. (2011). Anger among allies: Audre Lorde's 1981 keynote admonishing the National Women's Studies Association. *Quarterly Journal of Speech,* 97(3), 283-308.

Tracy, K. (2011). 'Reasonable hostility': Its usefulness and limitation as a norm for public hearings. *Informal Logic,* 31(3), 171-190.

"If You Are A Scientist You Cannot Stop Such A Thing": Scientific Assent and Dissent in the Manhattan Project

DAVID ERLAND ISAKSEN
University of South-Eastern Norway
david.e.isaksen@usn.no

In science, dissent is encouraged in the search for truth. Yet when it comes to some of the basic assumptions about science the scientific community is less tolerant. I will show how some of these assumptions were used by J. Robert Oppenheimer as premises for arguments that had the weight of science without having scientific validity. I will show how they were used to suppress dissent and justify the development and use of the first atomic bombs.

KEYWORDS: rhetoric of science, dissent, Oppenheimer, atomic weapons, nuclear physicists, indexing, Kenneth Burke, consummation, Los Alamos

1. INTRODUCTION

Teachers often use scientists as the prime examples of critical thinkers, mentioning Albert Einstein and Niels Bohr as examples of how one can make progress and approximate truths about the universe by questioning and testing one's assumptions. However, scientists are able to radically question the nature of reality only because they confine that questioning within the framework established by science: "It is specifically because science provides such a framework of rules and regulations to control and set bounds to paranoid thinking that a scientist can feel comfortable about taking the paranoid leaps" (Eiduson 106). When the questions go beyond what Thomas Kuhn calls the "puzzle form" of "normal science" scientists are often less able and willing to think critically. As Kuhn writes in *The Structure of Scientific Revolutions*, "A paradigm can . . . insulate the [scientific] community from those socially important problems that are not reducible to the puzzle form, because they cannot be stated in terms of the conceptual and instrumental tools the paradigm supplies" (37). And as argumentation research has shown, scientific experts may even be more prone to "overconfidence" and "polarization" if argument quality is "not sufficiently high in a domain" (Mercier 313).

As any god-term, "science" has been effectively used to justify many actions that are ethically highly questionable. Similar to religious orthodoxy, it is a tool that can be used to keep scientists in line and suppress dissent. As Michael Polanyi writes, "No one can become a scientist unless he presumes that the scientific doctrine and method are fundamentally sound and that their ultimate premises can be unquestioningly accepted" (45), and yet "the scientific doctrine" is not a closed canon and has taken various forms through the ages, at times making such doctrines as scientific racism and positivism interchangeable with "science" to scientists and lay people alike. The concepts, methods, and assumptions embodied by the term "science" differ from generation to generation, and yet scientists are often blind to this difference because of how their training and research experience reinforce a homeostatic view of scientific history (Kuhn 152-65). As a consequence, science, a model of reasoned debate, can become a tool to suppress dissent simply by labeling it anti-science. I will here present one such case where the choices of scientists deciding whether or not to complete the atomic bomb were defined not as political choices but rather as adherence to or disavowal of the basic tenets of science. The case I will discuss is the Manhattan Project, and in particular I will discuss the arguments and thinking of J. Robert Oppenheimer, the director of the Los Alamos Laboratory.

2. OPPENHEIMER'S SCIENTIFIC CREED

Oppenheimer better than most scientists integrated the rigor of science with his early sensibilities as an artist. Some of his closest friends early on thought his career would be as an "artist" or "writer" rather than as a scientist (Smith and Weiner 66).

Gradually, science and especially physics replaces literature as the medium for Oppenheimer's aesthetic expression and appreciation. He early on begins to call physics his "stern and uncompromising muse" (57), and later calls it an obsession (59), a fixation (63), and even claims in jest "my muse still craves blood" (72) and "I need physics more than friends" (135). His descriptions of math and physics resemble those usually used about works of art. He praises the "beauty and simplicity" of math, the language of theoretical physics (100), and states, "physics has a beauty which no other science can match, a rigor and austerity and depth" (155). He later refers to theories as "pretty" (168), calls an experimental result beautiful (180), refers to data as "beautiful" (198), and speaks of the nuclear bomb development as yielding "intellectual or technical satisfaction" (312).[1] Infamously, he also stated about the potential method for developing an atomic bomb that it was "technically sweet" (USAEC 266) and that "when you see something that is technically sweet, you go ahead and do it and you argue about what to do about it only after you have had your technical success. That is the way it was with the atomic bomb" (266). There is a clear artistic and aesthetic dimension to Oppenheimer's work in physics that also was a motivating factor to develop the atomic bomb.

I have found 9 texts by Oppenheimer that were particularly instructive and relevant to his vision of nuclear weapons. Most of these are short letters, but the last one is a comprehensive speech to the Association of Los Alamos Scientists outlining his clearest vision for what nuclear weapons mean for the world. These texts show some of the developments in Oppenheimer's thoughts about science and nuclear weapons and the arguments he used to make scientists complete the nuclear bomb work. I will analyze these using the method Kenneth Burke called "indexing" (LAPE; Isaksen), looking for the key terms and god-terms Oppenheimer uses to fuse his vision for nuclear weapons with his concept of science.

2.1 Peace through War

The first of these documents is a letter written to Frank Oppenheimer written March 12th, 1932, and it is the clearest statement from Oppenheimer outlining his life philosophy and ethics before he became involved with the Manhattan Project. In the letter, he speaks briefly about the excellences of physics and biology (his brother was choosing between them for his vocation) before moving on to speak about "the virtue of discipline" (155). Oppenheimer claims discipline is fundamentally good for the soul and that it is the key to achieving "detachment" and ultimately "peace." However, he claims that discipline cannot be achieved without another real (though ultimately minor) objective. Oppenheimer organizes his thoughts in this letter by a logic of means and ends. The means to achieve discipline, which should therefore be "greeted with profound gratitude" are study, duties to men and to the commonwealth, war, personal hardship, and even the need for subsistence. These are some of all the real objectives that can lead a person to the virtue of discipline (the next level in the hierarchy). Discipline has value because "it is good for the soul" and is able to bring about an even more favorable

---

[1] This shift goes parallell to a transition from Sigmund Freud to Bertrand Russell as Oppenheimer's metaphysical reference point of choice. Oppenheimer often refers to Freud in his early years, especially in connection with his fiction writing (Smith and Weiner 13, 24, 48), but later seems to hold Russell as his metaphysical guide (Smith and Weiner 24, 48, 54, 71, 111).

condition. Robert Oppenheimer describes this more favorable condition as "detachment" and "that detachment which preserves the world which it renounces." This detachment is described as an ability to "see the world without the gross distortion of personal desire," to "learn to preserve what is essential to our happiness in more and more adverse circumstances" and to "abandon with simplicity what would else have seemed to us indispensable." This again leads to the final goal of peace, serenity, charity, and a small measure of freedom from the accidents of incarnation. This peace and serenity is arrived at by accepting finally "more easily our earthly privation and its earthly horror." Thus, war leads to discipline, discipline leads to detachment, and detachment leads to peace. In other words, out of war and striving, humans can gain peace for themselves.

Freeman Dyson, fellow physicist and colleague of Robert Oppenheimer at the Princeton Institute for Advanced Study, writes in *Weapons and Hope* that these words "contain a key to the central core of Robert's nature, to the sudden transformation which changed him eleven years later from a bohemian professor to driving force of the bomb project at Los Alamos" (125). For Dyson, this philosophy or ascesis of peace through war seemed a remnant of the nationalist ideologies preceding WWI, which had been brought to life again in left-wing circles supporting the Loyalist side in the Spanish Civil War (125-31). In either case, it seems significant that Oppenheimer would include war as one of those things that lead to discipline and therefore should be greeted with profound gratitude. This is a snapshot of the mental framework Robert Oppenheimer brings to the emerging problem of nuclear weapons and world war.

2.2 From Scientific Adventurer to Obedient Soldier

In January 1939 Oppenheimer writes a letter when he has just learnt about nuclear fission and is reacting to that discovery. Glenn Seaborg says of the time, "I do not recall ever seeing Oppie so stimulated and so full of ideas" (Smith and Weiner 207). The sense of excitement is palpable throughout the letter.[2] Oppenheimer starts the letter saying "The U [uranium] business is unbelievable" (207) and describes the frenzy among the scientists as they conduct all kinds of experiments, creating the same reactions and seeing "unbelievable ionization" (207). All the physicists are fixated on the question of a possible explosion: "Many points are still unclear... most of all, are there many neutrons that come off during the splitting or from the excited pieces? If there are then a 10 cm cube of U deuteride *should be quite something*. What do you think? It

---

[2] Letter to William A. Fowler, 28th of January 1939.

is I think exciting ... in a good honest practical way" (208). He expresses a similar sentiment to George Uhlenbeck on February 5th, 1939: "I think it really not too improbable that a ten cm cube of uranium deuteride *might very well blow itself to hell*" (209). From the last statement it seems that the main interest in the chain reaction is not the possibility of making a nuclear reactor for power, but rather the possibility of creating an explosive nuclear reaction: an atomic bomb. The physicists sound almost giddy, like boys playing with firecrackers, excited about the potential for nuclear explosions with almost no sense of gloom or worry about what the consequences of them could be.

As the war breaks out in Europe and grinds on from 1939-1941, Oppenheimer starts to think more about potential wartime applications of nuclear weapons. As his friend William Fowler goes to work for the National Defense Research Committee, Oppenheimer writes with encouragement, "I expect that as time goes on you'll have more and more a feeling of confidence and conviction in the work you are doing... I have a lot more misgivings even than you ever had about what will come of all of this; but even so I think surely if I were asked to do a job I could do really well and that needed doing I'd not refuse" (215).[3]

That request came in May 1942, when Robert Oppenheimer was asked to become "Coordinator of Rapid Rupture" which became a part of the new Manhattan Engineer District when it was established the next month. His letters start focusing on calculations of potential nuclear reactions, with the dual threat that the active material may either not be powerful enough to be worth the effort (a fizzle) or may be so powerful that it could set off a chain reaction that would ignite the atmosphere and kill off all of humanity (227-234).[4]

As it becomes clear that a new laboratory will need to be set up for the effort, Oppenheimer's concerns expand to recruitment for the laboratory. Smith and Weiner note that "it often took an interview with Oppenheimer, in which he cautiously but eloquently described a project that would *end the war* and have *peacetime applications of untold benefit to mankind*, to persuade a man to uproot his family and *join the adventure in the New Mexico mountains*" (239). The three motives of ending the war, providing "peacetime applications of untold benefit to mankind" (presumably electricity from nuclear power), and joining in an adventure were the main arguments Oppenheimer used to recruit scientists for the project.

---

[3] Letter to William A. Fowler, spring of 1941.
[4] When Arthur Compton heard about that possibility, he thought, «Was there really any chance that an atomic bomb would trigger the explosion of the nitrogen in the atmosphere or the hydrogen in the ocean? This would be the ultimate catastrophe. Better to accept the slavery of the Nazis than to run a chance of drawing the final curtain on mankind!» (Rhodes 419).

After November 1942, Oppenheimer is increasingly concerned with cross sections (measuring the rates and possibilities for fission reactions) and what magnitude of explosion the project can deliver for the army. He insists the project "will be principally interested in energies of 5 MeV and above" (237) and states that "we should be wanton to strive for . . . a low goal" of only exceeding a 1,000 ton TNT equivalent (240). The key term for his correspondence during this time is purity (referring to the uranium and plutonium), with impurity as the worst quality. High purity of the radioactive elements = less worry about maximum speed, simplicity, reliability, energy release of over 10,000 tons of TNT, and the chance of predetonation reduced to the formula 0,5n% (240-2).[5] Increasingly, his language seems to mirror the lectures held later at Los Alamos in April 1943, published as *The Los Alamos Primer*.[6]

2.3 Ending All War through Nuclear Weapons

Niels Bohr comes to Los Alamos in the beginning of 1944 and he gives Oppenheimer a copy of his memo to Roosevelt in the summer of 1944, with a vision of using nuclear weapons as a means to end all war between nation states. There are many indications that he adopted and adapted that vocabulary with its thinking and arguments and used it to stifle dissent among the Los Alamos scientists. Oppenheimer says February 1946 that Niels Bohr had "helped us reach the conclusion" that international control of nuclear weapons and the end of all war was "not only a desirable solution" but also that "there were no other alternatives" (Smith and Weiner 322). Robert R. Wilson records that Oppenheimer used an adapted version of Bohr's vision to convince the scientists at a critical juncture to keep working on the atomic bomb.

Towards the end of 1944 it became clear to the scientists at Los Alamos that the Germans were not going to succeed in developing nuclear weapons and they would soon be conquered. The initial impetus and argument for initiating the program was now gone, and many scientists started to wonder in private and in small groups "What will this terrible weapon do to the world?" and how should it be used (Bird and Sherwin 284). Oppenheimer tried to discourage public discussion of the matter, citing concerns with the G-2 (military security) (283). Despite

---

[5] Letter to James B. Conant, November 30th, 1942.
[6] The lectures were held by Robert Serber, one of Robert Oppenheimer's former students at Berkeley, and they followed the same trajectory as Robert Oppenheimer's thoughts on the project up to that point, with purity and maximizing damage and efficiency as key concepts.

this, there seem to have been three or four public meetings discussing the ethics and potential impact of nuclear weapon development.[7]

Oppenheimer attended these meetings and used different arguments to persuade the scientists to continue developing the bombs. To one group he said they had "no right to a louder voice in determining the gadget's fate than any other citizen" (284). To another group he said that "although they were all destined to live in perpetual fear, the bomb might also end all war" (284). This second argument echoes Bohr's words. Wilson gives the most detailed explanation of the argument Oppenheimer used in the meeting Wilson organized on "The Impact of the Gadget on Civilization":

> The war ... should not end without the world knowing about this primordial new weapon. The worst outcome would be if the gadget remained a military secret. If that happened, then the next war would almost certainly be fought with atomic weapons. They had to forge ahead ... to the point where the gadget could be tested. He pointed out that the new United Nations was scheduled to hold its inaugural meeting in April 1945—and that it was important that the delegates begin their deliberations on the postwar world with the knowledge that mankind had invented these weapons of mass destruction. (285)

This vision or argument convinces the other scientists to complete the project,[8] but Oppenheimer is given a sobering wake-up call when he finds out that this vision is not shared widely in President Truman's administration (Smith and Weiner 301).[9] In letters from August to November 1945, Oppenheimer keeps reiterating the hope that the bomb "may serve as a real instrument in the establishment of peace" adding at one point, "That is almost the only thing right now that seems to matter"

---

[7] Louis Rosen, a junior physicist, remembers "a packed daytime colloquium held in the old theater," the chemist Joseph O. Hirschfelder remembers a "discussion held in Los Alamos' small wooden chapel" in "early 1945," and Robert R. Wilson organized a meeting on "The Impact of the Gadget on Civilization" in March 1945 (284). In addition to this, there was a later meeting in April or May discussing whether or not nuclear scientists should unionize that also touched on the impact of nuclear weapons on the world (Wilson 3).

[8] As Wilson states, "It was to be the end of war as we knew it, and this was a promise that was made. That is why I could continue on that project" (285).

[9] A meeting with Truman (who initially rejected Oppenheimer's ideas for international control of nuclear weapons) famously has Oppenheimer stating "I feel like I have blood on my hands" and Truman dismissing him as a "cry-baby scientist" (Bird and Sherwin 332).

(303).[10] In November 1945, Oppenheimer arrives at one of his most well-formulated and enduring statements about science, the development of the atomic bomb, and his vision for a nuclear future. He imitates Bohr in this statement but he also diverges from him in important ways. The statement is titled "Speech to the Association of Los Alamos Scientists" and was given November 2nd, 1945.

2.4 Weapons Development as an Organic Necessity

The speech is roughly 6,000 words long, and it can be roughly divided into four parts: (1) Setting the scene and explaining the immediate impact of the bomb, (2) explaining the nature of science, (3) the qualitative change the bomb has brought to war and the world, and (4) Oppenheimer's vision for the future along with some of the challenges of implementing it. One of these sections stands out among the rest: why does he make what seems like a digression to talk about the nature of science? The other three parts function perfectly well together and are unified by the theme of the bomb. I argue that the section about the nature of science makes up the moral and philosophical foundation for the rest of the dynamics in the text. According to Oppenheimer, the bomb came because of the nature of science, the future is being formed by science and should be structured to best nurture the growth of science. This subordination of almost all other things to the nature of science (either being caused by science or being deemed less valuable than science) indicates that this is the god-term in this text, and the structure of the text is dramatic catharsis where the logical implications of a god-term are gradually unfolded.

    For Oppenheimer, science is not just a method or an approach to the world, but it is also a moral philosophy and an amalgam of practices and core beliefs similar to that of a religion. He postulates these beliefs, behaviors, and practices in a kind of "scientist's creed" where people who do not follow these "stop being scientists" (317). Some of these are rather uncontroversial even today: "It is not possible to be a scientist unless you believe it is good to learn" (317), unless you "think it is of the highest value to share your knowledge . . . with anyone who is interested" (317), and unless you believe "it is good to find out how the world works and what the realities are" (317). To learn, to teach, and to understand, these are the core values of science (325). However, there are some tenets of Oppenheimer's "science" that sound less benign: If you are a scientist you believe it is good "to attain a gradually greater and greater control over nature" (325), believe that "the knowledge of the world, and the *power* this gives, is *a thing which is of intrinsic value* to humanity" (317), and

---

[10] Letter to Marcelle Bier, August 31st, 1945.

believe "it is good to turn over to mankind at large *the greatest possible power to control the world*" (317). In essence, following the logical implications of these claims, there is no technology or weapon, no matter how destructive, that scientists would not be morally obligated to develop and turn over "to mankind at large" (317) as long as these tools would also give mankind greater understanding of and control over nature.[11]

This becomes his justification for the Manhattan Project: after mentioning some of the justifications from different scientists who joined the project, Oppenheimer states, "But when you come right down to it the reason that we did this job is because it was an *organic necessity*. If you are a scientist you *cannot stop such a thing*" (317). And yet, even though Oppenheimer admits that because of the work of science both "the life of science" and "the life of the world" are threatened (322) he still states that scientists resist "anything which is an attempt to treat science of the future as though it were rather *a dangerous thing*, a thing that must be watched and managed" (317-8).

For Oppenheimer, science has a power, direction, authority, and value that is connected to the core virtues of *knowledge of* and *power over* nature, and these are "a thing of intrinsic value" (317). Science, as the god-term and central motive, produces knowledge and power. These two can possibly be collapsed into one since, as Francis Bacon stated, "knowledge is power" (Scientia potentia est). In either case, "science" provides the logic that makes it "an organic necessity" or consummatory drive for scientists to discover and develop knowledge of and power over nature and spread this to the rest of humanity.

This drive leads to shocking and groundbreaking discoveries that force humans "to re-consider the relations between science and common sense" (315-6). As Oppenheimer states:

> They forced on us the recognition that the fact that we were in the habit of talking a certain language and using certain concepts did not necessarily imply that there was anything in the real world to correspond to these. They forced us to be prepared for the inadequacy of the ways in which human beings attempted to deal with reality, for that reality. (316)]

He mentions relativity, the whole development of atomic theory, and Bohr's interpretation of it "in terms of complementarity" (315) as some examples of such discoveries.

However, with the development of the atomic bomb, science has gone a step further from merely abstract concepts to real world

---

[11] As Kenneth Burke writes in *The War of Words*, "Power itself is impersonal and pitiless" (246).

developments that provoke profound change and unrest in human society. Oppenheimer compares "the impact of the creation of the atomic bomb and atomic weapons" to "the times when physical science was growing in the days of the renaissance" when "the threat that science offered was felt so deeply," or "when the theories of evolution seemed a threat to the values by which men lived" (316). By pushing the limits of power and knowledge, science provokes radical shifts in society, and Oppenheimer sees the development of atomic weapons as one of the most profound of these.

Oppenheimer argues that the development of atomic weapons is not just a dramatic quantitative change (increased magnitude of destruction, relatively cheap, with shifted advantage of aggression/attack compared to defense) but it also constitutes a change in quality: "wars have changed" and "if these first bombs ... can destroy ten square miles, then that is *really quite something*" (318).[12] It signifies "a change in the nature of the world" where "wars have become intolerable" and mankind faces a "common problem," "peril that affects everyone," and a situation where "the life of science and the life of the world is threatened" (318-9). In essence, this development has created "a new situation" and "and new field" or "new opportunity for realizing preconditions" (319).

So far Oppenheimer has outlined a logical sequence from the nature of science and from scientists following its "organic necessity," but the next step consists of possible rather than necessary developments. According to Oppenheimer, this new situation creates "a possibility of realizing ... those changes which are needed if there is to be any peace" (319). He describes them as "very far-reaching changes" in "relations between nations" in "spirit," "law," "conception" and "feeling" (319) based on a "complete sense of community responsibility" (319).

One of the most fundamental changes, which Oppenheimer describes as "an enormous change in spirit" (320) concerns the most basic commitment of the American people to their ideals:
There are things which we hold very dear ... I would say that the word "democracy" perhaps stood for some of them as well as any other word. There are many parts in the world in which there is no democracy. There are other things which we hold dear, and which we rightly should. And when I speak of a new spirit in international affairs I mean that even to these deepest of things which we cherish, and for which Americans have been willing to die ... even in these deepest things, we realize that there is something more profound than that; namely, the common bond with

---

[12] He uses the same term to describe his excitement of what kind of explosion one could get from nuclear fission in his first letter describing the newly discovered phenomenon.

other men everywhere. It is only if you do that that this makes sense. (320)

It is clear that Oppenheimer is here preparing scientists that they may have to give up some of their democratic ideals, at least temporarily, in order to achieve security for the world. It is unclear in the text exactly what he is referring to when he warns that "only by a profound revision of what it is that constitutes a thing worth fighting for and a thing worth living for can this crisis be met" (322) or how far such a radical change would have to go.[13]

The final goal of all these changes goes beyond the control of nuclear weapons to "a world that is united, and a world where war cannot occur" (320). He elaborates on this vision and the role of scientists in realizing it in his article "The New Weapon: The Turn of the Screw," published in the book *One World or None: A Report to the Public on the Full Meaning of the Atomic Bomb*:

> Scientists are ... humanists; science is ... universally human. It is therefore natural for scientists to look at the new world of atomic energy and atomic weapons in a very broad light. And in this light the community of experience, of effort, and of values that prevails among scientists of different nations is comparable in significance with the community of interest existing for the men and the women of one nation. It is natural that they should supplement the fraternity of the peoples of one country with the fraternity of men of learning everywhere, with the value that these men put upon knowledge, and with the attempt – which is their heritage – to transcend the accidents of personal or national history in discovering more of the nature of the physical world. (63)

The god-term of this structure is science. Science is the driving force of change in human history, and its final state is peace (leading to more scientific cooperation, which again leads to greater knowledge and power).

---

[13] The least controversial reading of Oppenheimer here would be that he is simply arguing for restraint and humility on the part of the US, "because if you approach the problem and say, 'We know what is right and we would like to use the atomic bomb to persuade you to agree with us,' then you are in a very weak position and you will not succeed" (320). However, he may also be sharing the assumptions made by H. G. Wells, Leo Szilard, and Niels Bohr that world government and democracy will be (at least initially) incompatible. The global Atomic Development Agency Oppenheimer later proposes in the Acheson-Lilienthal Report of 1946 can hardly be classified as a democratic organization, even though it would have a mandate superseding the individual nation states.

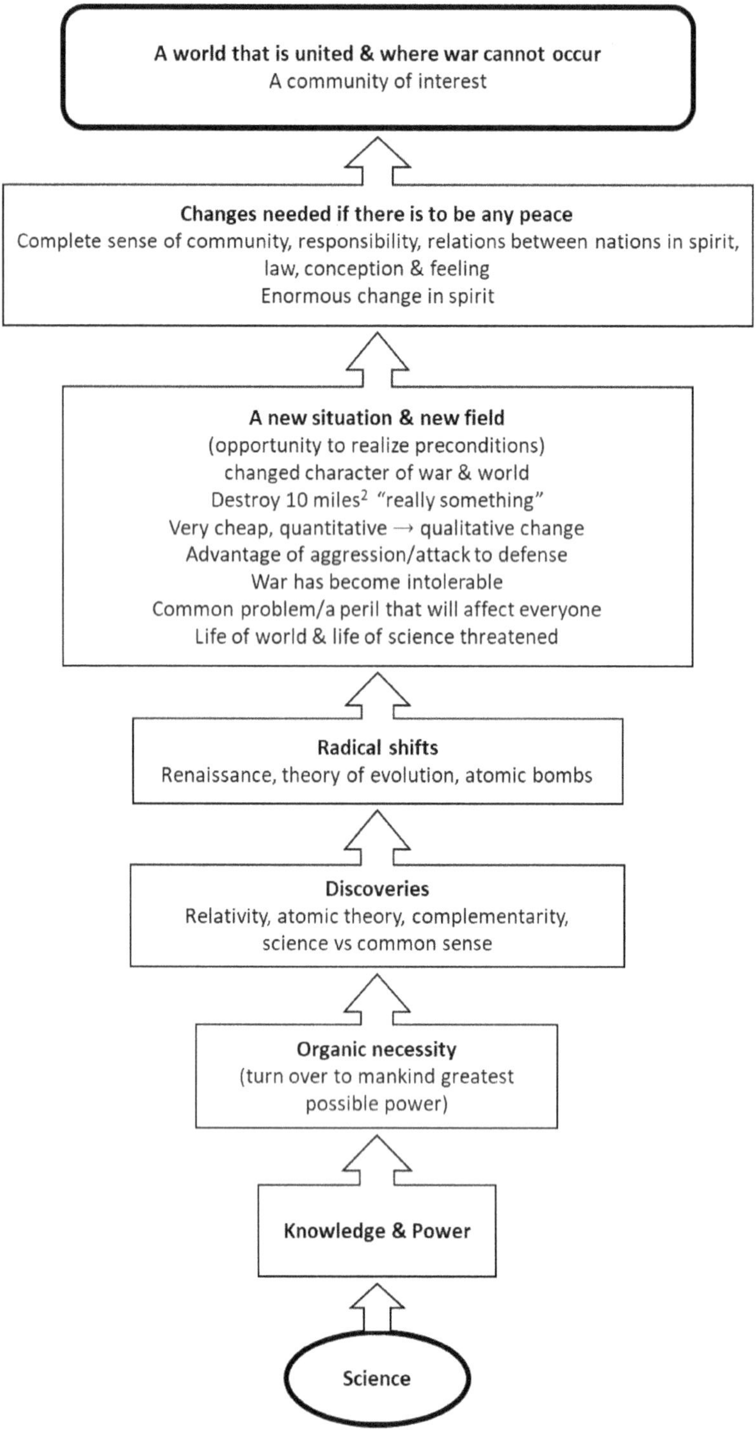

```
┌─────────────────────────────────────────────────┐
│   A world that is united & where war cannot occur │
│             A community of interest              │
└─────────────────────────────────────────────────┘
                        ⇧
┌─────────────────────────────────────────────────┐
│      Changes needed if there is to be any peace   │
│ Complete sense of community, responsibility, relations between nations in spirit, │
│              law, conception & feeling            │
│               Enormous change in spirit           │
└─────────────────────────────────────────────────┘
                        ⇧
┌─────────────────────────────────────────────────┐
│            A new situation & new field            │
│        (opportunity to realize preconditions)     │
│            changed character of war & world       │
│           Destroy 10 miles² "really something"    │
│       Very cheap, quantitative → qualitative change │
│         Advantage of aggression/attack to defense │
│                War has become intolerable         │
│     Common problem/a peril that will affect everyone │
│          Life of world & life of science threatened │
└─────────────────────────────────────────────────┘
                        ⇧
              ┌────────────────────────┐
              │     Radical shifts      │
              │ Renaissance, theory of evolution, atomic bombs │
              └────────────────────────┘
                        ⇧
              ┌────────────────────────┐
              │       Discoveries       │
              │ Relativity, atomic theory, complementarity, │
              │       science vs common sense    │
              └────────────────────────┘
                        ⇧
              ┌────────────────────────┐
              │    Organic necessity    │
              │  (turn over to mankind greatest │
              │        possible power)  │
              └────────────────────────┘
                        ⇧
              ┌────────────────────────┐
              │     Knowledge & Power   │
              └────────────────────────┘
                        ⇧
                   (  Science  )
```

Figure 1 – Hierarchy of terms for Oppenheimer's Los Alamos Speech

There are interesting similarities between the process Oppenheimer describes here and the one he describes to his brother in his letter from 1932. Both describe a transition from a state of turmoil (struggle/war vs. profound social change and shock) to a new condition (discipline vs. new situation). By choosing to use this new situation to purify oneself of the unnecessary (detachment which renounces the world it preserves vs. profound revision of what it is that constitutes a thing worth fighting and living for, such as democracy), one has the chance of obtaining the final goal of peace, which is the same in both texts. These similarities may indicate that Oppenheimer's mind had a preference for thinking in these patterns, transcending a situation that looks like a problem by appreciation (gratitude for struggle and war) and a form of ascesis ("learn to preserve what is essential to our happiness in more and more adverse circumstances" and to "abandon with simplicity what would else have seemed to us indispensable").

3. CONCLUSION

Throughout this text, Oppenheimer argues that no scientist can stop scientific or technological developments since this goes against all it means to be a scientist. Edward Teller would later borrow the same form of argument to insist on developing the hydrogen bomb and later "clean bombs" and neutron bombs:

> The spectacular developments of the last centuries, in science, in technology and in our own everyday life, have been produced by a spirit of adventure, by a fearless exploration of the unknown. When we talk about nuclear tests, we have in mind not only military preparedness but also the execution of experiments which will give us more insight into the forces of nature. Such insight has led and will lead to new possibilities of controlling nature. There are many specific political and military reasons why such experiments should not be abandoned. There also exists this very general reason—the tradition of exploring the unknown. It is possible to follow this tradition without running any serious risk that radioactivity, carelessly dispersed, will interfere with human life. (Teller and Latter 72)

There are echoes of the same argument among scientists pushing for human gene-editing and the development of weapons with artificial intelligence. Science is a wonderful tool for critical thinking, but we must also be able to think critically about all that this endeavor entails in any given generation, and the hierarchy of values it contains. Otherwise, science may also be used as a tool to stifle dissent and perpetuate harmful assumptions.

REFERENCES

Bird, K. & Sherwin, M. J. (2005). *American Prometheus: The triumph and tragedy of J. Robert Oppenheimer.* New York: Vintage Books.
Burke, K. (1955). Linguistic approach to problems of education. Ed. Henry B. Nelson. *Modern philosophies and education: The fifty-fourth yearbook of the National Society for the Study of Education.* Chicago: University of Chicago Press.
Burke, K. (2018). *The war of words.* Eds. Burke, A, Jensen, K, & Selzer, J. Berkeley: University of California Press.
Dyson, F. (1979). *Disturbing the Universe.* New York: Harper & Row.
Eiduson, B. T. (1962). *Scientists: their psychological world.* England: Basic Books
Isaksen, D. E. (2017). Indexing: Kenneth Burke's method of textual analysis. *KBJournal.* 12(2).
Kuhn, T. S. (2012). *The structure of scientific revolutions.* 4th edition. Chicago: University of Chicago Press.
Mercier, H. (2011). When experts argue: explaining the best and worst of reasoning. *Argumentation.* 25(1). 313-27.
Oppenheimer, Robert J. (1946). The new weapon: The turn of the screw. *One world or none: A report to the public on the full meaning of the atomic bomb.* The New Press, 2007.
Smith, A. K. & Weiner, C. (1995). *Robert Oppenheimer: letters and recollections.* Stanford, CA: Stanford University Press.
Palevsky, M. (2000). *Atomic Fragments: A Daughter's Questions.* Berkeley: University of California Press.
Polanyi, M. (1946). *Science, faith and society.* Chicago: University of Chicago Press.
Teller, E. & Latter, A. (1958). The compelling need for nuclear tests." *LIFE.* 44(8). 64-72.
United States Atomic Energy Commission (1954). In the matter of J. Robert Oppenheimer: Texts of principal documents and letters of personnel security board general manager commissioners. Washington, DC: US Government Printing Office.

Formal specifications for dialogue games in multi-party healthcare coaching

MATHIDLE JANIER
Université Grenoble Alpes, France
mathildejanier@hotmail.com

ALISON PEASE
Centre for Argument Technology, University of Dundee, UK
a.pease@dundee.ac.uk

MARK SNAITH
Centre for Argument Technology, University of Dundee, UK
m.snaith@dundee.ac.uk

DOMINIC DE FRANCO
Centre for Argument Technology, University of Dundee, UK
d.f.defranco@dundee.ac.uk

We present our analysis in terms of Inference Anchoring theory of a dataset of patient interviews, in the context of multi-party health coaching. For each dialogue game specification we first provide a general description of the game, followed by descriptions of the participants, and rules for: locutions, commitment, structure, termination, and outcome. We then implement these theoretical dialogue game specifications by taking their subsequent representation in a Dialogue Game Description Language.

KEYWORDS: formal dialogue game, health coaching, Dialogue Game Description Language

1. INTRODUCTION

To design dialogue games that allow for realistic interactions between patients and healthcare professionals in a virtual setting, it is first necessary to understand how such interactions might take place between patients and real healthcare professionals. By far the best way to understand these interactions is to examine them happening in real life;

this, however, is almost impossible to do. First, putting real patient consultations under observation risks changing the dynamic of those consultations, thus providing inaccurate data. Second, it is unusual for consultations to take place with more than one medical practitioner, and so finding such sessions in the first instance would prove a significant challenge. We therefore adopted a role-playing approach, in which real medical practitioners carried out a series of consultations with patients played by actors. Across the consultations, different actors played to carefully designed different personas, in consultation with different practitioners. In this paper we describe our analysis of the role plays, in terms of Inference Anchoring Theory (IAT) -- a philosophically grounded theory which has been developed to capture relationships between argument structures and dialogue structures (Budzynksa et. al., 2016).

We firstly use this analysis as the foundation for formal specifications for dialogue games in this context. We then implement these theoretical dialogue game specifications by taking their subsequent representation in a Dialogue Game Description Language (DGDL).

A total of 35 excerpts have been analysed in OVA+ (Janier et. al. 2014) using the IAT annotation scheme. These gave a total of 662 turns, out of 2179 total moves; around 31% of the total dialogues. In particular, a complete session has been annotated which gives a better insight into the shape and content of the Council of Coaches dialogues. The other analysed excerpts, taken from 5 different sessions, aim at being a representative sample of the wide variety of communication situations in couch dialogues. Since the topics tackled, the patients' character and the professionals' domain of expertise are different in every dialogue, the annotated data present a wide range of dialogical and argumentative dynamics which can help to refine and generalise the dialogue games. Our 35 annotated maps can be seen at http://corpora.aifdb.org/couch, with full argument analytics at http://analytics.arg-tech.org/overview.php?c=couch.

2. BACKGROUND

2.1 Patient Consultation Corpus

To design dialogue games that allow for realistic interactions between patients and their virtual coaches, it is first necessary to understand how such interactions might take place between patients and real medical practitioners. By far the best way to understand these interactions is to examine them happening in real life; this, however, is almost impossible to do. First, putting real patient consultations under observation risks changing the dynamic of those consultations, thus providing inaccurate data. Second, it is unusual for consultations to take place with more than one medical practitioner, and so finding such sessions in the first instance would prove a significant challenge.

We therefore adopted a role-playing approach, in which real medical practitioners carried out a series of consultations with patients played by actors. Across the consultations, different actors played to different personas (that we specified), in consultation with different practitioners.

The audio from each session was transcribed by a professional transcription service, then anonymised to remove the names of the medical practitioner ("patient" names did not need removed because they were fake to begin with).

Several different personas were devised for the actors to play to, which are summarised in **Error! Reference source not found.**. All personas describe patients that have recently been diagnosed with Type 2 diabetes. Note that while a gender is specified for the persona, this was not fixed: through only tweaking minor details, each persona was adaptable to be played by an actor of any gender. The sessions recorded are summarised in **Error! Reference source not found.**.

No	Gender	Age	Personality
1	Male	57	Know-it-all
2	Female	63	Anxious
3	Female	50	Unengaged
4	Male	67	Benchmark

Table 1: Patient personas

Session ID	Actor	Type of patient	Practitioners involved
S1	Male	Know-it-all	General practitioner, diabetes practitioner
S2	Male	Benchmark	General practitioner, diabetes practitioner
S3	Female 1	Unengaged	Podiatrist, general practitioner
S4	Female 1	Anxious	Podiatrist, general practitioner
S5	Female 1	Benchmark	Podiatrist, general practitioner
S6	Female 1	Know-it-all	Podiatrist, general practitioner
S7	Female 2	Unengaged	General practitioner, motivational interviewer, dietician
S8	Female 2	Know-it-all	Motivational interviewer, dietician
S9	Female 2	Benchmark	Motivational interviewer, dietician

Table 2: Sessions recorded

2.2 Inference Anchoring Theory

Inference Anchoring Theory (IAT) is a philosophically grounded theory which has been developed to capture relationships between argument structures and dialogue (Budzynska et. al., 2016). By taking into account the illocutionary force of utterances, IAT allows us to represent illocutionary structures which link locution nodes (L-nodes) to information nodes (I-nodes). Moreover, given that some speakers' communicative intentions cannot be determined without knowing the broader context of the dialogue that is, what an utterance is responding to – IAT assumes that it is only by taking into account the relation between L-nodes that some illocutionary forces can be inferred. As a consequence, these illocutionary structures are anchored in transition nodes (TA-nodes) and can target I-nodes or scheme nodes (S-nodes) (to elicit inference or conflict relations between propositions) (Budzynska et. al., 2016) IAT is therefore a framework developed for the analysis of dialogues in order to elicit argumentative structures.

By making the illocutionary forces of locutions apparent, the model allows us to identify argumentative dynamics which have been generated by dialogical moves. The IAT graphical representations of dialogical structures and the attached illocutionary and argumentative structures represent a valuable framework for fine-grained analyses of discourse.

This theory is very well suited to our goal of building a dialogue game from our corpus of patient interviews, since our corpus consists of natural language dialogue and IAT provides a way of linking dialogue argumentative dynamics via the analysis of speech acts.

Figure 1 shows an example of an IAT analysis taken from the Patient Consultation Corpus.

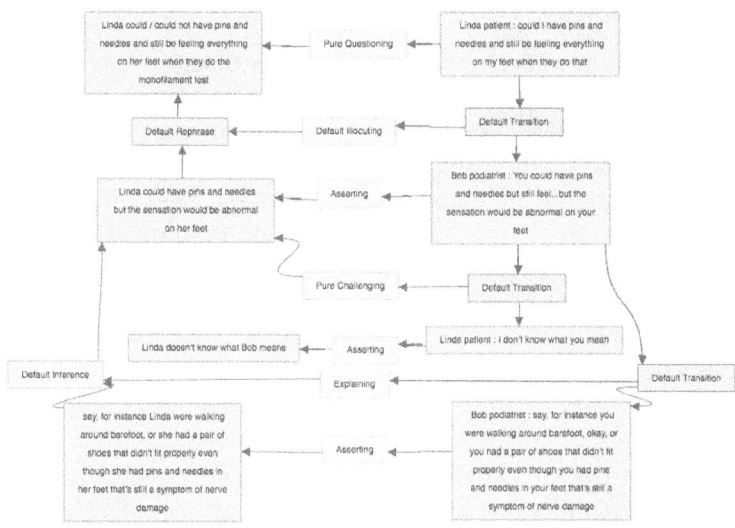

Figure 1: IAT analysis

3. DIALOGUE GAME SPECIFICATIONS

3.1 Game 1: Pre-interviews

The Pre-interview takes place before the patient is in the room. Its purpose is for the coaches to discuss how each of them may contribute and possibly what strategies might be effective in a particular case. This follows informal practice used in medical settings. The specification follows patterns found in the data collected from the patient interview sessions described in Section 3. Locution, commitment, structural, termination and outcome rules are shown in Tables 3 through 7 in the appendix, while a visualisation of the general structure of the game is in Figure 2.

The participants in a pre-interview dialogue consist of a set of at least two coaches (X), where a single coach (C) is designated the "Lead Coach" (LC). The Lead Coach is the coach who has the most familiarity with the patient and who can advise on which other experts should be present at the session and on strategies that might be useful, given the patient's personality and situation.

It is important to note that we do not specify a locution rule to permit players to argue or explain. As stated in (van Eemeren and Grootendorst, 1982) and (Budzynska et al., 2014a), 'arguing' is a complex illocutionary force which takes shape only by virtue of the interrelation between locutions: one can build an argument by asserting p and q and showing that there is an inference between p and q, e.g. "p because q". Hence, arguing is automatically created when support for a proposition is given and, in the pre-interviews game, PCh allows for triggering inference. Moreover, it has been shown that in some discursive contexts, AQ is more frequent than challenges to trigger argumentation (e.g. in debates, see (Yaskorska and Janier, 2015)) or in financial dialogues (Budzynska et al., 2014b)). Pure Challenging indeed has a low frequency in the COUCH corpus, this is explained by the fact that speakers do not necessarily wait to be challenged to support their opinion. However, formal dialectical systems' standards are followed here by including challenges which are, in the game, the only way for players to construct inference between propositions because parties cannot advance two propositions in a single turn.

Hamblin's view of speakers' commitments (Hamblin, 1971) is followed in our game: a speaker is committed to a statement if he personally utters the statement (CR1) or when he agrees with a statement uttered by an interlocutor (CR3). As in most formal dialogue systems (e.g. DC (Mackenzie, 1979), CB (Walton, 1984), PPD (Walton and Krabbe, 1995)), the pre-interviews game allows players to retract propositions: if a proposition is withdrawn, it is assumed that the players are no more in conflict about this proposition and consensus is reached on this particular proposition (CR2). Commitment rules in the pre-interviews game however differ from those in other dialogue games since propositions are added to a commitment store only if they have been asserted or agreed with. In many dialogue games, indeed, a stated proposition is added to all players' stores; if a player is not committed to this proposition, he has to explicitly withdraw it. In the pre-interviews game, on the other hand, a proposition is solely added in the store of the player who asserted (or agreed on) it. This is defined in CR1 and CR3. CR4 specifies that if a proposition p is disagreed with, then the opposite proposition (-p) is added to a store (see also (Wells and Reed, 2012)). This rule allows M to deploy a strategy: when :p is added to a player's commitment store after he disagreed with p, M is able to ask him whether his disagreement with p means that he is committed to :p. This is to ensure the relevance and consistency of dialogues: a player cannot simply disagree on p; he has to agree with :p, provide reasons for :p or withdraw :p.

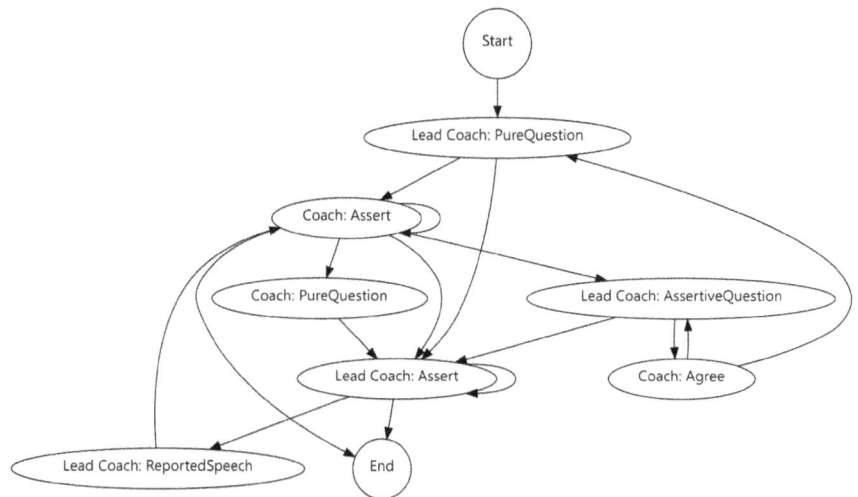

Figure 2: Visualisation of the pre-interview dialogue game

3.2 Game 2: Patient interview

The patient interview is the main consultation between the patient and multiple coaches, providing a broad framework for one or more coaches to engage in a consultation with a patient. Locution, commitment, structural, termination and outcome rules are provided in Tables 8-12 in the appendix.

The participants in a patient interview are a (possibly unit) set of coaches, and a patient. Note that there is no "Lead Coach" in this dialogue game – where there is more than one coach, all are given equal standing.

Due to the expressivity of the patient interview dialogue game, in all participants share the same set of locutions and (mostly) structural rules, any visualisation is highly complex and difficult to read. We therefore do not provide such a visualisation for this game.

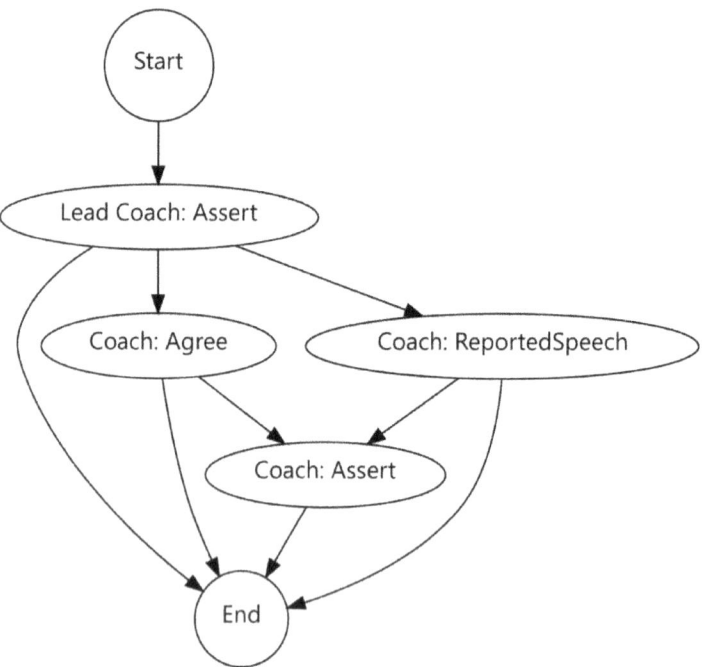

Figure 3: Visualisation of the post-interview dialogue game

3.3 Game 3: Post-interview

The Post-interview takes place after the patient interview. Its purpose is for the coaches to debrief and discuss how the session went. This follows informal practice used in medical settings. Locution, commitment, structural, termination and outcome rules are provided in Tables 13-16 in the appendix, while a visualisation of the general structure of the game is in Figure 2.

The participants in a Post-interview dialogue consist of a set of at least two coaches (X), where a single coach (C) is designated the "Lead Coach" (LC). The Lead Coach is the coach who has the most familiarity with the patient.

4. CONCLUSION

We have in this paper provided a set of specifications for dialogue games in multi-party health coaching. These are based on Inference Anchoring Theory analyses of a corpora of simulated consultations between various healthcare professionals and a patient.

Three dialogue games were provided: a pre-interview game, where the healthcare professionals discuss the patient's history; the interview game, which is the main consultation between the healthcare professionals and the patient; and the post-interview game, in which the healthcare professionals discuss what happened during the consultation (interview), and potential future steps.

Each dialogue game specification describes: the participants in the dialogue, and rules for locutions, commitment, structure (turn-taking), termination and outcome.

In future work, we will implement these game specifications computationally in Dialogue Game Description Language (DGDL for execution on the Dialogue Game Execution Platform (DGEP) (Wells and Reed, 2012).

ACKNOWLEDGEMENTS: This project has received funding from the European Union's Horizon 2020 research and innovation programme under grant number #769553. This result only reflects the authors' view and the EU is not responsible for any use that may be made of the information it contains. The authors are also grateful to the anonymous reviewers and the audience in our talk for their valuable feedback.

REFERENCES

Budzynska, K., Janier, M., Reed, C., & Saint-Dizier, P. (2016). Theoretical foundations for illocutionary structure parsing. *Argument & Computation, 7*(1), 91-108.
Budzynska, K., Janier, M., Reed, C., Saint-Dizier, P., & Stede, M. (2014a). A Model for Processing Illocutionary Structures and Argumentation in Debates. *Proceeedings of LREC 2014.*
Budzynska, K., Rocci, A., & Yaskorska, O. (2014b). Financial Dialogue Games: A Protocol for Earnings Conference Calls. *Proceedings of the Fifth International Conference on Computational Models of Arguemnt (COMMA 2014).*
Hamblin, C. L. (1971). Mathematical models of dialogue 1. *Theoria, 37*(2), 130-155.
Janier, M., Lawrence, J., & Reed, C. (2014). Ova+: An argument analysis interface. *Computational Models of Argument (COMMA).*
MacKenzie, J. D. (1979). Question-begging in non-cumulative systems. *Journal of philosophical logic, 8*(1), 117-133.
van Eemeren, F. H., & Grootendorst, R. (1982). *Regels voor redelijke discussies. Een bijdrage tot de theoretische analyse van argumentatie ter oplossing van geschillen [Rules for reasonable discussions: A contribution to the theoretical analysis of argumentation to resolve disputes].* Foris Publications Holland.
Walton, D. C., & Krabbe, E. C. (1995). *Commitment in dialogue: Basic concepts of interpersonal reasoning.* SUNY Press.
Walton, D. N. (1984). *Logical Dialogue Games and Fallacies.*
Wells, S., & Reed, C. (2012). A domain specific language for describing diverse systems of dialogue. *Journal of Applied Logic, 10*(4), 309-329.

Yaskorska, O., & Janier, M. (2015). Applying Inference Anchoring Theory for the analysis of dialogue structure in debate. *Proceedings of the First European Conference on Argumentation (ECA 2015)*.

APPENDIX

LR1	C can: 1. PQ(p) when he asks whether p is the case, i.e. if LC believes p 2. A(p) when he gives his opinion on p 4. PCh(p) when he seeks LC's ground for stating p 5. Agr(p) when he agrees on p
LR2	LC, in addition to those locutions available to all coaches, can: 1. AQ(p) when he seeks C's agreement on p 2. R(p) when he restates p (usually to summarise Patient's situation) 3. *ReportedSpeech(s, IllocutionaryForce(p))* when he reports that speaker s said proposition p with a specific Illocutionary Force. 4. *PatientSummaryConcluded* when the LC has concluded a patient summary

Table 3: Locution rules for pre-interviews dialogue game

CR1	Following a A(p), performed by C≠LC, p is added to CSc_i
CR2	Following a Agr(p), performed by C≠LC, p is added to CSc_i
CR3	Following a Disagr(p), performed by C≠LC, -p is added to CSc_i
CR4	Following a AQ(p), performed by C≠LC, p is added to CSc_i
CR5	Following a R(p), performed by C≠LC, p is added to CSc_i

Table 4: Commitment rules for pre-interviews dialogue game

SR1	LC moves first with *PQ(p)*, where p = "have a moment"
SR2	After LC OR C≠LC performs *PQ(p)*, C≠LC OR C must perform: 1. *A(p)*; or 2. *A(-p)*
SR3	After LC OR C≠LC performs *AQ(p)*, C≠LC OR C must perform: 1. *Agr(p)*; or 2. Dis*agr(p)*

SR4	After C≠LC performs *A(p₁),either*:
	    1. C can perform PQ(p), or
	    2. *LC can* perform a sequence of locutions asserting some finite number of propositions about S, many of which are rephrases (because he summarises S): *Assert(p_i)...*,
	where 1 ≤ <= i ≤ n for some n ∈ Natural Numbers (S=Situation) and then
	3. *LC* end the summary of the situation by saying: *PatientSummaryConcluded*

SR5	After LC asserts *PatientSummaryConcluded,* LC can perform: *ReportedSpeech*(P, *IllocutionaryForce*(p)) to report propositions p that the patient P has said in previous sessions
SR6	After *LC* performs *ReportedSpeech*(P, *IllocutionaryForce*(p)), C≠LC can perform: 1) a sequence of locutions asserting some finite number of propositions, with inferential structure between them: *A(p_i)...*, where 1 ≤ i ≤ n for some n ∈ Natural Numbers

SR7	After C≠LC performs *A(p_i)*... with inferential structure between them:
	C can perform PQ, where:
	    1. C≠LC performs PQ(s), where s=situation, or
	    2. LC performs PQ(p), where p=proposition

SR8	After C≠LC performs *A(p_i)*, LC performs: 1. AQ(p) where p = "see Patient P"

SR9	After LC performs AQ(p) where p = "see Patient P", all C≠LC must perform: 1. Agr(p)
SR10	After all C≠LC perform Agr(p) where p = "see Patient P", C≠LC can perform 1. PQ(p), or 2. AQ(p) where p is a strategy for dealing with the patient

Table 5: Structural rules for pre-interview dialogue game

T1	A dialogue terminates if any C≠LC performs: 1. *A(-p)*, where p = "have a moment" Or All C≠LC performs: 2. Agr(p), where p = "see Patient P" And, if C≠LC performs 3. if C≠LC performs PQ(p), then the LC performs Assert(p), or if 4. if C≠LC performs AQ(p), then the LC performs Agr(p) where p is a strategy for dealing with the patient.

Table 6: Termination rules for pre-interview dialogue game

Outcome	Conditions
Don't agree to see Patient P	any C≠LC performs *A(-p)*, where p = "have a moment"
Agree to see Patient P (no strategy for dealing with the patient)	All C in C/LC Agr(p) where p = "see Patient P"

| Agree to see Patient P (and strategy for dealing with the patient) | All C in C/LC Agr(p) where p = "see Patient P" |

Table 7: Outcome rules for pre-interview dialogue game

| LR1 | All participants can:
1. PQ(p) when they ask whether p is the case, i.e. if the hearer believes p
2. A(p) when they give their opinion on p
4. PCh(p) when they seek hearers' ground for stating p
5. Agr(p) when they agree on p
6. R(p) when they restate p (to exemplify, generalise, paraphrase, repeat, etc)
7. AQ(p) when they seeks the hearer's agreement on p
8. *ReportedSpeech*(s, *IllocutionaryForce*(p)) when they report that speaker s said proposition p
9. RQ(p) when they grammatically state a question, but in fact are just conveying that they do (or do) believe p and do not wait for the other participants to answer the question
10. Backchannel when they want the previous speaker to continue
11. Disagr(p) when they disagree on p |

Table 8: Locution rules for patient interview dialogue game

CR1	Following a A(p), performed by X, p is added to CSx_i
CR2	Following a Agr(p), performed by X, p is added to CSx_i
CR3	Following a Disagr(p), performed by X, -p is added to CSx_i
CR4	Following a AQ(p), performed by X, p is added to CSx_i
CR5	Following a R(p), performed by X, p is added to CSx_i
CR6	Following a RQ(p), performed by X, p is added to CSx_i

| CR7 | Following a Disagr(p), performed by X, -p is added to CSx_i |

Table 9: Commitment rules for patient interview dialogue game

SR1	[After greetings] The dialogue starts with C performing PQ(p) addressed to P
SR2	After X performs *PQ(p)*, the answerer must perform: 1. *Assert(p)*; or 2. *Assert(-p)*
SR3	After P performs Assert(p): 1. Any participant can Assert(q) where p and q form either a rephrasing structure or an inferential structure, or 2. Any participant can ReportSpeech(s,(IF(q)), or 3. Any participant can ReportSpeech(X,(A(p)), or 4. Any participant can AQ(p), or 5. Any participant can AQ(q), or 6. Any participant can RQ(p), or 7. Any participant can PQ(q), or 8. Any participant can PCh(p), 9. Any participant can Agr(p) 10. C can Disagr(p)
SR4	After C performs Assert(p): 1. Any participant can A(q) where p and q form either a rephrasing structure or an inferential structure, or 2. Any participant can ReportSpeech(s,(IF(q)), or 3. Any participant can ReportSpeech(X,(A(p)), or 4. Any participant can AQ(p), or 5. Any participant can AQ(q), or 6. Any participant can RQ(p), or 7. Any participant can PQ(q), or

	8. Any participant can PCh(p), 9. Any participant can Agr(p) 10. P can Disagr(p)
SR5	After P performs Assert(-p), 1. Any participant can A(q) where -p and q form either a rephrasing structure or an inferential structure, or 2. Any participant can ReportSpeech(s,(IF(q)), or 3. Any participant can ReportSpeech(X,(A(-p)), or 4. Any participant can AQ(-p), or 5. Any participant can AQ(q), or 6. Any participant can RQ(-p), or 7. Any participant can PQ(q), or 8. Any participant can PCh(-p), 9. Any participant can Agr(-p) 10. C can Disagr(p)
SR6	After C performs Assert(-p), 11. Any participant can A(q) where -p and q form either a rephrasing structure or an inferential structure, or 12. Any participant can ReportSpeech(s,(IF(q)), or 13. Any participant can ReportSpeech(X,(A(-p)), or 14. Any participant can AQ(-p), or 15. Any participant can AQ(q), or 16. Any participant can RQ(-p), or 17. Any participant can PQ(q), or 18. Any participant can PCh(-p), 19. Any participant can Agr(-p) 20. P can Disagr(-p)

SR7	After P performs ReportSpeech(s,(IF(p))), 1. Any participant can Assert(q) where p and q form either a rephrasing structure or an inferential structure, or 2. Any participant can ReportSpeech(s,(IF(q))), where p and q form an inferential structure, or 3. Any participant can AQ(p), or 4. Any participant can AQ(q), or 5. Any participant can RQ(p), or 6. Any participant can PQ(q), or 7. Any participant can PCh(p), 8. Any participant can Agr(p) 9. C can Disagr(p)
SR8	After C performs ReportSpeech(s,(IF(p))), 10. Any participant can A(q) where p and q form either a rephrasing structure or an inferential structure, or 11. Any participant can ReportSpeech(s,(IF(q))), where p and q form an inferential structure, or 12. Any participant can AQ(p), or 13. Any participant can AQ(q), or 14. Any participant can RQ(p), or 15. Any participant can PQ(q), or 16. Any participant can PCh(p), 17. Any participant can Agr(p) 18. P can Disagr(p)

SR9	After P performs AQ(p) addressed to C_i, C_i can: 1. C_i can Agr(p), or 2. C_i can Disagr(p), or 3. C_i can R(q) where q is a rephrase of p For i≠j, 1≤ i,j ≤ n where n is the number of coaches
SR10	After C performs AQ(p) addressed to P, P can: 1. Agr(p), or 2. Disagr(p)
SR11	After C performs AQ(p) addressed to C_i, C_I can: 1. Agr(p), or 2. R(q) where p is a rephrase of p
SR12	After X performs RQ(p), X can: 1. A(q) 2. PQ(q) 3. AQ(p) 4. AQ(q) 5. PCh(p) 6. R(q) where p and q form either a rephrasing structure or an inferential structure
SR13	After X performs PCh(p) addressed to C_i, 1. C_i can A(q) where p and q form an inferential structure, or 2. C_i can R(q) where p and q form a rephrasing structure addressed to X_i

SR14	After X performs Agr(p), any participant can: 1. A(q) where p and q form either a rephrasing structure or an inferential structure, or 2. ReportSpeech(s,(IF(q))), where p and q form a rephrasing structure or an inferential structure, or 3. AQ(q), or 4. PQ(q), or 5. Agr(p)
SR15	After P performs Disagr(p), 1. P can A(q) where -p and q form a rephrasing structure or an inferential structure 2. P can ReportSpeech(s,(IF(q))), where -p and q form a rephrasing structure or an inferential structure 3. P can PCh(p) 4. C can PCh(-p)
SR16	After C performs Disagr(p), 1. Any participant can PCh(-p) 2. C can A(q) where -p and q form a rephrasing structure or an inferential structure 3. C can R(q) where -p and q form a rephrasing structure or an inferential structure 4. C can AQ(-p) addressed to C_i 5. C can PQ(q) addressed to any other participant 6. C_i can Agr(-p)

Table 10: Structural rules for patient interview dialogue game

T1	A dialogue terminates if: 1. All participants agree on p, where p= "all issues have been raised and resolved"

Table 11: Termination rules for patient interview dialogue game

Outcome	Conditions
Plan of action and/or further session have not been agreed	P Agr(p) where p= "plan of action/further session
Plan of action and/or further session have been agreed	P Agr(p) where p= "plan of action/further session"

Table 12: Outcome rules for patient interview dialogue game

LR1	C can: 1. A(p) when he gives his opinion on p 2. Agr(p) when he agrees on p 3. *ReportedSpeech*(s, *IllocutionaryForce*(p)) when he reports that speaker s said proposition p 4. *ArgumentConcluded* when the C has concluded an argument

Table 13: Locution rules for post-interview dialogue game

CR1	Following a A(p), performed by C ∈ X, p is added to CSc_i
CR2	Following a Agr(p), performed by C ∈ X, p is added to CSc_i

Table 14: Commitment rules for post-interview dialogue game

SR1	LC moves first with: 1. a sequence of locutions asserting some finite number of propositions: *Assert(p_i)*..., where $1 \leq i \leq n$ for some n ∈ Natural Numbers with inferential structure between them, and then 2. *ArgumentConcluded*
SR2	After any coach performs *ArgumentConcluded*, any other coach can perform: 1. *Agr(p)*, or

		2. *ReportedSpeech*(P, *IllocutionaryForce*(q)) to report propositions q that the patient P has said in previous sessions
SR3		After *Agr(p)*, any coach can perform: 1. a sequence of locutions asserting some finite number of propositions: *Assert(q_i)...*, where $1 <= i <= n$ for some $n \in$ Natural Numbers with inferential structure between them, and then 2. *ArgumentConcluded*
SR4		After *ReportedSpeech*(P, *IllocutionaryForce*(p)), any coach can perform: 3. a sequence of locutions asserting some finite number of propositions: *Assert(q_i)...*, where $1 <= i <= n$ for some $n \in$ Natural Numbers with inferential structure between them, and then 4. *ArgumentConcluded*

Table 15: Structural rules for post-interview dialogue game

T1	A dialogue terminates if no-one performs a move.

Table 16: Termination rules for post-interview dialogue game

Outcome	Conditions
End of session	Post-interview is concluded

Table 17: Outcome rules for post-interview dialogue game

How courts should respond to the stories defendants tell: A Bayesian account of a Dutch ruling

HYLKE JELLEMA
Faculty of Law, University of Groningen, the Netherlands
h.jellema@rug.nl

> In criminal trials, a defendant sometimes provides an alternative explanation of the evidence. The Dutch Supreme court has set down a framework on how courts should respond to such explanations. Yet this framework is unclear. I offer an interpretation in terms of Bayesian probability theory.
>
> KEYWORDS: alternative explanations; Bayesianism; criminal law; justification; legal evidence; stories

1. INTRODUCTION

In criminal trials, the prosecution usually has a burden to prove that certain events happened which imply the guilt of the defendant. The defendant then often responds with an alternative account of the facts surrounding the alleged crime – one that implies his innocence, for example: "I did not kill her, I was just a bystander, it was someone else". Such accounts explain the facts by offering an alternative story of what happened.

How should fact finders – the jury or the judges who decide which version of the facts to accept - deal with such a story if they believe that it is false? Furthermore, does every explanation offered by the defendant require a reply? And what should this reply look like? In this paper I examine these questions in the light of a ruling of the Dutch Supreme Court. This ruling dealt with how courts should respond to the stories that defendants offer.

The Supreme Court's ruling was about a case that has become known as the Venray murder.[1] In this case the defendant was accused of having killed his wife. He offered an alternative explanation, according to which he came home and found his wife dead. According to Dutch criminal law, whenever a defendant offers such an explanation, the

---

[1] Dutch Supreme Court, March 16th 2010, ECLI:NL:HR:2010:BK3359. The name comes from the town where the victim and her husband lived.

court can only convict him if it provides a justification for rejecting this explanation in its ruling.[2] When the court of appeal ruled on the Venray case, it acquitted the defendant on the grounds that there was no evidence that refuted his explanation. In response, the Supreme Court decreed that while courts should ideally point to evidence that refutes the explanation, they can also reject alternative explanations even when there is no evidence that refutes it. In particular, the Supreme Court stated that courts can argue that the defendant's story "did not become plausible" or that it is "not credible". Finally, some explanations are so "highly improbable" that they require no explicit justification by the court to be rejected. The schema in figure 1 summarizes the ruling:

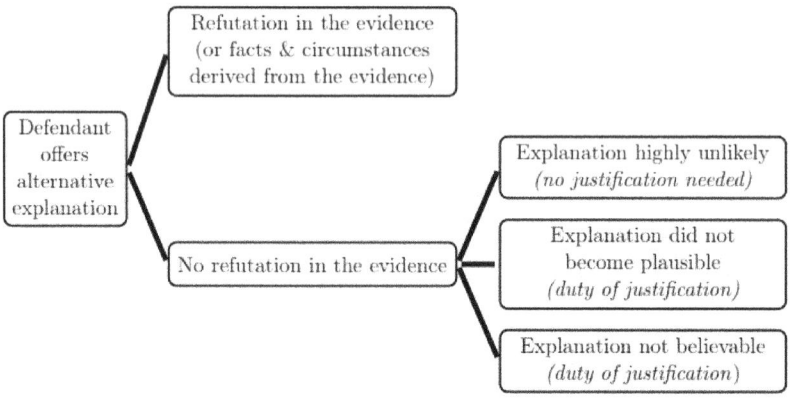

Figure 1 – The Venray ruling schematically

While this is an important ruling, it is also nebulous. The Supreme Court did not offer any explanation of the phrases it introduced (plausible, credible, highly unlikely), nor did it specify how these terms should be applied. As a result, both legal scholars and courts have been struggling to make sense of this ruling. In this paper I offer an interpretation of the ruling. My account is broadly Bayesian in that I use the language of Bayesian epistemology to clarify the necessary distinctions.

I start this paper in the next section by saying more about the Supreme Court's ruling and the problem it aims to solve, namely how courts should respond to explanations that are not eliminated by the evidence. After that, I offer an interpretation of each of the ways in which a court can reject an alternative explanation without referring to the evidence. I argue for the following interpretation:

---

[2] Article 359.2 Dutch code of criminal procedure.

The explanation did not become plausible: An alternative explanation fails to create a reasonable doubt its posterior probability is insufficiently high. When the defendant offers an explanation with a (very) low prior probability this explanation needs to become sufficiently probable in the course of the trial. If it does not, then the court can reject it.

The explanation is not credible: Some explanations can be probable when we look at them in isolation yet improbable when taking into account the credibility of the person who puts forward the story.

The explanation is highly improbable: Some explanations are not just improbable, but obviously improbable. The typical advantages of justifying decisions do not apply to these cases. Offering such a justification would take time and effort, thereby diminishing the efficiency with which decisions are reached in a criminal trial.

2. RESPONDING TO ALTERNATIVE EXPLANATIONS

The Supreme Court's ruling relates to the Venray murder case. In this case a man was accused of stabbing his wife to death. During the investigation, the man called upon his right to remain silent. He only offered an alternative explanation after one and a half years had passed. At that point, he knew the results of the forensic investigations. He then claimed that he had found his wife dead and hypothesized that criminals might have killed her because of an argument they had with him. As the court of appeal noted, this explanation fitted with the limited available evidence (blood stains and shoe prints) at least as well as the story that he killed his wife. However, the court did note that the defendant's story was somewhat hard to believe, especially since the defendant waited so long to come forward with it. Nonetheless, it did acquit him. The court reasoned that it could only convict if there was evidence that refuted the alternative explanation or if it was so implausible that it needed no explicit refutation. According to the court, neither was the case.

The court's position is understandable if we look at the Dutch proof standard for criminal cases. Proof standards determine when the defendant's guilt can be legally proven. In the Netherlands, the proof standard is that the court has to be convinced *based on the admissible evidence*.[3] So, it would seem that when the court has to explain why it is convinced of the guilt of the defendant, it should also do so by referring to the admissible evidence. In turn, when a conviction involves the court rejecting the defendant's story, this would require the court referring to some piece of evidence that refutes this alternative explanation. However, the Supreme Court did not share this reading – it ruled that

---

[3] Article 338, Dutch code of criminal procedure.

explanations can sometimes be rejected even when there is no evidence that contradicts it. It referred the case back to another court of appeal. The court of appeal then convicted the defendant of murder.[4]

So, the Supreme Court's ruling is about how courts should deal with cases in which the evidence does not refute a defendant's story. Before moving on to my interpretation of this ruling, I want to discuss both situations in which the court can point to evidence that refutes the defendant's story and situations in which the court can reject this explanation though no refuting evidence exists.

2.1 Refuting stories with evidence

When courts reject a defendant's explanation, they typically do so by referring to evidence that refutes this explanation. Evidence refutes a story insofar as it makes the story highly improbable. Courts should only reject a defendant's explanation if its probability is low, in order to avoid erroneous convictions. In Bayesian terms this means that the probability of the hypothesis (H, the explanation) conditional on the all the evidence in the case (E), ($P(H|E)$) is very low. This 'posterior probability' can be calculated with Bayes' formula:

$$P(H|E) = (P(E|H) * P(H))/P(E)$$

Whether the evidence in a case makes the hypothesis improbable therefore depends on the likelihood of the hypothesis, which refers to the probability of observing the evidence if we assume that the hypothesis (the explanation) is true, $P(E|H)$. A low likelihood means that we would not expect this evidence to occur if the hypothesis were true.

In cases with two competing explanations, such as the Venray case, we are often interested in the probability of one explanation compared to the other. For that purpose, we can rewrite Bayes' rule to its 'odds' version:

$$\frac{P(H_1|E)}{P(H_2|E)} = \frac{P(E|H_1)}{P(E|H_2)} \times \frac{P(H_1)}{P(H_2)}$$

Here H1 and H2 represent the hypotheses that either one or the other explanation is true. In this version of the formula, whether the evidence skews the prior ratio in favour of guilt or innocence depends on the 'likelihood ratio', $P(E|H1)/P(E|H2)$. When the likelihood ratio is lower

---

[4] Court of justice of Arnhem, October 15th 2012, ECLI:NL:GHARN:2012:BY0075.

than 1 it means that the evidence raises the probability of H1 whereas a likelihood ratio higher than 1 means that the probability of H2 is raised.

When reasoning about which story to accept, rejecting one story and accepting the other often means finding 'discriminating evidence', i.e. evidence that fits better with one story than another (Van Koppen, 2011, pp. 52-55). In Bayesian terms this means evidence where the likelihood ratio strongly favors one story over the other. Take again the example of the witness who testifies that he saw the husband kill the victim. This evidence discriminates between the two explanations because we would expect the evidence much more if the husband was the killer than if someone else was. This would mean that the likelihood ratio would be (much) greater than 1.

If the likelihood ratio is sufficiently much greater than 1, the probability of the story that the defendant killed the victim will be high and the probability that someone else killed her will be low. In such a case the court can point to the witness' statement as a reason why it convicts the defendant.

2.2 Refuting stories without evidence

Let us now turn to situations where the evidence does not refute the defendant's story, as in the Venray case. In that case the key question was which of two competing stories to accept, a situation that is best captured by the odds-version of Bayes rule. There the court noted that the evidence did not discriminate between these stories. Whether it was the husband who killed his wife or someone else, either way, we would expect to find the kind of evidence that was found (such as the shoe prints and the blood stains). This means that the likelihood ratio is close to 1. This means that the evidence did not significantly change the prior probability of either explanation.

Recall that on Bayes' rule, a low posterior probability of a hypothesis can depend either on a low likelihood, $P(E|H)$ or on a low prior probability, $P(H)$ of the hypothesis. So, we might assume that, if the evidence does not discriminate between explanations in terms of the likelihood, an alternative explanation's low posterior probability can only be because that explanation has a low prior probability. Is this how we should read the Supreme Court's ruling? To put it differently, when courts reject a defendant's explanation for being implausible, incredible or highly unlikely, is this always a judgment about that explanation's prior probability? And what should we then make of the distinction between these three terms?

In the following sections I will argue that prior probability only plays a key role in one of the three criteria that the Supreme Court mentioned, namely whether the explanation needs to "become plausible". I will look at this criterion in the next section. For the other two criteria we need different concepts, which I discuss in sections 4 and 5 respectively.

3. IMPLAUSIBLE EXPLANATIONS FAIL TO BECOME PROBABLE

Let us begin with the first term the Supreme Court introduces. According to the Supreme Court, courts can reject an explanation if it "did not become plausible" during the criminal proceedings.[5] The obvious question when interpreting this statement is why some explanations need to become plausible. The answer to this lies in the proof standard. As mentioned, in the Netherlands the proof standard states that the court should be convinced of the defendant's guilt based on the admissible evidence. However, in practice, many legal scholars believe that the standard is actually similar to that of common law countries - that guilt has to be proven beyond a reasonable doubt (Ter Haar & Meijer, 2018, 7.4; Nijboer 2017, pp. 73-74). So, if the defendant hopes to be acquitted by telling an alternative story, that story needs to be good enough to create a reasonable doubt about his guilt (assuming that the prosecution's case is in itself strong enough).

Suppose that the defendant's story is weak and that the prosecution's case is strong. This means that if no further evidence or arguments were to be adduced, the defendant would most likely lose the case and be found guilty beyond a reasonable doubt. So, if the defendant tells a story that initially seems weak, he risks losing the case if no new evidence confirms his story. The defendant may then have a burden to introduce new arguments or evidence that would make the court decide in his favor or he risks losing the case.

So, a defendant's explanation can be rejected if its posterior probability fails to become high enough during court proceedings. Recall that, on a Bayesian account, there are two reasons why an explanation can be improbable: due to the evidence in the case (likelihood) or due to its prior probability. So, one reason why an explanation can be improbable at the moment that the defendant offers it, is because it conflicts with (reliable) evidence that was already brought forward in court. However, in such cases the court can point to that evidence when justifying its decision to reject the explanation. Yet the Supreme Court's ruling is about cases in which courts cannot point to such evidence. So, it presumably describes situations in which an explanation is improbable due to its low prior probability. The lower the prior probability of an explanation is, the stronger the evidence has to be to make that explanation probable. If the evidence is not strong

---

[5] In Dutch "*niet aannemelijk geworden*", my translation.

enough (in terms of the likelihood) then the prior probability will not be raised sufficiently to create a reasonable doubt. In that case, the explanation has not become 'plausible'.

4. INCREDIBLE EXPLANATIONS ARE TOLD BY UNRELIABLE STORYTELLERS

Apart from arguing that the explanation is implausible, the Supreme Court also decreed that courts can reject explanations by arguing that they are "incredible".[6] How does this differ from an explanation that is implausible – i.e. improbable, either a priori or due to the evidence? When the term 'incredible' is used by Dutch courts, it typically refers to actions within the scenario that the defendant undertook or in how the defendant told the story (Lettinga, 2015). For instance, suppose that the defendant claims that he was a bystander of a murder but that he did not call the emergency services while he did spend time trying to hide possessions of the victim. Such a story would be implausible, in the way we just saw: it contains illogical elements and therefore has a low prior probability. However, it would also be incredible. The defendant would not come across as a reliable storyteller. Telling bad stories and lacking credibility as a storyteller often go hand in hand, but not always. Some stories fit well with the evidence and with our background beliefs, perhaps even better than the true explanation but are still improbable due to the lack of credibility of the defendant.

First, some stories are vague. For example, a defendant may claim that 'something else happened', without providing further details. People tend to find such stories difficult to believe because they lack relevant details (Pennington & Hastie, 1991). However, typically, the more general a claim the more probable it is. Suppose that a defendant offers the alibi "I saw someone else committing the murder." On its own this story should be much more probable than the story "I saw someone else, who had red hair and glasses, committing the murder." After all, there are many men but only a smaller subset of them have red hair and glasses. The claim that 'there was a man with red hair and glasses' therefore, by definition, implies that 'there was a man', and the former can therefore never be more probable than the latter.

But what we are assessing is not the statement 'there was a man' versus 'there was a man with red hair and glasses'. We are assessing whether this claim is credible given that the defendant tells it. On a Bayesian account, the credibility of the story depends on the answer to the following question: "given that a witness testifies to fact X, what is the probability of X?" (Goldman, 1999, 4.2–4.4). So, in the case where

---

[6] In Dutch *"ongeloofwaardig"*, my translation.

the defendant tells the general story, we are assessing the probability of 'there was a man' (H1) given that 'the defendant reports that "there was a man" (E1). When the defendant gives the more specific story, we are assessing the probability of 'there was a man with red hair and glasses' (H2) given that 'the defendant reports that "there was a man with red hair and glasses" (E2).

Suppose that the defendant was in a good position to observe the characteristics of the man. That means that, if he is truthfully reporting on his own experiences, we can reasonably expect him to testify to those characteristics. However, if the defendant then offers a vague story, we might become suspicious that he was lying by deliberately offering a vague story so that his story is not contradicted by the evidence. In other words, if we can reasonably assume that the defendant could tell a more specific story, which better explains the facts, then we have reason to doubt the credibility of his story.

A second important category of incredible stories are ad-hoc explanations. An ad-hoc explanation is an explanation that is made up to fit the available evidence but that is difficult or impossible to falsify (Lipton, 2003, p. 219). For instance, a guilty defendant could make up a somewhat believable, specific story intended to avoid falsification (Mackor, 2017). For instance, he can call upon his right to remain silent and only offer an explanation once all the evidence has been presented. This was what the defendant in the Venray case did. Such a story is not necessarily implausible or incoherent. On the contrary, false explanations of criminal evidence are sometimes more coherent (Vredeveldt et al., 2014) and better supported by the evidence (Gunn et al., 2016) than true explanations. This is because they can be tailored to the known facts. However, if we have good reasons to suspect that the defendant has fitted his story to the evidence, then this should lower our degree of belief that he is truthfully reporting on his own experiences.

5. HIGHLY IMPROBABLE EXPLANATIONS ARE OBVIOUSLY FALSE

When a court considers an explanation to be implausible or incredible it must generally justify why it does not believe the defendants explanation before convicting him. However, according to the Supreme Court, some explanations are so "highly improbable" that courts do not have a duty to respond to them.[7] Of the terms that the Supreme Court introduces in its ruling, this one is possibly the most nebulous. At first sight, the term would seem to refer to explanations that have a very low

---

[7] In Dutch "*zo onwaarschijnlijk is, dat zij geen uitdrukkelijke weerlegging behoeft*", my translation.

(posterior) probability. But this straightforward interpretation faces the difficulty that any alternative explanation that the court rejects is highly improbable. I mentioned earlier that (in practice) Dutch criminal law requires that a defendant's guilt must be proven beyond a reasonable doubt. This is a high standard for proof. In probabilistic terms, the standard is often taken to mean that the probability of guilt should be high enough (e.g. 95%) (Cheng, 2013, p. 1256). However, this means that the probability of any story consistent with guilt can be at most 5%. Furthermore, this 5% probability is a maximum, meaning that in many cases, the probability of the story that the defendant tells will be (far) lower. So, if all rejected alternative explanations are very improbable, what distinguishes those that are 'highly improbable' that they need not be addressed? We could, of course, answer this challenge by saying that while many explanations are improbable, some explanations are highly improbable, say less than 0.01%. Yet this still leaves us with the question why courts do not have to respond to such explanations. What makes highly improbable explanations special?

An answer to this question begins with a discussion about why courts usually should justify their decision to reject an alternative explanation. There are, broadly speaking, two purposes that such justification serves: making the explanation understandable and forcing the court to reflect on its reasoning. First, justification helps make the decision understandable for its audience, which includes the parties at trial, the legal community and society as a whole (Knigge, 1980; Dreissen, 2007, pp. 392-404). If the audience understands the arguments for the decision, then this makes the court's decision more legitimate for them. An explicit justification also allows courts of appeal, judicial scholars, experts and other interested parties to check whether the decision was correct and to point out possible flaws. Finally, by making the reasons for the decision understandable, parties might be less inclined to appeal the ruling. This would aid the efficiency of the criminal law system because courts of appeal would have to hear fewer cases (Buruma, 2005). The second reason why judges should justify their decision is that it forces courts to reflect on the arguments for their ruling. This in turn can help them avoid reasoning errors (see e.g. Dreissen, 2007, pp. 392-404). This is in line with psychological research that suggests that explaining one's decision-making process helps people make better decisions (Wilkenfeld & Lombrozo, 2015).

These benefits of justification also occur when courts justify why they reject an alternative explanation. In such cases the justification gives both the court and the audience insight into why that explanation is improbable enough not to create a reasonable doubt. However, there are cases in which this kind of insight is not required. In particular, some stories that defendants tell are so obviously improbable that we would gain little by arguing against them. For example, take a (real) case in which the defendant pleaded that he was not accountable for the child porn on his computer because his mind was controlled by aliens.[8] It seems fair to say that no reasonable audience would consider the 'alien' explanation remotely probable. Furthermore, a defendant who offers such an explanation would either be delusional or insincere. So, it is improbable that arguments would sway him. Hence, the court would (most likely) gain little by justifying why it rejects this alternative explanation, with respect to the parties, legal community and general audience's understanding of it. More generally, explaining does not help if the audience already believes that the explanation is improbable or if they believe that the explanation is probable, but this belief would most likely not be affected by further arguments from the court.

As I discussed above, justifying one's decisions is not just important for its outside relevance. It is also important to improve one's decision making by carefully considering one's reasoning and finding possible flaws in it. Why should spelling out the reasons against obviously improbable explanations not improve one's decision-making? My proposal is that the more difficult it is to see why an explanation is improbable, the more room for error there is. However, when an explanation's improbability is obvious, the reasoning required to understand its probability does not require much thought. Hence, there is less to be gained by carefully spelling out one's reasoning to see whether this reasoning is sound. For instance, the court does not have to carefully reflect on whether they might be making an error when they assume that mind controlling aliens do not exist.

So, there is little gain to justifying why we reject obviously improbable explanations. Yet spelling out such arguments does take time and effort and impedes the efficiency of decision making. So, with respect to obviously highly unlikely explanations, the costs of explicit justification will often outweigh the benefits.

It is not always obvious that an explanation is improbable. Seeing that an explanation is improbable often boils down to seeing that its elements do not cohere with one another, or that the explanation does not cohere with the evidence or with our general knowledge about the world. For instance, in criminal cases, the mere description of the evidence can sometimes be hundreds of pages long. Judging whether the evidence makes the explanation unlikely might therefore require seeing how numerous pieces of evidence cohere with one another. Similarly, an explanation can have a very low prior probability because of internal inconsistencies, without this being immediately obvious. Understanding that the explanation is highly improbable might then involve, for

---

[8] Court of Noord-Holland, November 24th 2014, ECLI:NL:RBNHO:2014:11709

instance, creating a time line of the story and seeing that the story implies that the defendant was in two places at the same time. This explanation then has a very low probability (perhaps even 0) but it is not immediately obvious that it does. This fact requires further explaining by the court.

6. CONCLUSION

When defendants plead for their innocence, they often do so by offering an alternative explanation of the evidence. When may courts reject such alternative explanations and when should they justify this decision to reject the defendant's story? In this paper, I discussed these questions in the context of the Dutch Supreme Court's ruling about the Venray murder case. What makes this ruling interesting is that it offers a nuanced approach to the above questions but it does so in nebulous terms, in need of further clarification. My goal was to make the ruling understandable by giving it a more precise interpretation, employing Bayesian probability theory. This interpretation ties the ruling to the goals of avoiding errors (in particular, courts should reject a defendant's explanation only when its posterior probability is sufficiently low), of making the decision to convict understandable to the parties and other audiences and of reaching these decisions efficiently.

At the heart of the Supreme Court's ruling is the idea that courts can reject a defendant's explanation even in cases where the evidence does not refute this explanation. While rejecting the story by referring to a smoking gun (i.e. refuting evidence) may be the ideal, other responses are possible too. The Supreme Court distinguishes three categories. First, some explanations can be rejected because they "did not become plausible." I argued that whether an explanation needs to become plausible during the criminal proceedings depends on its inherent plausibility at the time it is offered - its prior probability. If an explanation with a low prior probability does not become probable by means of the evidence, then the explanation fails to create a reasonable doubt. Second, explanations are "incredible". Whether an explanation offered by a defendant is probable does not just depend on whether it is a good story, it also depends on the credibility of the defendant. If an explanation is incredible, the defendant is not a believable storyteller, meaning that the posterior probability of his story is low because of the evidence of his unreliability. Finally, some explanations are so "highly improbable" that the court does not have a duty to respond to them. I argued that what distinguishes these explanations from explanations that the court should respond to is that their improbability is obvious. When an explanation is obviously improbable, the court would not serve the goals of making its decision understandable by offering a response. A duty to respond would then only reduce the efficiency of the decision process.

ACKNOWLEDGEMENTS: I would like to thank Henry Prakken and Anne Ruth Mackor for their extensive help and comments throughout the writing process of this article. I would also like to thank Pepa Mellema and Anne Kamphorst for commenting on previous versions of this paper. Finally, I would like to thank the audience for their helpful remarks. This work is supported by the Netherlands Organisation for Scientific Research (NWO) as part of the research programme with project number 160.280.142.

REFERENCES

Bex, F. J. (2011). *Arguments, stories and criminal evidence: A formal hybrid theory*. Dordrecht: Springer Science & Business Media.
Buruma, Y. (2005). Motiveren: Waarom? In A. Harteveld, D. H. Jong, & E. Stamhuis (Eds.), *Systeem in ontwikkeling. Liber amicorum G. Knigge*. Nijmegen: Wolf Legal Publishers.
Cheng, E. K. (2012). Reconceptualizing the burden of proof. *Yale LJ*, 122, 1254–1279.
Dreissen, W. H. B. (2007). *Bewijsmotivering in strafzaken*. Maastricht University.
Goldman, A. I. (1999). *Knowledge in a social world*. Oxford: Clarendon Press.
Gunn, L. J., Chapeau-Blondeau, F., McDonnell, M. D., Davis, B. R., Allison, A., & Abbott, D. (2016). Too good to be true: When overwhelming evidence fails to convince. *Proc. R. Soc. A*, 472, 20150748.
Kaye, D. (1979). The paradox of the gatecrasher and other stories. *Arizona State Law Journal*, 101–143.
Knigge, G. (1980). *Beslissen en motiveren (de artt. 348, 350, 358 en 359 Sv)*. Alphen aan den Rijn: Tjeenk Willink.
Lettinga, B. (2015). Recht doen aan alternatieve scenario's. *PROCES*, (1), 50-61.
Lipton, P. (2003). *Inference to the best explanation*. London: Routledge.
Mackor, A. R. (2017). Novel facts: The relevance of predictions in criminal law. *Strafblad*, (15), 145–156.
Nijboer, J. F., Mevis, P. A. M., Nan, J. S., & Verbaan, J. H. J. (2017). *Strafrechtelijk bewijsrecht*. Nijmegen: Ars Aequi Libri.
Pennington, N., & Hastie, R. (1991). A cognitive theory of juror decision making: The story model. *Cardozo L. Rev.*, 13, 519–557.
Ter Haar, R., & Meijer, G. M. (2018). De rechterlijke overtuiging. In *Elementair Formeel Strafrecht (Praktijkwijzer Strafrecht nr. 9)*. Deventer: Wolters Kluwer.
Twining, W. (1999). Necessary but dangerous? Generalizations and narrative in argumentation about 'facts' in criminal process. In M. Malsch & J. F.

Nijboer (Eds.), *Complex cases: Perspectives on the Netherlands criminal justice system* (pp. 69–98). Amsterdam: Thela Thesis.

Van Koppen, P. J. (2011). *Overtuigend bewijs: Indammen van rechterlijke dwalingen* [*Convincing evidence: Reducing the number of miscarriages of justice*]. Amsterdam: Nieuw Amsterdam.

Vredeveldt, A., van Koppen, P. J., & Granhag, P. A. (2014). The inconsistent suspect: A systematic review of different types of consistency in truth tellers and liars. In R. Bull (Ed.) *Investigative Interviewing* (pp. 183–207). New York: Springer.

Wilkenfeld, D. A., & Lombrozo, T. (2015). Inference to the best explanation (IBE) versus explaining for the best inference (EBI). *Science & Education*, 24(9–10), 1059–1077.

Index of Authors

Aikin, Scott F., 3
Alfano, Mark, 243
Alfino, Mark, 15

Bailin, Sharon, 41
Balg, Dominik, 27
Battersby, Mark, 41
Baumtrog, Michael D., 55
Bobrova, Angelina, 65
Bova, Antonio, 79
Busuulwa, Huthaifah, 93

Carozza, Linda, 107
Casey, John, 117
Cheng, Martha S., 143
Cohen, Daniel H., 117, 161
Cozma, Ana-Maria, 175
Csordás, Hédi Virág, 191

Danka, Istvan, 205
Dufour, Michel, 231
Dykes, Natalie, 217

Eemeren, Frans H. van, 261
Evert, Stefan, 217

Farine, Léa, 275
Feteris, Eveline, 287
Franco, Dominic de, 433

Garnier, Marie, 303

Garssen, Bart, 261
Gobbo, Federico, 315
Grefte, Job de, 327

Hansen, Hans V., 341
Heinrich, Philipp, 217
Henderson, Leah, 357
Herman, Thierry, 369
Hinton, Martin, 379
Hitchcock, David, 393

Innocenti, Beth, 407
Isaksen, David Erland, 419

Janier, Mathilde, 433
Jellema, Hylke, 453

Lagewaard, Thirza, 243

Novaes, Catarina Dutilh, 243

Pease, Alison, 433

Saint-Dizier, Patrick, 303
Sierra Catalán, Guillermo, 129
Snaith, Mark, 433
Stevens, Katharina, 161
Sullivan, Emily, 243

Wagemans, Jean H.M., 315

www.ingramcontent.com/pod-product-compliance
Lightning Source LLC
Chambersburg PA
CBHW050117170426
43197CB00011B/1622